AMARANTH
Biology, Chemistry, and Technology

AMARANTH
Biology, Chemistry, and Technology

Edited by

Octavio Paredes-López, Ph.D.
Professor of Food Biotechnology
Chairman
Department of Biotechnology and Biochemistry
Center of Research and Advanced Studies
National Polytechnic Institute
Irapuato, México

CRC Press
Taylor & Francis Group
Boca Raton London New York

CRC Press is an imprint of the
Taylor & Francis Group, an informa business

First published 1994 by CRC Press
Taylor & Francis Group
6000 Broken Sound Parkway NW, Suite 300
Boca Raton, FL 33487-2742

Reissued 2018 by CRC Press

Library of Congress Cataloging-in-Publication Data

Amaranth : biology, chemistry, and technology / edited by Octavio Paredes-Lopez.
p. cm.
Includes bibliographical references and index.
ISBN 0-8493-5374-2
1. Grain amaranths. 2. Amaranths. 3. Amaranth products.
I. Paredes-Lopez, Octavio.
SB191.A42A43 1994
583'.913--dc20 93-36812

A Library of Congress record exists under LC control number: 93036812

Publisher's Note
The publisher has gone to great lengths to ensure the quality of this reprint but points out that some imperfections in the original copies may be apparent.

Disclaimer
The publisher has made every effort to trace copyright holders and welcomes correspondence from those they have been unable to contact.

ISBN 13: 978-1-315-89050-0 (hbk)
ISBN 13: 978-1-351-06960-1 (ebk)

Visit the Taylor & Francis Web site at http://www.taylorandfrancis.com and the
CRC Press Web site at http://www.crcpress.com

This book
is dedicated
to the people
of Mocorito
located where the
Pacific Ocean,
Gulf of California,
and Sierra Madre
meet.

PREFACE

Amaranth:Biology, Chemistry, and Technology had its genesis during informal discussions with the bachelor and graduate students from The Center of Research and Advanced Studies (CINVESTAV-IPN) in Irapuato, Mexico, on the food potential of indigenous minor crops and on the role of different biological and chemical techniques, including modern biotechnology, to exploit in a rational way such potential. These discussions were continued with some of the scientists who attended the First International Congress on Amaranth held in 1991.

This book is meant to be a comprehensive set of reviews of our present knowledge on amaranth (*Amaranth*). It is a multiauthor work, with the advantage that each subject is explored by an expert in that field. Amaranth is a most remarkable crop, from which about 45 species are native to the Mesoamerican region and 10 others originated in Africa, Asia, and Europe. Amaranth was the foundation of the extensive North and South American ancient civilizations and was still important in the agriculture of more recent Indian cultures. However, this crop almost disappeared after the Spanish conquest. In view of the outstanding agronomic performance of the plant and of the high nutritional value of the grain, it is now becoming an important crop in various regions of the world. It should be highlighted that progress in utilization of amaranth may be directly related to the generation of scientific and technical information on its biological, physical, and chemical properties.

The book was written principally as a reference for students, researchers, and practitioners. It begins with a chapter on the use of tissue culture, molecular biology, and genetic engineering techniques for crop improvement; such approaches have brought fresh insights to the research studies on amaranth. The next chapter deals with classical genetics of wild and cultivated grain species, genetic stocks, and conservation of germplasm resources. The chapter on breeding of grain amaranth involves modern breeding techniques and breeding for specific characters (e.g., early maturity, insect and disease resistance, drought resistance). The development and physiological aspects are discussed in another chapter, emphasizing the information related to plant morphology and anatomy, seed dormancy and germination, environmental effects, photosynthesis, and water use efficiency.

Seed and leaf proteins are among the largest sources of human nutrition. The biochemical aspects of such proteins are analyzed in great detail in another chapter; their physicochemical and functional properties are also here considered. The following chapter deals with carbohydrates — mostly with starch isolation and characterization of functional attributes and the structure-function relationship. Amaranth grain has a higher lipid content than most cereal grains. The chapter on amaranth oil analyzes its composition and nutritional qualities.

One of the ancient ways of consuming amaranth grain by the Mesoamerican societies was by popping the seed in a hot plate. The structure of the seed before and after popping, the possible physicochemical mechanisms leading to popping, type of equipment, and uses of popped seeds are discussed in another chapter. The chapter on food uses of amaranth analyzes its potential and the numerous food products in which amaranth, both vegetable and grain, may be incorporated.

Amaranth grain has an attractive chemical composition and superior nutritive value when compared on an equal basis to other grains, whether cereals, food legumes, or oilseed crops. Its nutritive value, the effects of processing, and its supplementary and complementary alternatives are reviewed extensively in another chapter. The last chapter focuses on amaranth amenability to modern technology, its niches in world agriculture, and its future commerce associated with various crop development models.

Finally, thanks are extended to all of the authors for their excellent contributions to this book and to the editorial staff of CRC Press, Inc.

Octavio Paredes-López

THE EDITOR

Octavio Paredes-López, Ph.D., is Chairman of the Department of Biotechnology and Biochemistry and Professor of Food Biotechnology at the Center of Research and Advanced Studies, National Polytechnic Insitute (IPN), at Irapuato, Mexico.

Dr. Paredes-López received his B.Sc., biochemical engineering, and M.Sc. degrees, food science, from IPN in 1968 and 1975, respectively. He received another M.Sc. in biochemical engineering from the Czechoslovak Academy of Sciences in 1971, and his Ph.D in 1980 from the Department of Plant Sciences, University of Manitoba, Canada. He has received training and spent research stages at the University of Massachusetts, North Dakota State University, University of Florida, University of Maryland, Food and Drug Administration, ETH-Switzerland, and University of London. He has been Visiting Professor at the University of Manitoba.

Dr. Paredes-López is a member of the Institute of Food Technologists — USA, American Association of Cereal Chemists, International Society for Plant Molecular Biology, Mexican Society of Biotechnology, Mexican Society of Biochemistry and Mexican Academy of Sciences. He is serving on the Scientific Advisory Council to the President of Mexico, and is a member of the Editorial Board of the journal *Plant Foods for Human Nutrition* — The Netherlands and of *Biotecnologia* — Mexico.

He has received the National Bank of Mexico Award, the Nestlé Research Award, the Coca-Cola Award in Food Science and Technology, the Mexican Award in Chemistry, the NPI Distinguished Teaching Award, the NPI Distinguished Research Award, and the National Award in Science (the highest Mexican Award). He was distinguished in 1992 as Doctor Honoris Causa by the Autonomous University of Queretaro, Mexico. He has been the recipient of many research grants from the Mexican Council of Science and Technology, National Science Foundation — USA, European Economic Community and Organization of the American States.

Dr. Paredes-López is the author of more than 150 scientific and technical papers, several book chapters, various articles for the popular press on the role of science, and over 100 invited lectures at national and international meetings, and at universities and institutes. His current research interests relate to the use of molecular biology techniques involved in the improvement of the functional performance of plant proteins and of the food properties of some crops.

CONTRIBUTORS

Ana Paulina Barba de la Rosa, Ph.D.
Department of Biotechnology and Biochemistry
Centro de Investigación y de Estudios Avanzados
Irapuato, México

Robert Becker
United States Department of Agriculture
Agricultural Research Service
Western Regional Research Center
Albany, California

Luis Arturo Bello-Pérez, Ph.D.
Department of Biotechnology and Biochemistry
Centro de Investigacion y de Estudios Avanzados
Irapuato, México

William M. Breene, Ph.D.
Department of Food Science and Nutrition
University of Minnesota
St. Paul, Minnesota

Ricardo Bressani, Ph.D.
Instituto de Investigaciones
Universidad del Valle de Guatemala
Guatemala, Central America

Edmundo Brito, Ph.D.
Department of Food and Biotechnology
Division of Engineering
Faculty of Chemistry
National Autonomous University of Mexico
Alvaro Obregon, México, D.F.

Eduardo Espitia-Rangel, M.Sc.
Campo Experimenal Valle de México
Instituto Nacional de Investigaciones Forestales y Agropecuarias
Chapingo, Edo. de México, México

Holly Hauptli, Ph.D.
Potter Valley Amaranth
Potter Valley, California

Jaime Kigel, Ph.D.
Department of Agricultural Botany
Faculty of Agriculture
The Hebrew University of Jerusalem
Rehovot, Israel

Peter A. Kulakow, Ph.D.
The Land Institute
Salina, Kansas

James W. Lehmann, Ph.D.
Lehmann Crop Genetics
Rushmore, Minnesota

Mercedes G. López, Ph.D.
Department of Biotechnology and Biochemistry
Centro de Investigación y de Estudios Avanzados
Irapuato, México

Edmundo Lozoya-Gloria, Ph.D.
Department of Genetic Engineering
Centro de Investigación y de Estudios Avanzados
Irapuato, México

Octavio Paredes-López, Ph.D.
Professor of Food Biotechnology
Chairman
Department of Biotechnology and Biochemistry
Center of Research and Advanced Studies
National Polytechnic Institute
Irapuato, México

Kent A. Schnetzler, M.S.
Department of Food Science and Nutrition
University of Minnesota
St. Paul, Minnesota

Magdalena Segura-Nieto, Ph.D.
Department of Genetic Engineering
Centro de Investigacion y de Estudios Avanzados
Instituto Politecnico Nacional
Irapuato, México

Luis Raul Tovar, Ph.D.
Biosfera Investigación y Desarrollo Tecnologico
S.A. de C.V.
México, D.F., México

Maria de los Angeles Valdivia, M.Sc.
Department of Food and Biotechnology
Division of Engineering
Faculty of Chemistry
National Autonomous University of Mexico
México, D.F., México

TABLE OF CONTENTS

Chapter 1

Biotechnology for an Ancient Crop: Amaranth

E. Lozoya-Gloria

CONTENTS

I. INTRODUCTION

The potential of amaranth grain as food was recognized by old cultures in America; the Aztecs and Incas used it commonly until it was replaced, for reasons not well-known, by other vegetation such as corn and common beans after the Conquest. Although there are now some farmers dedicated to this crop, in many places it remains only as a familiar ornament in the backyard. Interestingly, the word "amaranth" means "everlasting" in Greek, which seems adequate for the history of this crop.[1]

For some years now, amaranth has been "rediscovered" as a useful and promising plant, mainly because it looks like an interesting alternative to alleviate the increasing need for food by some countries of the Third World. Ironically, amaranth has been used as food for a long time by some of those countries, although it appears that the fast expansion of the desertic zones and the huge demographic explosion have been too much for any crop, including amaranth.

Some important efforts have been made in order to return to the use of amaranth and to increase the potential of this crop. In 1981, grain amaranth was selected as a priority for a research grants program by the Board on Science and Technology for International Development (BOSTID) of the U.S. National Academy of Sciences/National Research Council.

The suggested goals for this program were:

- Determination of the environmental conditions under which amaranth could be a nutritious and economical crop
- Identification of the best varieties and growing strategies for different regions
- Development of techniques to harvest and use grain amaranth in local settings
- Development of a base of information to support the expanded production and use of amaranth

Promising results have been obtained so that the *Amaranth Newsletter* was established,[2] and an international society on amaranth was proposed during the *First International Amaranth Congress* (Oaxtepec, México, September 22 to 28, 1991).

Amaranth production in 1990 was over 100,000 ha in the Soviet Union and was used mainly as a forage crop. Among other principal producers of grain amaranth are China, México, Guatemala, Perú, India, and Kenya.[3]

Although many species of amaranth are considered as opportunistic weeds and eventually used as food, there are three species which are consumed by humans as grains (*Amaranthus hypochondriacus* L., *A. cruentus* L. and *A. caudatus* L.). The grains (or seeds) are considered as pseudocereal and in addition to their use for food purposes, the pigments of leaves and grains of some species have helped the preservation of these plants by using them as ornamentals.

However, what is the attraction of amaranth? Why is it considered a "new crop", in spite of its history? Among the principal attributes recognized to describe this plant as a "new crop" are[3]

- Has a high protein content (13 to 19 %)
- Has a high level of the essential amino acid lysine, in comparison to other grains like corn and rice
- Contains 1.5 to 3.0 times more oil than other grains which represents a high caloric content

0-8493-5374-2/94/$0.00+$.50

- Can successfully grow in adverse environmental conditions such as drought, high temperatures, and saline soils
- Is a dicotyledonous plant using an efficient type of photosynthesis known as C_4 carbon-fixation pathway
- Is a crop with multiple uses such as food, forage, silage, green manure, and animal feed
- Has a promising potential in industrial applications, from cosmetics to biodegradable plastics

Currently, a big and important amount of work is being carried out, mainly on breeding. There exist already three major germplasm collections at the:

- U.S. Department of Agriculture (USDA) Plant Introduction Center, Ames, IA
- University of Cuzco, Cuzco, Perú
- National Bureau of Plant Genetics Resources, Simla, India

Some agronomical and biochemical aspects of this crop are described in the following chapters of the present work. These include updated information about environmental conditions, fertility, and location requirements for the cultivation of amaranth. In addition, important factors related to food applications, nutritional properties, and economic and social considerations are discussed. The current status of this crop in México, Perú, Kenya, China, and the U.S. is also analyzed.[4]

Although all this information is very valuable and useful, there are other aspects of amaranth which have not been so deeply investigated. These pertain to the potential of amaranth in biotechnology and even if some work is already being carried out, it is necessary to increase research efforts toward these aspects.

This chapter summarizes and discusses research already done regarding tissue culture and molecular biology of amaranth, as well as the future prospects for this crop in modern biotechnology.

II. TISSUE CULTURE OF AMARANTH

The advantages of plant tissue cultures for biotechnological applications and research have been described at large in many books and journals. In general, this is the path of choice for achieving many controlled manipulations on plants resulting in the selection of new and useful traits, such as resistance to chemicals/pathogens, micropropagation of "hardy" plants, and studies about the physiology, biochemistry, and genetics of specific cultivars, among others.

Particularly, tissue culture of cereals was a challenge for some time until the discovery of the right conditions for the establishment of embryogenic calli of different Gramineae.[5]

The first reported studies about tissue cultures of amaranth[6] were done using *A. cruentus*, *A. hypochondriacus,* and *A. tricolor.* The authors reported that cell cultures from different explants of these species were easily established in several media (B5, MS, and White's) using combinations of various plant growth regulators, such as 2,4-dichlorophenoxyacetic acid (2,4-D), naphthaleneacetic acid (NAA), kinetin (K), benzyl adenine (BA), and zeatin (Z). Appropriate calli for suspension cultures were obtained from leaf discs and, in some cases, embryo-like structures were observed. Using high concentrations of cytokinins, it was possible to get shoot regeneration from hypocotyl-derived calli. Regarding protoplasts obtained from leaves of amaranth, the same authors reported that neither cell-wall regeneration nor cell division were observed under the used conditions.

Later on, and using a different species of amaranth (*A. paniculatus*), the full regeneration of complete plants was reported. This was done with calli from hypocotyl segments which were induced to shoots formation in K and NAA; then, these shoots were rooted and transferred to soil.[7]

A recent paper[8] using several varieties of four amaranth species (*A. caudatus*, three varieties; *A. cruentus*, four varieties; *A. hybridus*, two varieties; and *A. hypochondriacus*, eight varieties) reported on the successful (>95%) callus formation from hypocotyl and stem tissues in MS medium with a combination of either NAA plus benzylaminopurine (BAP) or 2,4-D plus K, for all assayed varieties. Callus growth was recorded, and results showed significant differences among varieties of the same species. In general, the NAA/BAP combination was more effective for callus growth in *A. cruentus* while 2,4-D/K was best for *A. hybridus*. A similar and scarce callus growth was observed for *A. hypochondriacus* and *A. caudatus* at any of the combinations of plant growth regulators used. Callus growth increased between 0.0215 to 0.1439 g fresh weight after 1 month of incubation, depending on the species and varieties. In contrast, only some of the calli of *A. hypochondriacus* and *A. caudatus* were more efficient at regenerating different organs such as shoots or roots. All regenerated shoots were able to form roots and, after transfer to pots, plantlets developed.

Another paper[9] reported on the use of inflorescences of *A. paniculatus* to produce secondary inflorescences and plantlets *in vitro*. The immature inflorescences were placed in MS with several concentrations of kinetin; after four weeks, various responses were observed. At low concentrations (0.5 mg K/L), tissues formed calli; as kinetin content was increased (3 to 6 mg K/L), the results were secondary inflorescences; then leaves, shoots and calli were formed at 8 to 12 mg K/L; and finally leaves appeared at the highest levels (10 to 15 mg K/L). The average percent of floral buds responding per inflorescence was up to 10% independently of the kinetin concentration. If the buds forming leaves and shoots were kept in the same media, they formed calli at the base, with subsequent elongation but without roots. After transfer of these tissues to media with 12 mg K/L and 15% coconut milk, the authors observed the formation of roots and shoots. These plantlets flowered after 20 days *in vitro* without a special photoperiod.

All these reports have shown that although it is easy to get tissue cultures of amaranth (mainly calli), they are still recalcitrant species for plant regeneration, and, so far, there is not a reproducible procedure for this goal. The genotype of the different varieties of one species plus the endogenous auxin/cytokinin balance may play an important role. After these results, it is clear that more work should be done in this direction. Considering the results obtained in cereals, plant regeneration of amaranth does not seem impossible, and even some hints could be used from similar work already done on Gramineae.[10]

III. MOLECULAR BIOLOGY OF AMARANTH

Molecular biology techniques have been the most powerful tool for the transfer of specific and new genetic information into living organisms. Currently, the procedures of plant molecular biology are already established for many species and are used for research and applied purposes. The abundance and variety of commercial kits for RNA and DNA isolation, DNA cloning, cDNA and genomic libraries construction, directed mutagenesis, nonradioactive hybridization, and even a kit for plant transformation using the reporter gene β-glucuronidase (GUS) and which includes all necessary plasmids and other materials for gene detection and genetic expression analysis, talks about the availability of these methods in addition to the profitable aspects of their commercialization.

Through biotechnology and molecular biology, plants, as well as other living organisms, can be used either as receptors of genes introduced by genetic transformation (which will give those plants some specific traits or advantages over their wild variants) or as a source of interesting genes to isolate, characterize, and engineer for their reintroduction into the same plant or into different plants or organisms.

Current techniques for plant genetic transformation range from the initial one based on the natural genetic transformation mediated by the Ti plasmid of *Agrobacterium tumefaciens* to the biolistic or gene gun procedure.

With the *Agrobacterium* system, it is possible to do several variations by using co-integrative or binary vectors.[11] There exist also many different and fairly well-characterized plasmids containing available restriction/cloning sites, the necessary promotor and 3'-end plant sequences for correct gene expression, replication origins for both *Escherichia coli* and *A. tumefaciens*, as well as different reporter and selectable (antibiotic resistance) genes. In this regard, some work involving amaranth is currently carried out at CINVESTAV-Irapuato, México.[12]

Other procedures for genetic transformation use plant protoplasts. Because of the nature of these cells which do not have cell walls, it is possible to achieve transformation through different techniques. These techniques include direct gene transfer using polyethyleneglycol (PEG)-mediated transformation, electroporation (i.e., introduction of DNA vectors by electric shock), and transformation via co-cultivation with *A. tumefaciens.*[13] By using the electroporation procedure, maize cells have been transformed.[14] Recently, another procedure for electroporation-mediated transformation of maize was reported[15] using callus tissue cultures instead of protoplasts, which represents an advantage since those tissues are more manageable and resistant than protoplasts.

In a more recent procedure named "biolistic",[16] high velocity microprojectiles are used to carry DNA past cell walls and membranes. The advantages of this technique are that it has been effective regardless of species or tissue type; it is rapid and simple; it should facilitate the direct transformation of totipotent tissues such as pollen, embryos, meristems, and cell cultures; and it seems also suitable for organelle transformation. Although it requires special equipment, currently it is commercially available.

However, a necessary procedure for the genetic transformation of plants is a good regeneration system from tissue cultures of the target plant. This is required for the appropriate selection of the individuals which not only possess but also express the introduced genetic information. Although there are some procedures where this requisite could be overcome, most plant genetic transformation procedures are based on selection for antibiotic resistance conferred by genes contained in the introduced molecular vector (e.g., plasmid, cosmid, phage).

In the case of amaranth, and according to the previous section, plant regeneration from different tissue cultures is not yet established in an efficient and reproducible way. Even the agroinfection, which allows for a viral infection mediated by *Agrobacterium* and does not require the tissue culture system,[17] is still beyond the current knowledge regarding amaranth processing.

Once this "bottleneck" is solved in the amaranth model, then the genetic transformation could be a reality, mainly because the diverse techniques for DNA introduction into plant cells could be assayed until the right one for amaranth is found.

This situation does not mean that amaranth is less interesting for molecular biology of plants because, independent of the problem of regeneration, it could be used right now as a source of interesting genes. A detailed description of some useful amaranth proteins is already included in the following chapters of this book and, in terms of molecular biology, we can propose some strategies for using this information to isolate the respective genes.

Albumins are among the most important seed proteins of amaranth. There are studies about their localization and some of their biochemical and physicochemical properties. They have been isolated and purified, and their amino acid composition is known. Among the albumin fractions of amaranth, there is a small group of methionine rich proteins (MRPs).[18] These proteins have a high methionine content (16 to 18 mol %) as well as a relatively high lysine content (6.1 to 6.4 mol %). Constituent polypeptides have been isolated, and the N-terminal amino acid sequences (first 25 amino acids) of two MRPs were determined.

Other interesting information has to do with antimicrobial peptides found in seeds of *A. caudatus*, where two of these peptides were identified and the complete sequence of 30 amino acids was also determined. These peptides are homologous to the cysteine/glycine-rich domain of chitin-binding proteins.[19] These amaranth peptides inhibit the growth of fungi to a much higher extent than other antifungal chitin-binding proteins. This property suggests that they may play a role in the defense of seeds against fungal attack.

Finally, there is a proteinase inhibitor peptide isolated from *A. hypochondriacus* which is a trypsin inhibitor named ATI. The sequence of the first 26 N-terminal amino acids was determined.[20] The purified inhibitor showed a typical trypsin-like inhibitor activity against proteases from fungi, storage insects, and germinated amaranth grains.

Considering the state of the art regarding these amaranth proteins, there are two possible strategies to isolate the respective genes. One is a classical approach including the raising of specific antibodies against each one of these purified proteins. Then the antibodies can be used to screen the protein or peptide synthesized in *E. coli* by using expression vectors in a cDNA library. This strategy was successfully used to isolate a gene encoding an albumin with a well-balanced amino acid composition and high level of essential amino acids from *A. hypochondriacus*.[21]

Another currently available strategy is the design and construction of degenerate oligonucleotides based on the flanking and already determined amino acid sequences of amaranth proteins and peptides. These site-directed oligonucleotides can be used to amplify the entire nucleotide sequence in between with the polymerase chain reaction (PCR).[22] PCR is based on the enzymatic amplification of a DNA fragment by using two oligonucleotides as primers which flank the defined target DNA sequence. A typical cycle of the reaction starts with the melting of the target DNA duplex, then continues with the hybridization of the PCR primers to the single-stranded target DNA, and finishes with the polymerization of the desired sequence. Thus, events are carried out in several steps of high and low temperature for the melting, hybridization, and polymerization of the respective molecules. Cycle repetition results in the exponential reproduction of a specific DNA fragment of defined sequence. The use of a thermostable DNA polymerase from the bacterium *Thermus aquaticus* allows for use in the reaction mix without adding new polymerase for each cycle. The goal is to use the amplified DNA fragment as an homologous probe to screen particular DNA sequences and ultimately to identify and isolate the corresponding gene from a DNA library.[23]

Because the N-terminal or the full amino acid sequence of MRPs, antimicrobial peptide, and proteinase inhibitor of amaranth are known, there are good prospects for these strategies. Once the respective genes are isolated, they can be used for transforming other plants to increase the nutritional properties or as a plant defense mechanism. Although the achievement of these objectives is still a bit far away and some other problems should be solved first, in principle it is possible to attain them.

Another interesting trait is the natural herbicide resistance of the weed *A. hybridus*. Some biotypes of this species are resistant to the herbicide atrazine; a single amino-acid mutation in the Q_B protein is responsible for this resistance. The Q_B protein is encoded by the chloroplast *psba* gene. After isolation and cloning of this gene, some modifications were done in order to express it as a nuclear gene.[24] By *Agrobacterium*-mediated transformation, it was possible to get transgenic atrazine-tolerant tobacco plants.

Finally, the characterization of genetic variability of amaranth species regarding specific traits such as early maturity, plant height, resistance to lodging, pest resistance, male sterility, seed size, drought resistance, and others, could, in principle, be addressed at the molecular level by using the restriction fragment length polymorphisms (RFLPs) and/or the random amplified polymorphic DNA (RAPDs) methods.[25]

With RFLPs, differences in length of DNA fragments and number of sites for different restriction enzymes are determined. In this technique, it is necessary to isolate genomic DNA from leaves; after digestion of samples of the DNA with different restriction enzymes, the resulting fragments are separated electrophoretically. The separated fragments are then denatured and blotted onto a nylon or nitrocellulose membrane. Next, radioactive or nonradioactively labeled single-strand DNA probes are hybridized to the denatured genomic DNA fragments. After washing and autoradiography or antibody detection of hybrids, the result is a specific pattern of discrete bands for each individual DNA. The development of RFLP molecular markers on amaranth chromosomes that are linked to traits of interest is currently in progress at Utah State University.[3]

For the RAPDs technique, the genomic DNA is first used as a substrate in PCR to amplify one or several sequences. Then the resulting fragments are separated by electrophoresis and the amplified sequences are observed directly on the gel after nucleic acid staining. The distances between the PCR primers may be specific for each DNA assayed, giving a particular pattern of bands which reflect a genetic polymorphism analogous to RFLP. PCR primers could be either targeted to known sequences or amplifying unknown sequences.

IV. PERSPECTIVES

The previous sections deal with the most direct approaches and examples of biotechnology in relation with amaranth, but there is still a lot more to do in this field. It is clear that more extensive research should be carried out regarding the regeneration of amaranth plants from tissue cultures.

Perhaps a similar work as in maize should be conducted in terms of identifying the right callus type of amaranth suitable for transformation and regeneration processes. In the maize model, the attainment of highly embryogenic, pale, friable, and fast growing callus tissues (named as type II) was considered as a requisite for transformation. This type of callus is genotype dependent and obtained at low frequency. In contrast, callus type I — which is embryogenic, compact, nodular, organized, readily obtained at high frequency, and in a wide variety of maize genotypes — was not successfully transformed until recently[15] by electroporation.

Other factors which may have some influence for developing the biotechnology of amaranth, besides the suitability of the tissue type, are the presence of antibiotic substances. This influence could be due to the concentration effect of a particular substance or to the antibiotic effect itself. In relation to these effects, it is known that oxalates can reach levels up to 7% of the dry weight in leaves of vegetable amaranth, and a close correlation between nitrate and oxalate levels in amaranth seedlings and cell suspensions has been described.[26] Up until now, the effect of these concentrations of oxalate on *Agrobacterium* strains has not been studied. Similarly, although the antimicrobial peptides of amaranth seem to have an effect only on fungi, the possibility of some deletereous effect on *Agrobacterium* by this or by other similar peptides cannot be discarded. Being a dicotyledonous plant, amaranth is potentially transformable with *A. tumefaciens*. However, the presence of these factors in amaranth tissue cultures could be negative for the survival of the bacteria and hence for the transformation process. In that case, the use of other transformation procedures should be considered.

On the other hand, currently, amaranth can be used as an excellent source of useful genes not only through the procedures already described[21] but also by using the PCR approach for amaranth peptides and other proteins of known sequence.

Seed storage proteins include mainly prolamins and glutelins in cereals. Prolamins are soluble in aqueous ethanol, and their extraction can be improved by adding reducing agents while glutelins are the most insoluble protein fractions from the seed meal. In amaranth, the amount of prolamins is low, but the glutelins are the second most important fraction. Both kinds of proteins have been isolated from amaranth, and the amino acid compositions have been determined. A full description and results on these proteins will be presented in Chapter 5 of this volume.

Many cDNAs encoding cereal prolamins and glutelins have been cloned and sequenced, showing that the particular structures of these proteins contain unique and repetitive amino acid domains.[27] Consensus motifs of the repeats have been described for sulfur-rich and sulfur-poor high molecular weight (HMW) prolamins, zeins (maize prolamins), and others. With this information, degenerate oligonucleotides designed according to these sequences can be used to amplify by PCR a similar DNA fragment contained in amaranth DNA, thus allowing for the preparation of homologous probes. Finally, these probes could be used to screen for the respective gene in genomic and/or cDNA amaranth libraries.

All current and upcoming information, together with the ever expanding possibilities that molecular biology offers, could make the future and potential of amaranth impressive. Hopefully, researchers will give the biotechnology of the ancient and "everlasting" crop amaranth the opportunity to emerge as an important science.

REFERENCES

1. **Kauffman, C.,** Realizing the potential of grain amaranth, *Food Rev. Int.,* 8(1), 5, 1992.
2. **Bale, J.R. and Kauffman, C.S.,** Editor's overview, *Food Rev. Int.,* 8(1), 1, 1992.
3. **Komen, J.,** Grains for the tropical regions, Biotechnology and Development Monitor, publication of the Ministry of Foreign Affairs and the University of Amsterdam, The Netherlands, No. 10, March 1992, 3.
4. **Kauffman, C.,** The status of grain amaranth for the 1990s, *Food Rev. Int.,* 8(1), 165, 1992.
5. **Vasil, V. and Vasil, I.K.,** Induction and maintenance of embryogenic callus cultures of Gramineae, in *Cell Culture and Somatic Cell Genetics of Plants,* Vol. 1, Vasil, Indra K., Ed., Academic Press, New York, 1984, 36.
6. **Flores, H.E. and Galston, A.W.,** *In vitro* culture of grain and vegetable amaranths (*Amaranthus* spp.), *Am. J. Bot.* 69, 1049, 1982.
7. **Bragga, S., Venkateswarlu, K., and Sopory, S.K.,** *In vitro* regeneration of plants from hypocotyl segments of *Amaranthus paniculatus, Plant Cell Reports,* 6, 183, 1987.
8. **Bennici, A., Schiff, S., and Bovelli, R.,** *In vitro* culture of species and varieties of four *Amaranthus* L. species, *Euphytica* 62, 181, 1992.
9. **Arya, I.D., Chakravarty, T.N., and Sopory, S.K.,** Development of secondary inflorescences and *in vitro* plantlets from inflorescence cultures of *Amaranthus paniculatus, Plant Cell Reports,* 12, 286, 1993.
10. **Hibberd, K.A.,** Induction, selection and characterization of mutants in maize cell cultures, in *Cell Culture and Somatic Cell Genetics of Plants,* Vol. I, Vasil, Indra K., Ed., Academic Press, New York, 1984, 571.
11. **Herrera-Estrella, L. and Simpson, J.,** Foreign gene expression in plants, in: *Plant Molecular Biology, A Practical Approach,* Shaw, C.H., Ed., IRL Press, Oxford, 1988, 131.
12. **Jofre-Garfias, A., Cabrera-Ponce, J.L., Simpson, J., and Herrera- Estrella, L.,** Genetic transformation of amaranth (*Amaranthus hypochondriacus*) mediated by *Agrobacterium tumefaciens,* in preparation.
13. **Shillito, R.D. and Saul, M.W.,** Protoplast isolation and transformation, in *Plant Molecular Biology, A Practical Approach,* Shaw, C.H., Ed., IRL Press, Oxford, 1988, 161.
14. **Fromm, M.E., Taylor, L.P., and Walbot, V.,** Stable transformation of maize after gene transfer by electroporation, *Nature,* 319, 791, 1986.
15. **D'Halluin, K., Bonne, E., Bossut, M., De Beuckeleer, M., and Leemans, J.,** Transgenic maize plants by tissue electroporation, *Plant Cell,* 4, 1495, 1992.
16. **Sanford, J.C.,** Biolistic plant transformation, *Physiol. Plantarum,* 79, 206, 1990.
17. **Grimsley, N.H., Ramos, C., Hein, T., and Hohn, B.,** Meristematic tissues of maize plants are most susceptible to agroinfection with maize streak virus, *Biotechnology,* 6, 185, 1988.
18. **Segura-Nieto, M., Rodríguez-Nester, C., and Olguín-Martínez, L.,** Identification of methionine rich proteins from *Amaranthus hypochondriacus* seed, presented at the *Int. Symp. Biotechnol. Crop Improvement in Latin America* (CQI 10), November 1 to 7, 1992.

19. **Broekaert, W.F., Mariën, W., Terras, F.R.G., De Bolle, M.F.C., Proost, P., Van Damme, J., Dillen, C.M., Rees, S., Vanderleyden, J., and Cammue, B.P.A.,** Antimicrobial peptides from *Amaranthus caudatus* seeds with sequence homology to the cysteine/glycine-rich domain of chitin-binding proteins, *J. Biochem.*, 31, 4308, 1992.
20. **Valdés, S., Chagolla, A., Martínez, N., Mendiola, E., Segura, M., and Blanco, A.,** Study of a protease inhibitor from amaranth (PO-17T), presented at the *VII PAABS Congress* and *XIX Congr. Mexican Biochem. Soc.*, Ixtapa, Gro., México, September 27 to October 2, 1992.
21. **Raina, A. and Datta, A.,** Molecular cloning of a gene encoding a seed-specific protein with nutritionally balanced amino acid composition from *Amaranthus*, *Proc. Natl. Acad. Sci. U.S.A.*, 89, 11774, 1992.
22. **Saiki, R.K., Gelfand, H.D., Stoffel, S., Scharf, S.J., Higuchi, R., Horn, G.T., Mullis, K.B., and Erlich, H.A.,** Primer-directed enzymatic amplification of DNA with a thermostable DNA polymerase, *Science,* 239, 487, 1988.
23. **Dehmer, K.J., Graner, A., and Wenzel, G.,** Screening for defined DNA sequences in minimal amounts of barley tissue by PCR, *Plant Breed.*, 107, 70, 1991.
24. **Cheung, A.Y., Bogorad, L., Montagu, M.V., and Schell, J.,** Relocating a gene for herbicide tolerance: a chloroplast gene is converted into a nuclear gene, *Proc. Natl. Acad. Sci. U.S.A.*, 85, 391, 1988.
25. **Pérez de la Vega, M.,** Biochemical characterization of populations, in: *Plant Breeding: Principles and Prospects,* Hayward, M.D., Bosemark, N.O., and Romagosa, I., Eds., Chapman & Hall, London, 1993, 184.
26. **Teutonico, R.A. and Knorr, D.,** Non-destructive method for determination of water-soluble oxalate in cultured *Amaranth tricolor* cells, *J. Agric. Food Chem.*, 33, 60, 1985.
27. **Utsumi, S.,** Plant food protein engineering, *Adv. Food Nutr. Res.*, 36, 89, 1992.

Chapter 2

Genetic Characterization of Grain Amaranth

Peter A. Kulakow and Holly Hauptli

CONTENTS

I. INTRODUCTION

The world has been experiencing a resurgence of interest in grain amaranth as a "new" crop in the last few decades. However, amaranth is an ancient crop that was already under cultivation 5000 to 7000 years ago.[1] We have inherited a richly varied germplasm and cultural traditions that rival many major food crops, while relatively little is known about the genetic structure of the grain amaranths. What is meant by genetic structure may be summarized as the amount and kinds of genetic variation, its' geographic distribution, and natural (or selected) reassortment. Ancient amaranth cultivation and harvest traditions continue as new efforts in breeding, plant physiology, culture, and application begin around the world.

The amaranth germplasm grown in Mexico, Central and South America, Asia, and Africa probably represents a remnant of the crop gene pool under more extensive cultivation several hundred years ago. The future success of genetic improvement for grain amaranth will reflect the value and respect we give to the past and present farmers responsible for propagating grain amaranth. Rapid selection of few improved varieties and replacement of old cultivars could lead to short term successes, but it could also lead to a dangerously narrow genetic base under cultivation. The long term prospects for grain amaranth breeding will depend on the availability of a diverse germplasm to respond to changing needs. An initial imperative for grain amaranth breeding must be the preservation and study of the remaining genetic diversity in the crop. We felt such a study would begin with a description of the existing genetic variation, how it is distributed globally, and how it may have been modified through selection.

Some important features to consider in breeding grain amaranth and preserving our rich heritage are reviewed here. Most of the ideas discussed here will present more questions than answers for amaranth breeders. As international efforts with grain amaranth grow, we will begin to acquire an understanding of how best to breed and conserve grain amaranth genetic resources.

0-8493-5374-2/94/$0.00+$.50

II. GENETIC VARIABILITY

We must consider the total pool of genetic variation available to breed grain amaranth to get a clear picture of the potential for genetic improvement, especially since amaranth has a very limited history of recent crop improvement. There are at least five levels of the grain amaranth gene pool corresponding to more and more highly selected genetic materials. Between each level there are cases of recent or ancient hybridization that blur the boundaries between groups. Beginning with the least highly selected genetic materials, there is the diverse group of wild and weedy relatives of grain amaranths with potential for transfer of genes through hybridization. The second level is the group of three cultivated grain species. The third level corresponds to races or grain types within each cultivated species. The fourth level includes individual landrace populations. The final level corresponds to genetic stocks, breeding lines, and cultivars that have been selected in recent years as part of amaranth breeding programs. Each of these groups will be discussed briefly.

A. WILD AND WEEDY RELATIVES

Although grain amaranth evolved from its wild and weedy progenitors thousands of years ago, many weedy amaranths have retained the ability to hybridize with the cultivated grain amaranths. These hybrids have been used to study sex expression in dioecious and monoecious species of amaranth.[2] Ancient hybridization between cultivated grain amaranth species and weed species has been hypothesized in the evolution of unique "hybrid" populations in the Sacramento-San Joaquin Delta of California.[3] Introgressive hybridization between cultivated and weedy amaranths may have played a role in the evolution of the grain species.[4-5] Most of the hybrids between wild or weedy species and the cultivated amaranths result in high levels of sterility. However, there is evidence that genes from *Amaranthus retroflexus* can be readily transferred to *A. cruentus* through backcrossing.[6] Apparent recent hybrids between *A. retroflexus* and *A. caudatus* have been seen in Peru,[7] indicating there is current interaction between the weedy and domesticated species groups. The wild and weedy *Amaranthus* species capable of hybridizing with the grain amaranths together would comprise the tertiary gene pool of grain amaranth.[8]

Many wild and weedy *Amaranthus* species possess unique adaptations attractive to breeders such as salt or drought tolerance. Weedy species have other desirable characteristics such as early flowering. Kulakow and Jain used backcrossing to introduce a gene governing early flowering from *A. retroflexus* to *A. cruentus* through four backcross generations.[6] Indehiscent utricles have been found in rare populations of weedy and crop-weed hybrid populations which may aid breeders in alleviating shattering problems.[9] At least one weedy species (*A. pumila*) has been observed that bears seed approximately three times the weight of average domesticated amaranth grain.[10] This species may represent a good source of genes for increased seed size. Like other crops, the grain amaranths have been cultivated for more than their primary "intended" purpose. For instance, the nonbranching, sometimes white-seeded grain amaranth species are grown in many parts of Africa for use as a potherb rather than grain crop.[11] Carlsson has observed good yields of liquid protein concentrate and suggests it be grown for that application as well.[12] By a similar token, wild *A. australis* may be a good source of genes for overall biomass production, as it is reported to grow 9 m tall.[10]

In many fields of cultivated amaranth, naturally occurring crop/weed hybrids are common. The weedy amaranth species present a classical dichotomy between the possibility of unintentional introduction of weedy genes into crop varieties on the one hand and a treasure trove of desirable traits to choose from on the other. This situation is not uncommon among the major crops. One of many crop examples is that of the sunflower situation in California, a major region for the breeding of new varieties and production of hybrid sunflower seed. Although very weedy forms of *Helianthus annuus* can be a real weed problem for the same crossability reasons in sunflower seed fields, they have also been used to donate many useful characteristics and improve breeding populations.[13]

B. CULTIVATED GRAIN SPECIES

The cultivated grain amaranths differ from their wild and weedy relatives by having pale, rather than black, seed coats and are superior for many cooking uses. While most cultivated populations have pale seeds, some populations have black seeds, especially those used for ornamental or vegetable purposes. Some polymorphisms for seed coat color may be derived from previous hybridization events with weedy species.

The grain amaranths have unisexual flowers borne in glomerules (branched flower clusters) described as dicasial cymes which collectively form the characteristic large flowering panicle. One glomerule has varying numbers of male and female flowers. The cultivated species are distinguished by characteristics of the female flowers according to the taxonomic key developed by Sauer.[1,14] The cultivated species also tend to differ by morphological features of the large terminal panicle that is characteristic of all the cultivated grain amaranths. As morphology is highly plastic to changes in growing conditions and landrace populations within a species vary in conformation of the panicle, morphological features of the panicles are less reliable species indicators than is female flower morphology.

The cultivated species originated in the New World although they are presently distributed through much of Asia and Africa. Currently, grain amaranth is most widely cultivated in the Himalayan foothills of Asia. Sauer developed two hypotheses concerning the origin of grain amaranth.[1] In one scenario, the three cultivated species were independently domesticated from weedy species in three regions: *A. cruentus* from *A. hybridus* in Central America, *A. hypochondriacus* from *A. powellii* in Mexico, and *A. caudatus* from *A. quitensis* in South America. The alternative hypothesis proposed a single origin of the pale seeded amaranths where *A. cruentus* arose from *A. hybridus* in Central America. Movement of the early domesticate north into Mexico and south into South America, followed by subsequent hybridization with local weed populations, lead to the selection of *A. hypochondriacus* and *A. caudatus* . Sauer favored the single origin hypothesis partly because of the apparent rarity of pale seeded mutants among black seeded amaranths, despite extensive contact with humans. Three types of data have addressed the origin question in grain amaranth since Sauer's investigations: patterns of genetic diversity in germplasm collections of the cultivated and related weedy species, hybridization studies, and inheritance patterns of seed coat color. Hauptli and Jain studied geographic patterns of allozyme diversity in cultivated and weedy amaranths.[5] They found the cultivated amaranth species to be more closely related to each other than to their hypothesized weedy progenitor species. With the exception of the *A. caudatus-A. quitensis* pair, this supported a single origin hypothesis. The relationship of *A. quitensis* to the other cultivated and weedy species has been questioned. *A. quitensis* populations likely represent *A. caudatus* that has escaped from cultivation or a semidomesticated variety of *A. hybridus*.[4] Some evidence of recent introgression and/or segregation of crop-weed hybrids between *A. caudatus* and *A. retroflexus* was seen in the form of rare individuals in crop populations with crop allozyme alleles, except for a single homozygous weedy allele at one locus.[15] More extensive surveys of variation in weedy populations in the likely regions of origin are needed to clarify their evolutionary, as well as more recent, relationship to related crop species.

Hybridization results from crossing landrace populations of the three cultivated species have given variable results on the presence of hybrid barriers within and between grain amaranth species. This is further complicated by limited reports of variable chromosome numbers between species with $2n = 32$ for *A. caudatus* and *A. hypochondriacus* and $2n = 34$ for *A. cruentus*.[16] Successful crosses with at least partial fertility have been made between each pair of the cultivated species except *A. cruentus* and *A. caudatus*; however, numerous cases of cross incompatibility have been observed.[17-19] Gupta and Gudu made interspecific hybridizations among the major grain species to reveal genetic relationships in evolutionary study of the group.[19] They, too, found reciprocal crosses of *A. caudatus* and *A. cruentus* incompatible as hybrids. Although hybrid seeds were produced, all F_1 plants died at the seedling stage. The most striking cases of hybrid weakness concern abnormal F_1 hybrids with virus-like symptoms derived from crosses between particular populations of *A. caudatus* and *A. hypochondriacus*[20-22] and certain interpopulation crosses within *A. caudatus*.[4] In our work, while abnormal *A. hypochondriacus* × *A. caudatus* F_1 plants were highly sterile, it was possible to obtain backcrosses to the parental lines. Segregation ratios indicated two genes determine this type of hybrid weakness.[22]

Hybridization studies have not conclusively supported one theory of the origin of grain amaranth. Despite varying crossability of different populations, both within *A. caudatus* and between different cultivated grain species, it is possible to transfer genes among all three cultivated species. Not only may genome size differences affect cross compatability, but differences in the environments in which hybridizations are made and grown may also play a significant role in determining observed cross-compatiblity.[19] Further surveys of hybrid barriers using more complete samples of amaranth landraces and populations of the likely progenitor species grown in diverse environments are needed to better address the question of species relations and evolution.

The dark vs. pale seed coat difference between wild and cultivated amaranths is determined by at least two genes. Within *A. hypochondriacus* and *A. caudatus*, black seed coat color was governed by two genes.[21] In a cross between *A. cruentus* and *A. retroflexus*, two genes also determined dark vs. pale seed coat color although *A. retroflexus* has not been directly implicated in the origin of grain amaranth.[22] Multiple gene determination of seed coat color would lend support to a single origin of pale seeds in the cultivated amaranths since the probability of recovering pale seeded mutants would decrease as the number of genes determining seed coat color increased. Further crosses and allelism tests using several populations of each cultivated species and the related weedy species would provide useful insight into the evolution of seed coat color in amaranth.

Interspecific crosses can be useful for increasing the genetic variance especially for characteristics not readily available within a species. There are at least two reasons to be cautious about relying too heavily on such wide crosses. First, considerable progress may be achieved by hybridization and selection among landraces within each species. Second, wide crosses may introduce hybrid sterility problems. Some of this may be due to the aforementioned cytogenetic differentiation between the species, with 2n = 32 for *A. hypochondriacus* and *A. cauatus* while *A. cruentus* is reported to bear 2n = 34. While hybrids can be made between certain populations of each species, substantial hybrid sterility is observed in many crosses. Residual hybrid sterility could impose new problems that may be avoided by utilizing parents within species when possible.

Even though considerable genetic erosion is presumed to have occurred during the centuries when amaranth cultivation fell into disfavor in the New World centers of their origin, genetic variation within the domesticated grain amaranths is still vast. As an indication of how much variation remains, significant responses were achieved by applying mass selection within landrace populations for high and low outcrossing rates,[23] time of flowering, plant height, seed yield per plant, harvest index, and yield to height ratio.[24] Good responses to selection were also observed in an interpopulation cross in *A. cruentus* for days to first anthesis and leaf length.[25] Realized heritabilities calculated from these studies varied from lower values for seed yield (0.09) and leaf length (0.08 to 0.19) to higher values for plant height (0.22 to 0.49) and days to first anthesis (0.35 to 0.66). Two cycles of divergent selection for days to first antheisis resulted in early and late selection lines differing by 20.5 days to first flowering. This illustrated that rapid and large selection gains are possible.[26] Of course, these values are dependent upon the particular populations and environments sampled in the studies, but the magnitude of such responses and significant heritabilities indicate a source of quantitative genetic variation available for selection.

Discoveries are being made continuously about amaranth's physiological attributes and genetic variation for them. For instance, 10 years ago, it was known that protein levels in amaranth grain were generally higher at 15 to 18% than protein levels of graminaceous grain crops.[27-28] It was also known that the amino acid complement of that protein was complementary to proteins from cereals, as well as from legumes, as amaranth's limiting amino acid was not lysine, tryptophan, methionine, nor cysteine.[28] Quite recently, however, more detailed work has progressed on components of amaranth seed to uncover additional nutritional properties not readily obvious before now. Some of these newfound characteristics are unique to amaranth; others match similar characteristics in other crops.

Okuno and Sakaguchi[29] reported the existence of a simply inherited trait in populations of *A. hypochondriacus* resulting in exclusive amylopectin production and virtually no amylose production in amaranth seed starch. This glutinous perisperm was shown to be simply inherited[30] and the gene named "glutinous" (Gl, gl). The enzymatic mode of gene action of gl in amaranth is directly analogous to the corresponding "waxy" genes in sorghum, rice, barley, and maize.[30] This is interesting partly because it is the first report of the occurrence of such a gene in a dicotyledonous plant. Also, the amaranth perisperm is diploid tissue, whereas the starchy endosperm in monocots that display this trait are triploid in nature.

High levels of a rare form of vitamin E (alpha, gamma, and delta tocotrienols, or T3s) have been found in grain amaranth seed oil. These compounds were first isolated from cereals, notably high protein barley,[31] and have been shown to reduce overall cholesterol production and the production of LDL cholesterol in diets of rats and pullets.[32] More recent reports show tocotrienol levels in amaranth are generally high and reduce cholesterol biosynthesis in chickens.[32] Some preliminary information shows gross genetic variation as differential expression of these three forms of T3 in different landraces.[33]

The squalene present in high levels in amaranth grain has also been implicated as hypocholesterimic by inhibiting HMG CoA reductase, the key regulatory enzyme for cholesterol biosynthesis.[34] Work in the area of genetic variation for the already high squalene expression in amaranth germplasm is in progress.[35] In addition, it has recently been shown in rat feeding trials that fiber from whole amaranth grain seems to combine the cholesterol-lowering effects of soluble fiber with colon-cancer reduction benefits of insoluble fibers.[36] Considering the relatively high levels of T3s, squalene, and unusual fiber, amaranth may complement the cereals not only in amino acid score, but also provide complementary vitamin E isomers and cancer preventative fiber as well.

Amaranth seed proteins are now being described biochemically in more refined ways than before.[37] Most amaranth protein is composed of a globulin fraction that shares approximate weight, size and apparent multimeric dissociation characteristics with both of the two major legume storage proteins.[38] It is likely that amino acid and/or DNA sequence of these proteins will also be elucidated. When this information is available, a bit more light may be shed on the evolutionary time of origin for these legume proteins shared by a handful of other domesticated and undomesticated taxa.[39] Certainly more will be known about evolutionary relationships within the genus. Perhaps one day the functional utility of different amaranth globulin genotypes for use in particular food or industrial applications will be as well known as the relationships in wheat between particular high molecular weight glutenin genotypes and bread baking quality in bread wheat,[40] or low molecular weight glutenin genotypes and pasta quality in durum wheat.[41]

Gene-cytoplasmic male sterility is used extensively with some crops, notably maize, for the production of hybrid planting seed.[42] It has been observed in amaranth as well and should be investigated further to ascertain whether or not the levels of heterosis found in amaranth are great enough to warrant the expense that would be required to establish a hybrid breeding program. In certain cases, amaranth inbred lines and hybrids have shown significant inbreeding depression or heterosis, indicating nonadditive variation is important in determining seed yield and other traits in amaranth. In comparing F_1 and F_2 generations from several interpopulation crosses between two *A. cruentus* populations, mean inbreeding depression was -5.6% for days to first anthesis, 4.8% for leaf length, 9.0% for plant height at maturity, and 24.4% for panicle weight.[26] In a six parent diallel in *A. hypochondriacus,* nonadditive genetic variation was an important factor for days to flowering, plant height, pollen fertility, seed size, and grain weight per plant.[43] These observations suggest the usefulness of developing breeding methods and varieties that take advantage of hybrid vigor as would synthetic varieties. Interspecific hybrids between *A. hypochondriacus* and *A. caudatus* and between *A. cruentus* and *A. retroflexus* showed tremendous hybrid vigor for plant height and vegetative vigor. The vegetative vigor of interspecific hybrids in amaranth could have potential use in forage production systems, with some hybrids yielding over 150% more than their high parent.[44] While the importance of heterosis in grain amaranth needs further investigation, the observation of hybrid vigor suggests that nonadditive genetic variation may be important for selection progress in population improvement, as well as for the performance of varieties.

C. RACES OR GRAIN TYPES

Many morphological features of the cultivated grain amaranths overlap between species, but there are some emerging multiple character patterns that are useful for defining races within each species based on geographic origin and morphology. Workers at the Rodale Research Center made the first attempt to group germplasm accessions based on common morphological features.[45] Description of the grain types was considered preliminary and based on collective experience in observing a large number of accessions over many growing seasons. Grain types have been used to group germplasm collections for early agronomic studies and breeding efforts. Reliably described amaranth races will have great value for genetic conservation and breeding and perhaps be useful for eventually designating core collections.[46]

Three groups of characters have been used to classify grain types. A number of other agronomic characters and genetic markers tend to be characteristics of certain grain types. The first group of characters includes flowering time, plant height, and time of maturity. These traits often go together in defining a grain type due to likely physiological and genetic correlations. The environmental responses that induce flowering in landrace populations may commonly represent adaptations to different day-lengths and climates that show geographic patterns. The second character group is vegetative branching pattern. The third group includes panicle characteristics such as the number of female flowers per glomerule. In a small sample of Mexican, Guatemalan, and African grain types of *A. cruentus*, Mexican grain types had the highest average numbers of female flowers per glomerule and African grain types had the lowest numbers.[21]

Within each cultivated species based on Sauer's taxonomic key,[14] there are several grain types as defined by Kauffman and Reider.[45] For example, within *A. cruentus*, collections Kauffman and Reider described as African, Mexican, and Guatemalan grain types have certain morphological and phenological traits in common. Some populations do not fit any of the described races. These populations either represent new races or they exhibit variability due to past hybridization between races. Over the years, the Rodale group's racial classification scheme has proven to be useful. A review of the 15 designated morphological groups is given by Kauffman.[47] Although necessarily limited because all of the material was grown for evaluation in a single location, this classification scheme demarcates landrace phenotypes quite well. Landraces differentiated by this classification scheme have been used successfully for crop evalution in Thailand[48,49] Mexico,[50] and Kenya.[51]

Additional races are likely to be discovered as new germplasm is evaluated in different environments (i.e., *A. caudatus* germplasm from South America) or as multivariate techniques are used more often to describe germplasm collections.[52] Racial groups may represent complexes with specific adaptations. Races may also show repeatable heterotic patterns in interracial hybrids. Races will help define incipient hybrid barriers that have been described in many hybridization studies.[5,18] Future efforts to describe races of grain amaranth based on morphological and biochemical characteristics would be very useful for guiding selection of genetic materials in a breeding program.

D. LANDRACE POPULATIONS

Landraces of grain amaranth can be defined as distinct populations of amaranth cultivated in different localities under different environmental and cultural conditions throughout the range of amaranth production. Collections of landrace populations comprise most of the germplasm available for introducing amaranth to new localities. Since the late 1970s, extensive germplasm collections have been made in many regions of amaranth cultivation including Bolivia, Ecuador, Peru, Guatemala, Mexico, India, and Nepal.[53-59] Descriptions of the genetic structure of these landraces is useful for determining how best to conserve amaranth germplasm and for selecting plant materials to initiate breeding programs. Such descriptions include factors such as the breeding system, patterns of genetic variation within and among families, and inheritance patterns for various qualitative and quantitative traits.

Hauptli and Jain studied the genetic structure of six landrace populations from the New World by examining variation for allozyme markers, morphological markers, and quantitative traits.[60] There was significant genetic variation both within and among landrace populations for numerous traits. Using this data and a survey of zymograms from more Central and South American landrace populations, Hauptli concluded that grain amaranth germplasm from New World centers of origin is genetically structured in the form of many distinct and polymorphic populations composed of largely homozygous individuals.[15] Adjacent populations are often fixed for different enzyme alleles at the same locus or exhibit very different allele frequencies at polymorphic loci. A breakdown of genetic diversity in the crop species shows about half of the total diversity is due to between-species differences, with the remainder equally partitioned to between population and within population components.

The observed low heterozygosity implies predominant inbreeding. Increased apparent inbreeding may be the result of selection under human cultivation. Since one plant may produce hundreds of thousands of seeds, the strict selection of a single plant to reseed the next season's crop presents an extremely narrow genetic bottleneck. Current New World farming practices virtually always include interplanting, often with maize, which might reduce cross-pollination as well.[15] For these reasons, these landrace populations may have been selected for low outcrossing. The presence of some largely outcrossed individuals may be due to heterotic advantage, as discussed here earlier. These populations may have originated under divergent selection pressures for high seed set and heterosis vs. domestication under small effective population size often leading toward inbreeding.

Genetic structure of Old World populations of the crop are similar. Vaidya[52] studied ten populations of *A. hypochondriacus* from two geographic regions of India. Although these populations showed no variation for allozyme markers,[61] they did show relatively high levels of heterozygosity for marker genes compared to the New World amaranths. There was significant genetic variance both among populations and among families within populations for all nine metric traits studied by Vaidya.[52] This situation of plants from New or Old World amaranths exhibiting largely homozygous biochemical genotypes but bearing apparently heterozygous, variable genotypes for marker and metric traits[23,52] is anomalous. More precise biochemical analysis of genetic variation (i.e., restriction fragment length polymorphism or RFLP survey techniques) may show better correspondence between molecular and morphological data.

Study of the genetic structure of landraces of grain amaranth is perhaps the most fruitful area for basic genetic research to further amaranth breeding. The extent to which landraces within each species have diverged can give clues of their value in breeding. Genetic diversity of some populations may represent previous interspecific hybridization between the grain species or with related weedy species. While extensive germplasm collections have been made, the genetic structure of these collections has only begun to be explored.

E. GENETIC STOCKS, BREEDING LINES, AND CULTIVARS

Grain amaranth is an ideal plant for identifying and studyng major genes governing plant development. Large quantities of seed can be produced by self- or cross-pollination, large progenies can be easily raised under greenhouse and field environments. Numerous differences distinguish individuals for traits from allozymes and pigments to flowering time and plant architecture. During a 5-year period, we identified over 20 major genes and discovered variation for other traits with likely simple inheritance.[21,22] A concerted effort to study qualitative and quantitative genetic variation in amaranth will likely lead to insights into the genetic changes involved in domestication and evolution of landraces over a range of environments. These studies would give amaranth breeders useful genetic stocks and techniques to rapidly achieve a variety of breeding objectives. Products of biochemical and genetic engineering approaches will be integrated into amaranth breeding through the generation of new genetic stocks.[62] Grain amaranth breeding shows tremendous promise for a successful long-term selection effort combining the use of major genes with population improvement methods. We will briefly review what is known about some of the simply inherited traits in amaranth.

1. Pigmentation

The amaranths show a diversity of betacyanin pigmentation patterns; six genes have been identified that control differences in the presence vs. absence of red or orange pigments (2 genes), the timing of pigment production (2 genes), pigment distribution on different plant parts (1 gene), and the intensity of pigment production (1 gene).[21] Numerous other pigmentation differences have not been studied. Pigmentation differences have many practical and esthetic uses. They serve as useful genetic markers for variety identification, genome mapping, estimation of natural outcrossing, and control of hybridization in breeding.

2. Leaf Markings

Amaranth landraces and related weedy species differ in numerous leaf marking patterns that include oval spots (1 gene) and "V"-shaped marks (2 genes).[21,63,64] These markings are particularly prevalent in certain landraces of *A. hypochondriacus*.

3. Seed Traits

As discussed earlier, the grain amaranths are distinguished from wild and weedy relatives by having pale rather than black seeds. Within certain landraces and hybrids there are numerous seed coat colors including black; dark burgundy; dark, medium, and light brown; yellow; and various white or cream colors. Polymorphisms for seed coat color within particular landraces have shown segregation for one or two genes[21,64] with indications of multiple alleles or additional loci.[21] Some amaranth landraces, particularly within *A. caudatus*, have seeds with pink embryos visible through the seed coat. In these seeds, the betacyanin pigment is produced during embryo development rather than during seed germination. This early pigment production is determined by one gene although its intensity is modified by other genes.[64] This is the first trait known in amaranth where the seedling genotype can be determined on the mother plant. This could be useful for detecting hybridization in some populations. Glutinous vs. nonglutinous perisperm is determined by one gene with glutinous seeds having an opaque phenotype and nonglutinous seeds having a translucent phenotype. The allele determining glutinous perisperm is recessive.[29] A final seed trait in amaranth is shatter resistance. In most cultivated landraces, seed shattering occurs when an abscission layer develops in the utricle at seed maturity. Brenner and Hauptli reviewed variation for shatter resistance in amaranth[9]. A single recessive gene determines the absence of this abscission layer in one *A. hypochondriacus* landrace.[21]

4. Panicle Structure

Grain amaranth species and landraces differ in the structure of the large terminal panicle. Within *A. caudatus*, two of these major differences are controlled by a few genes. The determinant inflorescence trait characteristic of *A. caudatus* var. *edulis* is governed by a single recessive gene where the allele for the common indeterminant panicle type is dominant over the determinant allele.[64] In near isogenic (F_6) lines segregating for this trait, seed from determinant plants was larger than seed from indeterminant plants, indicating that use of determinant varieties may increase seed size. The second panicle trait, drooping panicle orientation of many *A. caudatus* landraces, is determined by two major genes with other minor genes modifying panicle orientation.[64] Drooping panicles were significantly longer, with more lateral inflorescence branches compared to erect panicles in the same segregating progenies. While drooping panicles are not suited for mechanical harvest, they could result in higher seed yield per plant in some populations. Drooping panicles may have utility in more labor intensive cropping systems.

5. Dwarfing

Several germplasm evaluation programs have identified variation for short stature.[53,63] While many short stature lines in grain amaranth achieve a dwarf habit by early flowering, other sources are dwarfed through reduced internode length with a similar leaf number compared to tall plants. A reduced internode type dwarf in *A. caudatus* was determined by a single recessive gene.[64]

6. Flowering

Time of flowering and time of plant maturity are important adaptive traits in grain amaranth. In temperate growing regions, there has been a need to select earlier flowering varieties from populations that evolved at lower latitudes and have short day-length requirements which delay flowering at higher latitudes. Vaidya and Hauptli found almost continuous variation in photoperiod response within several populations.[52,60] Collections of *A. retroflexus*, a weedy species, showed very early flowering under both short day-length environments (8 hours) and long day-length environments (16 hours). Crop-weed hybrids and backcrosses between *A. retroflexus* and an accession of *A. cruentus* from Africa indicated that a single gene governs the very early flowering trait in *A. retroflexus* with dominance of the allele determining earliness.[6] As many as three genes plus polygenic modifiers were needed to explain other flowering variability in the cross.[6] Isolation of a single gene determining very early flowering may be useful for developing very short season varieties or for facilitating rapid cycling of crosses in a breeding program with six generations possible in a year. Further genetic studies of day-length responses in grain amaranth germplasm is needed to uncover other major genes involved in determination of flowering.

7. Male Sterility

As is the case in many other of the major world crop species, amaranth male sterility has been found at low frequencies in some populations. Although amaranth plants are monoecious, inflorescences contain both male and female flowers, both of which are quite small, in the range of 2 to 4 mm across. This makes identification of individual male sterile plants in the field laborious, unlike the situation with maize.[65] Male sterility has been identified in four landrace populations of *A. hypochondriacus* from India[66] and one from Nepal.[67] Analysis of segregating progenies within landraces showed monogenic or digenic determination of male sterility. However, numerous crosses within and between landraces of particular Indian landraces supported a gene-cytoplasmic mode of inheritance.[66] These sources of male sterility may be useful for facilitating recombination in population improvement schemes or in the utilization of backcrosses in varietal development. They also may contribute to development of synthetic or hybrid varieties. Some heterotic effect has been seen in the form of increased biomass yields in a diallel cross comparing specific and general combining abilities in amaranth.[44] However, it is not known whether heterosis is applicable to other traits, occurs overall at a high level, or is widespread enough in grain amaranth to warrant the effort required to develop hybrid varieties.

8. Breeding Lines and Cultivars

Recent breeding efforts in several countries have led to selection of improved breeding lines and cultivars for distribution to farmers and plant breeders. These lines will serve as initial standards to compare new selections. The most widely distributed breeding lines have originated from the amaranth breeding program at the Rodale Research Center in Pennsylvania from 1977 to 1990. These lines have been released both as early generation hybrids between diverse landraces and as advanced uniform selections. The most widely grown amaranth lines in the U.S. come from the Rodale program. Named cultivars have been released in several countries: three from Perú, two from the U.S., and one from India.[53,68,69]

III. BREEDING SYSTEM

The cultivated amaranths have a mixed mating system with variable levels of self- and cross-pollination. Both the genotype and the environment influence outcrossing rates in quantitative as well as qualitative ways. Mean outcrossing rates for single landrace populations varied from 5.8 to 28.8% percent for *A. caudatus*, 3.5 to 14.0% for *A. hypochondriacus*, and 31% for *A. cruentus*.[70] Estimates for *A. caudatus* were based on two allozyme loci, and other estimates were based on the red-green seedling marker. Within a population, outcrossing rates of individual plants have varied from complete selfing to complete outcrossing.[23,71] Outcrossing rates in an F_2 generation of an interpopulation cross between two *A. cruentus* accessions were estimated for 3 consecutive years at Davis, CA and resulted in mean outcrossing rates of 21, 24, and 34%.[71] Agong and Ayiecho observed intraspecific outcrossing rates of 10.4 and 10.9%

among populations of *A. hypochondriacus* and *A. cruentus*, respectively.[72] Interspecific hybridization between those two species was much lower at 1.4%.

These wide differences in outcrossing rate (approximately 6 to 34%) are presumed to be largely due to variability in environment and mode of pollination. The large aerial inflorescences of amaranth make it seem logical to assume that wind is the main pollen vector when cross-pollination occurs. However, many insects (mainly of the order *Hymenoptera*) have been observed foraging for pollen on copiously pollen-laden inflorescences.[7] It is not understood to what extent these insects have effected cross-pollination.

Using bidirectional mass selection in a population of *A. cruentus* varying for the simply inherited seedling pigment marker R/r, Hauptli was able to select for high and low outcrossing lines to demonstrate genetic variability in breeding system.[15,23] Outcrossing rates in the S_0 generation ranged from 0 to 100% and were strongly negatively correlated with the ratio of male to female flowers within an inflorescence. In the S_0 generation, mean outcrossing rate in the high outcrossing group was 93%; mean rate for the low outcrossing group was 2%. Although the selected populations did not reach the distribution of the parental population values, there was evidence of response to mass selection. Mean rates in the S_1 generation were 12% (±6%) for the low selection and 20%(±9%) for the high selection. Differences in outcrossing rates and male fertility between S_1 high selected and low selected outcrossing groups were significantly different.[15] Not only were the absolute values of male and female flowers different than taxonomic description of *A. cruentus*, but aberrant flower morphology, position, and apparent fertility or sterility was variable as well.[23]

These differences in flower sex ratio in glomerules may partly explain some of the wide variation in outcrossing rate seen across populations and species and the high apparent levels of inbreeding found in some of the material observed. In a preliminary study, Hauptli[15] found a wide range of mean numbers of male and female flowers per glomerule in landrace materials grown in a common garden in California. Flower sex ratio, coupled with the variable abundance of insect pollinators working the flowers, all show that amaranth breeding system is a very complex trait that, although genetically controlled in part, may also be dependent upon environmental factors and is difficult to describe precisely.

Amaranth compares well with described mating systems for some other crops. Breeding system characteristics of many legumes have been reported to be under polygenic control and intermediate in outcrossing expression.[73,74] Others with a mixed mating system and similar outcrossing rates are cotton ($t = 0.2$ to 0.3)[75] and sorghum ($t = 0.05$ to 0.4).[76]

For plant breeders, the mixed mating system in grain amaranth offers both opportunities and challenges. A wide choice of breeding methods are available using both controlled and natural self- or cross-pollination. Until breeding system components of particular breeding stocks and environments are considered, it must be assumed that natural outcrossing rates can vary over a wide range. Knowledge of outcrossing rates and pollen dispersal patterns in particular environments will be crucial to designing isolation requirements needed to retain germplasm and varietal purity.

IV. BREEDING OBJECTIVES

Although it has not been completely described, this vast array of variation is already being applied to amaranth for varietal improvement. Breeding objectives and priorities will vary with each breeding program and environment; however, several common objectives have emerged.[53,77] For mechanized agriculture using monoculture cropping systems, common objectives include shortened time of flowering to fit local conditions, shortened stature, reduced branching, placement of the inflorescence above the leaf canopy, synchronous dry-down at maturity either after a frost in temperate zones or before a frost in subtropical zones, shatter resistance coupled with suitability for mechanical threshing, stalk strength, lodging resistance, and high seed yield potential. For more labor intensive systems, breeding objectives that contrast with mechanized systems have been suggested.[78] Important differences in breeding objectives include selection for tall plant height, high yield potential per plant, late maturity, suitability for hand threshing, and compatibility with other species in an intercrop. Many of the available landraces were selected in more labor intensive multiple cropping systems and fit this ideotype.

Many breeding objectives may be common to most programs although specific priorities will vary. Some of these objectives include selection for large seed size, seedling vigor, and resistance to insect pests and diseases. A few diseases have been reported as problems on grain amaranth, such as damping off of seedlings and crown and root rot associated with insect tunneling at flowering. Lygus bugs, *Lygus lineolaris*, are the most important insect pest of amaranth.[79] Selection for pale seed coats (white, gold, or pink) and against black seeds would be common to most programs emphasizing use of the grain. Selection for specific nutritional and functional components such as protein, oil, or starch fractions may be important to specific breeding programs especially as industrial applications of amaranth are marketed. Cooking characteristics such as popping quality and taste are also important breeding considerations. Many of these important quality considerations have been reviewed. [80,81] Other objectives likely to be important in specific environments include drought tolerance,[82] aluminum tolerance,[83] salt tolerance, heat tolerance, germination/seedling tolerance to low spring temperatures, increased crop residue after harvest for soil erosion control from wind, and forage quality. A less obvious breeding consideration is selection for hybridization barriers with locally occurring weedy amaranths to facilitate quality seed production. Certainly new breeding objectives will emerge as grain amaranth is more widely cultivated. While it is not difficult to make an extensive list of crop weaknesses and desirable attributes of a grain amaranth variety, successful breeding programs will need to define a specific set of high priority objectives.

V. CONSERVATION OF GERMPLASM RESOURCES

The grain amaranths could have been considered staple grain crops in the major centers of their origin in Latin America. For example, they are mentioned in the Codex Mendoza[84] as tribute paid to Aztec kings 500 years ago at the time of the European conquest of the New World. However, in the course of the following four centuries, the value of the crop was not as strongly recognized, and its cultivation was gradually supplanted by crops demanded by colonial interests and international commerce, rendering amaranth less popular. The crop was outrightly discouraged or prohibited in some places, notably Mexico, because of its place in the religious ritual of local peoples.[1] Unknown quantities of germplasm resource have been lost during this quiescence in large scale amaranth cultivation.

Although interest is presently being renewed in the grain amaranths, genetic erosion may still be inadvertantly promoted by the introduction of improved varieties. Breeding programs at public institutions (universities and government agricultural research centers) have released improved grain amaranth varieties in Mexico, Perú, India, and the U.S. For these reasons, the first imperative for preservation of the crop is seed collection of all species, domesticated as well as weedy species. Of course, it is best to gather as much as possible from the crop's centers of diversity, but weedy collections outside these areas may be useful as well.[10] Opportunities for *in situ* preservation of grain amaranth germplasm should also be explored.

Researchers in many countries have begun to assemble amaranth germplasm of all species for preservation and study in the last 15 years. Since 1977, several thousand accessions of the cultivated grain amaranths have been assembled. Quantitative descriptions of these collections are beginning to emerge from evaluations in diverse locations such as Bolivia,[54] India,[53] Thailand,[85] the USA,[18,45] Russia,[68] and other countries.

Concurrent with collection and preservation efforts, germplasm that has been collected must be evaluated for presence of specific genes and/or characteristics. Inheritance of observed variation must be confirmed. This variation must be catalogued in accessions as well as geographical regions. It was discussed in the sections above on breeding system and landrace genetic variation that races, or better yet, populations, should be evaluated in more than one environment. Another issue that needs to be considered is adherence to a uniform list of descriptors. Common descriptors would facilitate comparison of evaluations in different environments. Development of a descriptor list is limited with a "new" crop, as new characteristics are discovered faster than the evaluations can even be completed and as procedures for the efficient, rapid screening of germplasm collections lag behind. New terms describing and perhaps simplifying complex physiological variables may be added to such a list. For instance, stem girth may be a simpler measurement to make in amaranth for indication of relative overall biomass accumulation.[44] Photoperiod response may be more informative a descriptor than "flowering time" or "time to flowering".[6] Molecular markers will also serve as descriptors to be integrated with morphological and physiological characteristics.

Amaranth plant breeders will likely maintain and describe their own germplasm collections, in addition to governmental or other public preservation programs. While this should aid conservation efforts, more may be done by breeders to save amaranth germplasm. Strategies to breed varieties for specific environments (rather than a single or few widely adapted varieties) while concurrently conserving valuable germplasm should be instituted now, early in the amaranth genetic development as a major crop in the world arena. Emphasis should be placed on incorporating as much genetic diversity as possible in varietals through the use of the variation in breeding population materials, multilines, and synthetic varieties. If breeding programs with a local history of amaranth culture emphasize the use of local genetic diversity while improving new varieties, a kind of *in situ* genetic preservation may occur. If heterotic advantage is found to be so high that the development of hybrid varieties becomes inevitable, effort should be placed on locating as many different sources of sterile cytoplasm and restorers as possible.

If all of these aspects of genetic diversity are kept at the forefront, perhaps the genetic development of amaranth as a major world crop can proceed, avoiding some of the disasters caused by genetic uniformity in other crops. Such an example of disaster because of genetic uniformity pertinent to our mention of sources of sterile cytoplasm is the corn epiphytotic of 1970 due to instant susceptibility of most of the popular hybrid maize varieties to the Southern Corn Leaf Blight organism.[87] Because nearly all U.S. corn hybrids grown then were produced using the same cytoplasm, the corn crop presented a very narrow genetic base. When the blight organism became virile to that cytoplasm, most of the 1970 crop was decimated, and there were few alternatives for corn breeders.

Examples of crop failures due to mutations in disease organisms that overcame a host crop's single somatic gene resistance abound.[88] The use of quantitatively inherited resistances,[89] and breeding more for tolerance rather than resistance to insect and disease pests,[90] can avoid these problems. Similarly, the use of mixed cultivars and composites may avert these problems. Hopefully, progress in grain amaranth genetic development will begin by accentuating the use of genetic diversity, thereby preserving amaranth germplasm while evading genetic uniformity problems like those that have been suffered by other crops.

VI. CONCLUSIONS

We have shown the grain amaranths to be extremely valuable for present as well as future generations of human agriculture. Every aspect of the crop's evolution, physiological characteristics, breeding system, and potential development parallels one or more of the world present day globally important major food crops. At the same time, amaranth germplasm also possesses unique characteristics that complement well other major crop products. For these reasons, it is indeed time to welcome this "new-old" crop into the family of world crops and place more importance on the conservation of its existing genetic resources around the world.

REFERENCES

1. **Sauer, J. D.,** The grain amaranths: a survey of their history and classification, *Ann. Mo. Bo. Gard.*, 37, 561, 1950.
2. **Murray, M. J.,** The genetics of sex determination in the family Amaranthaceae, *Genetics*, 25, 409, 1940.
3. **Tucker, J. M. and Sauer, J. D.,** Aberrant *Amaranthus* populations of the Sacramento-San Joaquin Delta, California, *Madrono*, 14, 252, 1958.
4. **Coons, M. P.,** Relationships of *Amaranthus caudatus*, *Econ. Bot.*, 36, 129, 1982.
5. **Hauptli, H. and Jain, S. K.,** Allozyme variation and evolutionary relationships of grain amaranths (*Amaranthus* spp.), *Theor. Appl. Genet.*, 69, 153, 1984b.
6. **Kulakow, P. A. and Jain, S. K.,** The inheritance of flowering in *Amaranthus* species, *J. Genetics*, 64, 85, 1985.
7. **Hauptli, H.,** unpublished data, 1979.
8. **Harlan, J. R.,** *Crops and Man*, American Society of Agronomy, Madison, WI, 1985.
9. **Brenner, D. and Hauptli, H.,** Seed shattering control with indehiscent utricles in grain amaranths, *Legacy*, 3, 2, 1990.
10. **Brenner, D.,** Potentially useful chartacteristics from wild *Amaranthus* species, poster presented at *Assoc. Adv. New Industrial Crops*, Peoria, October 2 to 6, 1989.
11. **Grubben, J. H.,** *The Cultivation of Amaranth as a Tropical Leaf Vegetable*, Communication 67, Royal Tropical Institute, Amsterdam, 1976, 206.
12. **Carlsson, R.,** Amaranth species, and related species for leaf protein concentrate, in *Proc. 1st Amaranth Sem.*, Rodale Press, Emmaus, PA, July 29, 1977, 83.
13. **Beard, B. H.,** Registration of Helianthus germplasm Pools III and IV, *Crop Sci.*, 22, 1276, 1984.
14. **Sauer, J. D.,** The grain amaranths and their relatives: a revised taxonomic and geographic survey, *Ann. Mo. Bot. Gard.*, 54, 103, 1967.

15. **Hauptli, H.,** *Genetic Variation for Allozyme, Morphological, and Mating System Characteristics in Grain Amaranths,* Ph.D. dissertation, University of California, Davis, 1983.
16. **Pal, M., Pandey, R. M., and Khoshoo, T. N.,** Evolution and improvement of cultivated amaranths. IX. Cytogenetic relationship between the two basic chromosome numbers, *J. Hered.*, 73, 353, 1982.
17. **Pal, M. and Khoshoo, T. N.,** Evolutionary Studies in World Crops, In *Diversity and Change in Indian Sub-Continent,* Hutchinson, J. B., Ed., Cambridge University Press, London, 1974, 129.
18. **Jain, S. K., Kulakow, P. A., and Peters, I.,** Genetics and breeding of grain amaranth: some research issues and findings, in *Proc. 3rd Amaranth Conf.*, Rodale Press, Emmaus, PA, 1986, 174.
19. **Gupta, V. K. and Gudu, S.,** Interspecific hybrids and possible phylogenetic relations in grain amaranths, *Euphytica,* 52, 33, 1991.
20. **Pal, M. and Khoshoo, T. N.,** Evolution and improvement of cultivated amaranths. V. Inviability, weakness, and sterility in hybrids, *J. Hered.*, 63, 78, 1972.
21. **Kulakow, P. A., Hauptli, H., and Jain, S. K.,** Genetics of grain amaranths. I. The inheritance of six color characteristics, *J. Hered.*, 76, 27, 1985.
22. **Kulakow, P. A.,** unpublished data, 1985.
23. **Hauptli, H. and Jain, S. K.,** Genetic variation in outcrossing rate and correlated floral traits in a population of grain amaranth (*Amaranthus cruentus* L.), *Genetica*, 66, 21, 1985.
24. **Vaidya, K. R. and Jain, S. K.,** Response to mass selection for plant height and grain yield in amaranth (*Amaranthus* spp.), *Z. Pflanzenzuecht.*, 89, 1, 1986.
25. **Ayiecho, P. O.,** *Quantitative Studies in Two Grain Amaranth Populations Using Two Selection Methods*, Ph.D. dissertation, University of California, Davis, 1986.
26. **Kulakow, P. A. and Jain, S. K.,** Genetics of grain amaranths. IV. Variation and early generation response to selection in *Amaranthus cruentus* L., *Theor. Appl. Genetics,* 74, 113, 1987.
27. **Saunders, R. M. and Becker, R.,** *Amaranth:* a potential food and feed resource, *Advances in Cereal Science and Technology*, 6, 357, Amer. Assoc. Cer. Chem. Pilot Knob, 1984.
28. **Downton, W. J. S.,** *Amaranthus edulis*: a high lysine grain amaranth, *World Crops*, 25, 20, 1973.
29. **Okuno, K., and Sakaguchi, S.,** Inheritance of starch characteristics in perisperm of *Amaranthus hypochondriacus, J. Hered.*, 73, 467, 1982.
30. **Okuno, K.,** Expression of mutant genes specifying starch synthesis in cereal grains, *Gamma Field Symp.,* 24, 39, 1985.
31. **Qureshi, A. A., Burger, W. C., Peterson, D. M., and Elson, C. E.,** The structure of an inhibitor of cholestorol biosynthesis isolated from barley, *J. Biochem.,* 261, 10544, 1986.
32. **Qureshi, A. A. and Lehmann, J. W.,** Cholesterol regulation by grain amaranth and its oil in 6-week old female chickens, *Cereal Foods World*, 36, 710, 1991.
33. **Lehmann, J. W., Qureshi, A. A., and Putnam, D. H.,** Genetic variability of vitamin E isomers in grain amaranths, *Agronomy Abstracts*, Agronomy Society of America, 1990, 98.
34. **Becker, R., Wheeler, E. L., Lorenz, K., Stafford, A. E., Grosjean, O. K., Betschart, A. A., and Saunders, R. M.,** A compositional study of amaranth grain, *J. Food Sci.*, 46, 1175, 1989.
35. **Rayas-Duarte, P. and Jacob, R. A.,** Study on the squalene content in amaranth grains, presented at the Amaranth Inst. 6th Annu. Meet., Fargo, ND, August 21 to 22, 1992.
36. **Danz, R. A. and Lupton, J. R.,** Physiological effects of dietary amaranth (*Amaranthus cruentus*) on rats, *Cereal Foods World,* in press, 1992.
37. **Segura-Nieto, M., Barba, de la Rosa, A. P., and Paredes-López, O.,** Biochemistry of amaranth proteins, *Amaranth: Biology, Chemistry, and Technology,* Paredes-López, O., Ed., CRC Press, Boca Raton, 1994, chap. 5.
38. **Barba de la Rosa, A., Paredes-López, O., and Gueguen, J.,** Characterization of amaranth globulins by ultracentrifugation and chromatographic techniques, *J. Agric. Food Chem.*, 40, 937, 1992.
39. **Gepts, P.,** 1992. Genetic diversity of seed storage proteins in plants, in *Plant Population Genetics, Breeding, and Genetic Resources*, Brown, A. H. D., Clegg, M. T., Kahler, A. L., and Weir, B. S., Eds., Sinauer Associates, Sunderland, MA, 1989, 64.
40. **Carillo, J. M., Rousset, M., Qualset, C. O., and Kasarda, D. D.,** Use of recombinant inbred lines of wheat for study of associations of high- molecular-weight glutenin subunit alleles to quantitative traits, I. Grain yield and quality prediction tests, *Theor. Appl. Gen.,* 79, 321, 1990.
41. **Feillet, P., Ait-Mouh, O., Kobrehel, K., and Autran, J.,** The role of Low Molecular Weight glutenin proteins in the determination of cooking quality of pasta products, *Cereal Chem.*, 66, 26, 1989.
42. **Craig, W. F.,** Production of hybrid corn seed, in: *Corn and Corn Improvement*, Sprague, G. F., Ed., American Society of Agronomy, Madison, WI, 1976, 671.

43. **Pandey, R. M.,** Genetic studies of yield contributing traits in *Amaranthus, Theor. Appl. Genet.,* 68, 121, 1984.
44. **Lehmann, J. W., Clark, R. L., and Frey, K. J.,** Biomass heterosis and combining ability in interspecific and intraspecific matings of grain amaranths, *Crop Sci.,* 31, 1111, 1991.
45. **Kauffman, C. S. and Reider, C.,** Rodale amaranth germplasm collection. *Rodale Res. Rep. NC-84/2*, Rodale Press, Emmaus, PA, 1984.
46. **Marshall, D. R.,** Crop genetic resources: current and emerging issues, in: *Plant Population Genetics, Breeding, and Genetic Resources,* Brown, A. H. D., Clegg, M. T., Kahler, A. L., and Weir, B. S., Eds., Sinauer Associates, Sunderland, MA, 1989, 367.
47. **Kauffman, C. S.,** The status of grain amaranth for the 1990s, *Food Rev. Int.,* 8, 165, 1992.
48. **Duriyaprapan, S., Buranasilpin, P., Tanpanich, S., Chitnawarsarn, S., and Watanakul, J.,** Preliminary evaluation and screening of introduced amaranth in Thailand, *Food Rev. Int.,* 8, 125, 1992.
49. **Senthong, C., Julsrigival, S., Tiyawalee, D., and Wivutvongvana, P.,** Germplasm screening of grain amaranth in Chiang Mai, Thailand, *Food Rev. Int.,* 8, 143, 1992.
50. **Espitia, E.,** Amaranth germplasm development and agronomic studies in Mexico, *Food Rev. Int.,* 8, 71, 1992.
51. **Gupta, V. K. and Thimba, D.,** Grain amaranth: a promising crop for marginal areas of Kenya, *Food Rev. Int.,* 8, 51, 1992.
52. **Vaidya, K. R.,** Genetic Variation in Landrace Populations of Indian Amaranths, *Ph.D. dissertation, University of California,* Davis, 1984.
53. **Joshi, B. D. and Rana, R. S.,** Grain Amaranths: The Future Food Crop, *National Bureau of Plant Genetic Resources.* Simla, 1991.
54. **Avila, G. L., Alvarez, P. Z., and Castellon, S. T.,** Introductory steps for improvement of "millmi" (*Amaranthus caudatus*) at the Phytoechogenetic Center of Pairumani, Bolivia, *Amaranth Newsletter,* 6, September 1989.
55. **Hauptli, H., Lutz, R. L., and Jain, S. K.,** Germplasm exploration in Central and South America, in *Proc. 2nd Amaranth Conf.,* Rodale Press, Emmaus, PA, 1980, 117.
56. **Jain, S. K., Vaidya, K. R., and Joshi, B. D.,** Collection and evaluation of Indian grain amaranths, in *Proc. 2nd Amaranth Conf.,* Rodale Press, Emmaus, PA, 1980, 123.
57. **Peralta, E. I.,** The grain amaranth status in Ecuador, *Amaranth Newsletter,* 3, June 1985.
58. **Perez, C. A. A. and Salan, M. G.,** Genetic resources in Guatemala: *Amaranthus* sp., *Amaranth Newsletter,* 5, June 1988.
59. **Kalinowski, L. S.,** The small giant, *Amaranth Newsletter,* 1, June 1983.
60. **Hauptli, H. and Jain, S. K.,** Genetic structure of landrace populations of the new world grain amaranths, *Euphytica,* 33, 875, 1984a.
61. **Jain, S. K., Wu, L., and Vaidya, K. R.,** Levels of morphological and allozyme variation in Indian amaranths: a striking contrast, *J. Hered.,* 71, 283, 1980.
62. **Flores, H. E.,** Biotechnology and amaranth, in *Amaranth: Perspectives on Production, Processing and Marketing,* Minnesota Extension Service, St. Paul, 1990, 143.
63. **Gupta, V. K. and Gudu, S.,** Inheritance of some morphological traits in grain amaranthus, *Euphytica,* 46, 79, 1990.
64. **Kulakow, P. A.,** Genetics of grain amaranths. II. The inheritance of determinance, panicle orientation, dwarfism, and embryo color in *A. caudatus, J. Hered.,* 78, 293, 1987.
65. **Neuffer, M.G., Jones, L., and Zuber, M. S.,** *The Mutants of Maize,* Crop Science Society of America, Madison, WI, 1968.
66. **Peters, I. and Jain, S. K.,** Genetics of grain amaranths. III. Gene- cytoplasmic male-sterility, *J. Hered.,* 78, 251, 1987.
67. **Gudu, S. and Gupta, V. K.,** Male-sterility in the grain amaranth (*A. hypochondriacus* ex-Nepal) variety 'Jumla', *Euphytica,* 37, 23, 1988.
68. **Kauffman, C. S. and Weber, L. E.,** Grain amaranth, in *Advances in New Crops,* Janick, J. and Simon, J. E., Eds, Timber Press, Portland, OR, 1990, 127.
69. **Schulz-Schaeffer, J., Baldridge, D. E., Bowman, H. F., Stallknecht, G. F., and Larson, R. A.,** Registration of 'Amont' grain amaranth, *Crop Sci.,* 31, 482, 1991.
70. **Jain, S. K., Hauptli, H., and Vaidya, K. R.,** Outcrossing rate in grain amaranths, *J. Hered.,* 73, 71, 1982.
71. **Kulakow, P. A. and Jain, S. K.,** Genetics of grain amaranths. VI. A three-year study of within population variation in outcrossing rates, *Amaranth Newsletter,* 5, Dec. 1988.

72. **Agong, S. G. and Ayiecho, P. O.,** The rate of outcrossing in grain amaranths, *Plant Breed.*, 107, 156, 1991.
73. **Harding, J. and Tucker, C. L.,** Quantitative studies on mating systems. I. Evidence for non-randomness of outcrossing in *Phaseolus lunatus, Heredity,* 19, 361, 1964.
74. **Green, A. G., Brown, A. H. D., and Oram, R. N.,** Determination of outcrossing rate in a breeding population of *Lupinus albus* L. (white lupin), *Z. Pflanzenzuecht.*, 84, 181, 1980.
75. **Stephens, S. G. and Finkner, M. D.,** Natural crossing in cotton, *Econ. Bot.*, 7, 257, 1953.
76. **Ricelli-Mattei, M.,** Estimation of Genetic Parameters Related to the Mating System in Populations of Sorghum, *Sorghum bicolor* (L.) Moench, *Ph.D. Dissertation,* University of California, Davis,1968.
77. **Weber, L. E. and Kauffman, C. S.,** Plant breeding and seed production, in *Amaranth: Perspectives on Production, Processing and Marketing*, Minnesota Extension Service, St. Paul, 1990, 115.
78. **Maldonado, U. A. and Estrada, A. L.,** Amaranth genetic variation and utilization in Mexico, in *Proc. 3rd Amaranth Conf.*, Rodale Press, Emmaus, PA, 1986, 120.
79. **Wilson, R. L.,** Insect and disease pests of amaranth, in *Amaranth: Perspectives on Production, Processing and Marketing*, Minnesota Extension Service, St. Paul, 1990, 163.
80. **Bressani, R,** Grain amaranth. Its chemical composition and nutritive value, in *Amaranth: Perspectives on Production, Processing and Marketing*. Minn. Extension Serv., St. Paul, 1990, 19.
81. **Saunders, R. M. and Becker, R.,** *Amaranthus*: A potential food and feed resource, in *Advances in Cereal Science and Technology*, VI, Pomeranz, Y., Ed., 1989, 357.
82. **Cavagnaro, J. B. and Jain, S. K.,** Effect of water stress in four *Amaranthus* species, *Amaranth Newsletter*, 3, September 1985.
83. **Campbell, T. A. and Foy, C. D.,** Selection of grain *Amaranthus* species for tolerance to excess aluminum in an acid soil, *J. Plant Nutr.*, 10, 249, 1987.
84. **Anawalt, P. R. and Berdan, F. F.,** The Codex Mendoza, *Sci. Am.*, 266, 70, 1992.
85. **Duriyaprapan, S.,** Study on grain amaranth production in Thailand, report submitted to NAS, *Thailand Inst. of Sci. and Tech. Res.*, 1986.
86. **Girenko, M. M., Borodkin, A. S., and Voskresenskaya, V. V.,** Variation in the main economically useful characters in amaranth, *Sborni Nauchnykh Trudovpo Prikladnoi Botanike, Genetike i Selektsii,* 118, 59, 1990.
87. **Sill, W. H.,** *Plant Protection: An Integrated Interdiscinplinary Approach*, Iowa State University Press, Ames, 297 pp.
88. **Day, P. R.,** Crop resistance to pests and pathogens, in *Pest Control Strategies for the Future*, Division of Biology Agriculture, National Research Council, National Academy of Science, Washington, D.C., 1972.
89. **Allard, R. W.,** Future directions in plant population genetics, evolution, and breeding, in: *Plant Population Genetics, Breeding, and Genetic Resources*, Brown, A. H. D., Clegg, M. T., Kahler, A. L., and Weir, B. S., Eds., Sinauer Associates, Sunderland, MA, 1989, 64.
90. **Gould, F.,** Evolutionary biology and genetically engineered crops, *Bioscience*, 38, 26, 1988.

Chapter 3

Breeding of Grain Amaranth

Eduardo Espitia-Rangel

CONTENTS

I. INTRODUCTION

Amaranth is a new crop with the potential of broadening the diversity of commercially grown crops and so make an important contribution to world food supplies.[1] Amaranth is an old crop, grown by pre-Columbian civilizations in America; it was replaced by crops introduced from the old continent and continued to decline in popularity over the years.[1,2,3] Even today, some traditional farmers have continued to abandon amaranth as they devote more of their land to high-yielding modern crops.[1,2] Now there is a renewed interest in the cultivation of this under-exploited crop as its potential as a new food, its nutritional benefits, and its agricultural advantages become more widely known.[1,4] Amaranth is resistant to drought and requires less water than many other staple crops. This characteristic makes it a prospective dryland crop for farmers in rainfed areas. In irrigated areas amaranth provides an alternative for farmers who seek to reduce irrigation costs and soil salinization.

0-8493-5374-2/94/$0.00+$.50

The amaranth germplasm which farmers continue to cultivate in Mexico, Central and South America, and Asia probably represents a remnant of the crop diversity which existed under more extensive cultivation in pre-Hispanic times. A portion of this remaining germplasm has been assembled for preservation and study in many countries. The success of genetic improvement of amaranth will probably depend on our appreciation of the peasants responsible for preserving and propagating the crop in the past and present. Selection of a few improved varieties and replacement of old landraces may lead to short term success. The long term prospects for amaranth breeding will depend on the availability and knowledge of its genetic diversity.[5,6,7]

II. ORIGIN AND EARLY DEVELOPMENT

The origin of the genus *Amaranthus* was probably in America. This assumption is reasonable considering that 60 species are native to the New World and about 15 to the Old World. Most of them are widely distributed throughout the temperate and tropical regions of the world in cosmopolitan weed or cultivated forms.[8-10] The origin of various species of cultivated amaranths is not easy to trace because wild ancestors are pantropical cosmopolitan weeds. Amaranth is presently cultivated in both the Old and New Worlds, and the regions of origin and domestication of the amaranth are still controversial; however, several lines of evidence have been presented.[8] That the three grain amaranth species were domesticated prehistorically in the highlands of tropical and subtropical America.[8,10]

The grain amaranths were cultivated from 5,000 to 7,000 years ago in the New World, although at present they are distributed through much of Asia and Africa.[11] The three cultivated species *A. cruentus, A. hypochondriacus,* and *A. caudatus* have centers of diversity in Mexico and Central America, Mexico, and Andean South America, respectively. Currently, *A. hypochondriacus* is most widely cultivated in Nepal and India.[8-10] Archeological evidence shows that amaranth was collected by prehistoric people in the New World. After that the native groups learned to cultivate the wild and weed amaranth as a grain. Artificial selection directed the domesticates toward larger plant size with huge inflorescences and more seed production, although the size of the individual seeds was not increased. This selection also produced bright red ornate forms; the red coloration presumably had magical and ceremonial connotations, and some native groups grow amaranth primarily as a source of colorants.[3,8,12]

A crucial step in the evolution of the domesticated grain amaranth was the selection of mutant forms in which the normal black seed changed to white. This mutation resulted in a better flavored grain with superior popping quality and also helped the prehistoric Indian domesticators to limit crossing between their crop and the weed amaranth by eliminating dark seeds from their planting stock, thus favoring divergent selection of the domesticates.[12]

Maize and amaranth intercropping system is still being practiced in the Sierra Madre Oriental in Puebla state in Mexico. This system has been preserved by the Totonacas, the native group of this area. When the prehistoric peasants cultivated the wild and weed amaranth together with maize, this forced the amaranth plant to grow straight in order to compete with maize; thus, the characteristic of erectness evolved in amaranth. Moreover, the pre-Hispanic agricultural system did not include tillage, thus causing the amaranth to become a vigorous plant. Some of today's cultivated forms of *A. hypochondriacus* can reach a height of 3 m.

After the conquest, amaranth cultivation suffered a catastrophic decline that was partially due to the general disruption of native cultures; however, it was also due to deliberate suppression of amaranth as a symbol of sinful heathenism.[12] The crop has persisted among remote native groups and even in some places close to Mexico City where remaining genetic variability was preserved.

III. MODERN BREEDING OBJECTIVES

A series of characteristics has been developed to describe the desired model plant or ideotype of grain amaranth.[7] Nevertheless, this ideotype will vary with each breeding program, environment, and requirements of the food industry. In traditional production areas, hand threshing is a strong limitation; On the other hand, mechanized production is essential for increasing area of amaranth production. For mechanized agriculture, the improved varieties will be shorter, with reduced branching, resistance to lodging, inflorescence above the leaf canopy and synchronous dry-down of plant at maturity.[1,2,6,7,9] For traditional agricultural systems the breeding objectives would include selection for tall plant, high yield per plant, middle to late maturity, and suitability for hand threshing.[13]

Other breeding objectives may be common to most programs, although specific priorities will vary. Some of these objectives include selection for large seed size, seedling vigor, and resistance to insect pests and diseases. A few diseases have been reported as problems on grain amaranth, such as Stem Black Spot (*Macrophoma* sp.), and Damping Off.[2,14] Lygus bugs (*Lygus lineolaris*), and the Stem Borers (*Sciara* sp.) are the most important insect pests of amaranth.[2,15] Selection for specific nutritional and functional components such as protein, oil, and starch fractions may be important to specific breeding programs especially for industrial applications of amaranth. Cooking characteristics such as popping quality and taste are also important, especially in breeding programs that work for traditional utilization of amaranth.[16] Other important characteristics could be drought tolerance, salt tolerance, heat tolerance, and high biomass production for use as forage.[10,17,18] New breeding objectives will emerge as amaranth is more widely cultivated.

IV. GENETIC VARIABILITY

The total variation available for breeding grain amaranth must be considered to get a clear picture of the potential for genetic improvement. There are at least five levels of the grain amaranth gene pool corresponding to more and more highly selected genetic materials. The least selected genetic material is the diverse groups of wild and weedy relatives of the grain amaranths, which can transfer genes through hybridization. The second level is the group of the cultivated grain species. The third level corresponds to races, grain types, or morphological groups within each cultivated species. The fourth level includes individual landrace populations. The final level corresponds to genetic stocks, breeding lines, and cultivars that have been selected in recent years as part of amaranth breeding programs.[5] Some of these groups will be discussed here.

A. CULTIVATED AND WILD SPECIES

There are three species of grain amaranth: *A. hypochondriacus*, *A. cruentus*, and *A. caudatus*. Associated with these are three weedy species: *A. powelli*, *A. hybridus*, and *A. quitensis*. The cultivated species originated in the New World and now are distributed throughout Asia and Africa; only one weedy species, *A. hybridus* is now widely spread over almost the whole world.[5,21] *A. quitensis* is considered a geographic variant of *A. hybridus*.[22] Some authors recognize a fourth cultivated species, *A. mantegazzianus* Pass or *A. edulis* Speg, also from South America, but genetic studies do not seem to warrant the recognition of this group as a separate cultivated species.[21-23]

The grain amaranth species are characterized by compound monoecious inflorescence and pentamerous flowers with circumcessile utricules. The basic units of the inflorescence are dicacial cymes, usually called glomerules, each ordinarily consisting of an initial staminate flower and an indefinite number of pistillate flowers. The glomerules are crowded on a leafless axis to form complex inflorescence, technically Thyrses, which are generally called spikes.[3,8] The wild and weedy amaranth species are profusely branched and have a small terminal inflorescence. Most of them are pioneer annuals of naturally open habitats. They thrive by constant colonization of sites with disturbed soil, full sun, and little competition. Long natural selection for such conditions preadapted certain of the species for success in habitats drastically disturbed by man. Thus, amaranths are best known as nexious weeds.[3,8,23,24] Many wild and weedy *Amaranthus* species have unique adaptative characteristics such as early flowering.[26]

B. GRAIN RACES

There is a broad range of variability within each grain amaranth species. Since the early studies on amaranth, attempts at grouping different cultivated forms were made.[3] Recently, amaranth germplasm was grouped based on common morphological features.[27] Description of the grain types was considered preliminary and based on morphological traits such as plant height, main inflorescence size, branching pattern, number of pistillate flowers per glomerule, and on phenological traits like flowering time and time to maturity. Those amaranth grain types represent complexes adapted to different localities under different environmental and cultural conditions. They had been developing through artificial and natural selection in primitive agricultural systems.[5,10,28,29] Within each cultivated species there are several grain types. For example, within *A. cruentus*, the grain types Mexican, Guatemalan, and African were defined. The grain types of *A. hypochondriacus* are Aztec, Mercado, Mixteco, Nepal, and Spike. The grain types South American and Edulis were described in *A. caudatus*. Within *A. hybridus*, the grain types Prima and Sangorache were determined. Some populations do not fit any of the grain types, since they represent new grain types or exhibit variability due to hybridization among races.[5]

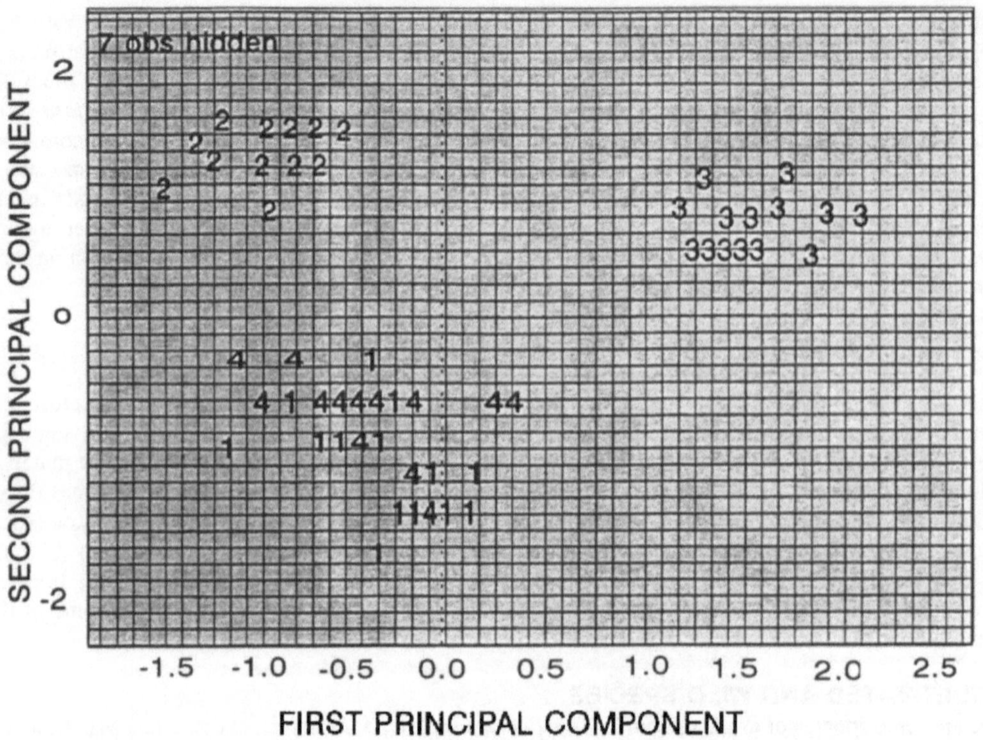

Figure 1 Graph of first two axis from a principal component analysis based on the means of 25 traits of grain amaranth. 1 = Mercado grain race, 2 = Mexican grain race, 3 = Aztec grain race, and 4 = Nepal grain race.

The grain types or morphological groups represent a part of the spectrum of genetic variation within each species. The accessions within a morphological group may actually be landraces; these are domesticated individuals of sexual reproduction with characteristics that permit them to be distinguished from other groups. Consequently, the most proper designation for the grain types would be races.[5-7] Each race has a defined distribution and has been developed under different agroclimatic conditions that made the evolution for each race different. Figure 1 is a graph of the first two components from a principal component analysis based on means from 25 traits of 60 grain amaranth landraces. The first principal component contains 63.3% of the total variation and is associated with the highest manifestation of traits such as days to flowering, plant height at flowering, number of leaves, stem diameter, plant height at maturity, inflorescence length, leaves dry matter, and stem dry matter. The second principal component contains 23.7% and is associated with the highest manifestation of traits such as seed filling period, days to maturity, and inflorescence dry matter; it is also associated with low manifestation of harvest index and grain yield. On this graph, the landraces are clustered into well-defined groups, which correspond to the three most important grain races from Mexico and one from Asia. Both Aztec and Mexican grain races are perfectly separated; the former shows high values of both the first and second principal components, while the last shows low value of the first principal component and high value of the second principal component. On the other hand, Mercado and Nepal grain races show low values of two principal components and their landraces have the earliest and shortest plants. This is evidence that the grain races are well-defined groups with their own distinctive traits. However, efforts to describe all the grain races based on morphological, cytological, and biochemical characteristics are needed to clarify the race classification. It might be useful to rename or reclassify some of the present races, which are described below.[6,29]

1. African Grain Race

This race is used as a leafy vegetable in West Africa. However, there are accessions useful as breeding parents for such traits as early maturity and shorter plant stature in the development of grain amaranth varieties.[28] This race has a height of 0.6 to 2.0 m; it branches at low densities. The apical inflorescence does not bear the majority of the grain yield. The main inflorescence has from 31 to 104 panicles; each panicle has from 5 to 10 branches. The glomerule has an average of 15 pistillate flowers. The bract is shorter than the utricule. The seed is small and almost all the landraces are black seeded. This race is not very sensitive to day-length.[2,27-30]

2. Aztec Grain Race

The Aztec grain race originated in the highlands of Tlaxcala, Puebla, Mexico, and Distrito Federal in Mexico where it is used only as grain crop. This race includes the largest plants of the genera; they can reach 3 m in height. Primarily unbranched, only some minor branching can be noted. The main inflorescence can reach 0.6 to 1.2 m and has a range from 80 to 150 erect panicles; each one with 5 to 13 branches. The glomerule has an average of 30 pistillate flowers. The bract is longer than the utricule, and this makes the inflorescence prickly. Seed shattering is slight in this race. The seed color is white or dark brown, and the size is medium to large. This race is sensitive to the photoperiod, and it matures in 160 to 180 days at Chapingo, Mexico, that is, at 19° 17' latitude north; in Kutztown, PA, at 40° latitude north, it does not mature.[2,27-30]

3. Edulis Grain Race

The place of origin of this race is Northwest Argentina; it is grown as a grain crop in the provinces of Tucuman, Jujuy, and Salta. This race is 1.5 to 2.0 m tall and shows branching at low densities. It has a distinctive determinate inflorescence that can be erect or decumbent with medium to large white seeds. This race is susceptible to stalk rot diseases.[22,27,28,31]

4. Guatemalan Grain Race

Originating in Guatemala and the southeast of Mexico, this race is cultivated as a grain crop; some semicultivated forms are used as ornamentals or eaten as a vegetable. This race has a height at maturity of 1.5 to 2.0 m and branching on upper stem. It has a small main inflorescence and often there are three or more seed heads borne at the top of the plant. The main inflorescence has from 46 to 81 semidrooped panicles, each one with 4 to 14 branches. The glomerules have an average of 32 pistillate flowers. The bracts are shorter than the utricule. Seed shattering is from moderate to severe. The seeds are tan, black, or dark brown and small to medium in size. This race matures early to mid-season, slightly earlier than the Mexican grain race. It is a little sensitive to day-length,[7,28-30]

5. Mercado Grain Race

This grain race includes some of the most valuable germplasm for development of improved varieties. Mercado grain race has excellent culinary qualities and the potential for high yields. The place of origin of this race is Mexico, primarily the warm regions of the states of Morelos and Puebla, where it is grown as a grain crop. The plant can reach a height at maturity of 1.5 to 2.2 m with branching on the upper stem. The main inflorescence has from 42 to 75 erect panicles, each one with 3 to 9 branches. The glomerules have an average of 44 pistillate flowers, and the bracts are shorter than the utricule. Seed shattering is light to moderate in this race. The seeds are white, golden, and brown and medium to large in size. This race is sensitive to the photoperiod matures in 135 days at Chapingo, Mex. and 15 days later at Kutztown, PA.[2,7,27-30]

6. Mexican Grain Race

The Mexican grain race has the best potential as a grain producer in the U.S. It originated in Mexico — primarily in the warm regions of Morelos, Puebla, and Guerrero — places in which it is now cultivated as a grain crop. This race can reach a height of 1.5 to 2.5 m; it has lateral inflorescences on the upper stem and one dominant apical inflorescence, which has 32 to 56 semidrooped panicles, each one having from 1 to 3 branches. The glomerules have an average of 50 pistillate flowers. The bracts are shorter than the utricule, tending to be shorter than the bracts of the Mercado grain race. The seeds are white with a few brown ones and are from medium to large sized. This race is sensitive to the duration of the photoperiod, and it matures in 80 to 145 days.[2,7,27-30]

7. Mixteco Grain Race

This race is used as a grain crop and originated in the states of Oaxaca and Michoacan in Mexico. This race is very similar to the Aztec grain race, but it takes longer to mature, from 180 to 220 days. This race can reach 3 m in height. The inflorescences have from 100 to 170 erect panicles, each one with 5 to 27 branches. The glomerules are formed by an average of 18 pistillate flowers. The bracts are longer than the utricule. The seeds are white or dark brown and the seed size is from small to medium. This race is sensitive to the length of the photoperiod.[2,7,27-30]

8. Nepal Grain Race

This race has a broad genetic variability. Now cultivated as a grain crop, it originated in Asia, mainly India and Nepal. Landraces of this race have a high yield potential at low latitudes; they have shown good performance in dry regions of Mexico and Kenya. This race can reach a height of 0.8 to 2.2 m and branches at low densities. The main inflorescence has from 36 to 56 erect panicles, each one with 1 to 4 branches. The glomerules have an average of 47 pistillate flowers. The bracts are generally longer than the utricule, but in some landraces they are shorter. Seed shattering is slight to moderate. This race is sensible to length of photoperiod, and it matures in 80 to 150 days.[2,7,27,28,30]

9. Prima Grain Race

The landraces that fall into this race are short with one massive inflorescence that arises from leaf axils near the ground level. This race originated in North and Central America. The inflorescences are spiny, since the bracts are long. The seeds are black and are small to medium in size. These traits are in marked contrast to the small axilary inflorescences and small seed size of weedy accessions of *A. hybridus*.[2,7,27-30]

10. Sangorache Grain Race

This race originated in South America, predominantly Ecuador, where it is used for dyes. This race is a product of introgressive hybridization between the weed and crop species. At maturity, this race can reach 1.5 to 2.5 m in height. It branches at low densities. The landraces of this race have black seeds. This race matures late.[22,27,28,31]

11. South American Grain Race

This race originated in the Andean region of South America, primarily Perú and Bolivia, but extended from North Chile and Argentina to Ecuador, where it is cultivated as a grain crop. The plants of this race are from 1.2 to 2.8 m in height, and it can branch at low densities. The inflorescence can be erect or decumbent with from 45 to 106 panicles, each having 1 to 5 branches. The glomerules have an average of 59 pistillate flowers. The bracts are shorter than the utricule. Seed shattering is moderate. The seed is white, dark golden, brown, and pink. The seed size is medium to large. This race is sensitive to the length of the photoperiod, and it matures in 100 to 160 days.[22,27,28,31]

12. Spike Grain Race

Originally from India and Nepal, the landraces of this race are segregates from the Nepal grain race. However, the plants of the Spike race are more branched and tend to lodge since they have a thin main stem which cannot support the large inflorescence. This race includes the shortest plants of the *A. hypochondriacus* species with a height of 0.7 to 1.7 m. The inflorescence has a long fingered panicle and tends to be prickly, since the bracts are longer than the utricule. The seeds are mainly black with a few white seeds, and the seed size is medium to large.[2] Landraces from this race show characteristics especially useful in dryland farming and mechanized agricultural systems. This race is sensitive to the length of the photoperiod and matures early at low latitudes.[2,7,27-30,32]

C. QUANTITATIVE GENETIC VARIATION

Knowledge of the genetic variability present in the original materials is of great importance in defining the most suitable breeding methods; the greater the genetic variability in the base genetic pool, the greater the possibility of obtaining the desired genotype.[33,34] When comparing the absolute value of genetic variance of two or more populations, it is possible to obtain erroneous conclusions. This can be avoided by comparing the genetic variation coefficient, which is a means of measuring the genetic variability that is not affected by the measuring units used.[35] In India, a study of the genetic variation of native amaranth varieties from different agroecological regions was carried out. Genetic variation from 3.1 to 24.7 for such traits as plant height, number of leaves per plant, leaf length, leaf width, inflorescence length, days to maturity, and yield per plant was found.[35]

The concept of heritability originated when trying to define whether the differences observed among individuals had their origin in the genetic differences of those individuals, or if they resulted from environmental effects.[36,37] In 20 genotypes of *A. hypochondriacus* it was found that there were high broad sense heritability values (0.60 to 0.76) for plant height and inflorescence length; intermediate values (0.26 to 0.47) for yield, days to maturity and number of leaves; and low values (0.01 to 0.13) for length and width of leaves.[35] In other studies, heritability values for plant height of 0.22 in *A. cruentus* and 0.47 in *A. hypochondriacus* were found while the value for yield was 0.09 in both species. In the latter species, heritability values from 0.35 to 0.66 for days to flowering, values of 0.21, 0.02 0.70, and 0.33 were reported for plant height, inflorescence length, harvest index, and yield per plant, respectively.[38-41]

In Mexico, a study was conducted to understand the genetic variability present among and within the grain races Mercado, Mexican, Aztec, and Nepal. It was found that the variation of genetic origin was mainly due to races as a source of variation, and only a small part was due to genotypes within races. That is, the races are integrated by approximately similar populations with a capacity for reproducing those traits that make them different from other races. For instance, the Mercado and Mexican grain races had their origin in an agroecological zone with a warm climate; on the other hand, the Aztec grain race originated in a temperate region, while the Nepal grain race covers a wider climatic range in India and Nepal. Those facts permit one to suppose that during the evolutionary process the races have been slowly adapted to different climatic conditions. In Table 1, the genetic variability coefficients of general and within-race analysis is presented. In general, those traits that showed a higher genetic variability were plant height at flowering, number of leaves, stem dry matter, seed dry matter, biological yield, seed-straw ratio, and seed filling index, with genetic variability coefficients from 31.36 to 66.32.[6]

The genetic variability coefficients for the 15 genotypes belonging to each race (Table 1) indicate that the Mercado grain race showed the highest genetic variability for 5 traits; the Mexican grain race turned out to have the highest genetic variability in 5 other traits; the Aztec grain race was the highest only in 3 traits; and the highest genetic variability for the remaining 12 traits corresponded to Nepal grain race. This last race represents the total variability of the species *A. hypochondriacus* that can be present in India and Nepal, covering a range of altitude from 900 to 2400 m above the sea level (a.s.l.). The genetic variability of this species that may be present in Mexico has been grouped in the Mercado and Aztec grain races, which are grown at 1400 to 2 400 m a.s.l. and 200 to 2800 m a.s.l., respectively. This explains why the Nepal grain race showed the highest genetic variability in a greater number of traits. It must be mentioned that several traits of agronomic interest (such as days to flowering, days to maturity, plant height at maturity, and yield), presented the highest degree of genetic variability within the Nepal grain race, indicating that this race may provide a good source of genetic pool for selecting grain amaranth.[6]

In Table 1 the broad sense heritability values for 25 traits are presented on a general level and by races. It can be observed that at genotypes within race level the heritability values are lower since the genetic variation of genotypes within each race is low. The Mercado grain race showed the highest heritability values in three traits; the Mexican grain race presented the highest heritability values in seven traits; the Aztec grain race also showed the highest heritability values in seven traits; while the Nepal grain race presented the highest heritability values for the other eight traits. Each trait can be more easily improved within the race that reaches the highest heritability values, since its genetic variability is mostly due to genetic effects compared to other races. For example, yield and some of its components (such as days to flowering, number of leaves, and stem diameter) will be more easily improved in the Nepal grain race; plant height and inflorescence length in the Mercado grain race; days to physiological maturity and harvest index in the Mexican grain race; and number of leaves and dry matter for each organ in the Aztec grain race. Generally speaking, those traits that can be more easily improved are days to flowering, plant height at flowering, number of leaves, seed filling period, days to physiological maturity, and plant height at maturity.[6]

D. TRAITS FOR INDIRECT SELECTION OF GRAIN YIELD

The meaning of genetic and phenotypic correlation is of great use to the plant breeder, mainly when selection is directed toward modifying several traits simultaneously, when trying to select for traits of difficult quantification or when low heritable traits are selected.[42,43] In a study carried out in India using six varieties of *A. hypochondriacus* and all of its possible crosses, it was found out that the genetic correlation was higher than the corresponding phenotypic correlation, while yield presented a positive association with inflorescence length and plant height.[44] In Mexico, positive association among grain yield and stem diameter, leaf length, days to flowering, and plant height was found.[45] In other work in this country, the relation of grain yield and its components was studied and analyzed at general level and within races. In Table 2 the genetic and phenotypic correlation values of grain yield and other traits are presented. Upon analyzing each race, changes in sense and degree of association among the grain yield

Table 1 **General and within-race genetic variance coefficient (GVC) and heritability in broad sense (H^2) values for 25 traits in amaranth. Chapingo, Mexico**

Traits	General		Mercado		Mexican		Nepal		Aztec	
	GVC	H^2	GVC	H^2	GVC	H^2	GVC	H^2	GVC	H^2
Days to flowering	15.61	0.96	2.36	0.40	1.52	0.21	0.53	0.06	4.50	0.64
Plant height at flowering	35.13	0.95	5.07	0.29	1.16	0.02	1.61	0.12	6.21	0.32
Leaves number	39.18	0.70	4.35	0.09	—	—	4.13	0.18	6.34	0.07
Middle leaf length	9.83	0.53	5.93	0.39	5.55	0.41	2.67	0.17	4.89	0.29
Middle leaf width	12.90	0.60	4.79	0.26	6.80	0.45	2.29	0.10	5.22	0.31
Middle leaf petiole length	10.44	0.20	4.66	0.07	3.58	0.04	13.29	0.29	5.30	0.05
Middle leaf dry matter	26.31	0.25	—	—	9.62	0.08	24.99	0.20	—	—
Middle leaf area	20.02	0.56	9.55	0.27	11.93	0.47	4.44	0.13	9.61	0.27
Seed filling period	9.81	0.71	1.63	0.14	4.89	0.53	—	—	1.72	0.12
Stem diameter	22.88	0.61	—	—	—	—	3.11	0.07	4.87	0.08
Days to maturity	8.55	0.88	1.42	0.54	2.14	0.80	0.27	0.09	2.18	0.70
Plant height at maturity	23.93	0.92	5.59	0.53	2.56	0.15	2.78	0.32	5.23	0.36
Inflorescence length	13.78	0.39	4.03	0.13	—	—	2.68	0.07	—	—
Leaves dry matter	28.15	0.42	14.02	0.10	2.90	0.01	26.27	0.54	5.87	0.03
Stem dry matter	66.32	0.76	17.56	0.15	—	—	13.20	0.28	13.72	0.06
Inflorescence dry matter	22.42	0.38	8.30	0.07	6.20	0.08	10.50	0.20	13.15	0.16
Seed dry matter	41.07	0.51	9.07	0.05	—	—	10.52	0.16	13.68	0.12
Total dry matter	31.36	0.66	9.97	0.12	0.75	0.001	6.66	0.16	13.17	0.20
Harvest index	23.90	0.43	2.50	0.01	18.18	0.20	5.56	0.06	2.74	0.01
Hectoliter weight	1.00	0.08	—	—	—	—	—	—	1.39	0.12
Yield	29.01	0.43	12.14	0.14	13.04	0.07	3.55	0.02	19.40	0.27
Economic yield rate	28.45	0.42	11.43	0.14	16.55	0.10	—	—	16.66	0.22
Biological yield rate	23.49	0.54	10.58	0.15	3.19	0.02	6.30	0.13	11.55	0.23
Seed-straw ratio	37.38	0.44	0	0	18.02	0.08	7.67	0.05	0	0
Seed filling index	32.62	0.49	32.00	0.17	19.22	0.11	5.02	0.04	18.07	0.23

[a] No estimated values due to negative variance.

and the rest of variables are observed; moreover, the correlation values are lower than those of the general analysis. Considering this, as with the other genetic parameters, correlation must be interpreted for each race without extrapolating one from the other.[6]

In both Mercado and Nepal grain races, highly significant positive correlation was observed for most of the variables studied: for harvest index, economic yield rate, biological yield rate and seed filling index in the Mexican grain race, and for economic yield rate, and seed filling index in the Aztec grain race. This means that those plants within each race that yield the most will be the ones that give the best manifestation of the desired traits, as well as others. However, from the agronomic point of view, several of such traits are not desirable. Correlation values greater than 1.0 in several cases within races can also be observed; this might be due to small, within-race genotypic variance values, which are even smaller than those within each race.[6]

As it was stated above, several traits show acceptable values in the general analysis for those genetic parameters which help in choosing the most adequate variables for indirect selection of the grain yield. In a breeding program for amaranth, those directly measured traits will be the most manageable since a great number of individuals or progenies must be observed. Hence, minute measurements which require too much care would be impractical, especially nowadays when economic resources are scarce. In this way, those traits of little practical evaluation, such as those that require dividing the plant in small pieces to determine its dry weight, would be inadequate for indirect selection or multiple selection schemes, such as selection indexes.[6]

In the within-race analysis, considering that the value of genetic parameters is more modest than the general analysis, the grain yield would be indirectly improved by selection of traits, such as days to flowering, plant height at flowering, length of the middle leaf, width of the middle leaf, days to physiological maturity, plant height at maturity, and seed filling index in the Mercado grain race; width

Table 2 **General and within-race genetic(r_G) and phenotypic(r_P) correlation values among yield and 24 amaranth traits. Chapingo, Mexico**

	General		Mercado		Mexican		Nepal		Aztec	
Traits	r_P	r_G	r_P	r_G	r_P	r_G	r_P	r_G	r_P	r_G
Days to flowering	0.29*	0.46*	0.23	1.12*	–0.07	0.21	0.09	1.64*	0.41	0.97*
Plant height at flowering	0.33*	0.48*	0.40	1.09*	0.02	0.50	0.14	–0.64*	0.12	0.01
Leaves number	0.36*	0.53*	0.20	1.12*	0.78*	—	0.13	0.48	0.26	1.00*
Middle leaf length	0.09	0.08	0.37	1.15*	0.14	–0.37	0.16	–0.10	0.30	0.78*
Middle leaf width	0.22	–0.60*	0.22	1.10*	0.15	–0.78*	0.12	–0.99*	0.33	0.78*
Middle leaf petiole length	0.04	–0.34*	0.11	0.77*	0.07	–0.03	–0.06	–0.41	0.08	0.31
Middle leaf dry matter	0.09	–0.33*	0.17	—	0.04	–0.82*	–0.14	–0.92*	0.10	—
Middle leaf area	0.10	–0.37*	0.36	1.20*	0.64*	–0.53	0.12	–0.60	0.27	0.84*
Seed filling period	0.34*	–0.63*	–0.01	–0.03	–0.16	–1.15*	–0.07	—	–0.11	0.03
Stem diameter	0.38*	0.64*	0.27	—	0.25	—	0.12	0.09	0.27	0.50
Days to maturity	0.07	0.13	0.22	1.02*	–0.25	1.24*	0.04	0.63*	0.36	1.04*
Plant height at maturity	0.34*	0.49*	0.34	0.93*	0.18	–0.03	0.21	0.76*	0.28	0.70*
Inflorescence length	0.39*	0.74*	0.20	1.07*	0.08	—	0.07	–0.77*	0.20	—
Leaves dry matter	0.29*	0.53*	0.34	0.95*	0.10	–0.76*	0.12	0.04	0.32	1.55*
Stem dry matter	0.36*	0.53*	0.24	0.84*	0.14	—	0.26	1.02*	0.37	1.44*
Inflorescence dry matter	0.07	0.10	0.22	0.55	0.10	–0.75*	0.15	0.10	0.31	0.18
Seed dry matter	0.56*	0.88*	0.39	1.58*	0.37	—	0.28	–0.38	0.42	1.22*
Total dry matter	0.37*	0.53*	0.38	0.01	0.24	0.25	0.31	0.48	0.49	1.03*
Harvest index	0.50*	0.86*	0.28	1.58*	0.33	1.62*	0.16	–0.46	0.10	0.89*
Hectoliter weight	0.19	0.90*	0.08	—	0.08	—	–0.05	—	0.01	0.19
Economic yield rate	0.41*	0.95*	0.99*	1.00*	0.98*	1.01*	1.01*	—	0.99*	1.00*
Biological yield rate	0.41*	0.64*	0.38	0.82*	0.28	1.00*	0.31	0.44	0.15	0.96*
Seed-straw ratio	0.44*	0.78*	0.31	—	0.26	0.69*	0.14	–1.12*	0.04	—
Seed filling index	0.97*	0.97*	0.98*	0.99*	0.85*	0.99*	1.00*	1.10*	0.98*	1.00*

* $p < 0.05$.

of the middle leaf, seed filling period, stem diameter, days to physiological maturity, and seed filling index in Mexican grain race; petiole length of middle leaf, plant height at maturity in Aztec grain race; and days to flowering, plant height at flowering, length of the middle leaf, width of the middle leaf, days to physiological maturity, and plant height at maturity in the Nepal grain race.[6]

V. BREEDING SYSTEM

The cultivated amaranths have a mixed mating system with varying levels of cross pollination. Outcrossing rates vary significantly with the genotype and the environment.[5,6] A bidirectional experiment based on male/ female flower ratio per plant suggested that breeding system is under genetic control.[9] The outcrossing rates for a single landrace population varied from 5.8 to 28.8% for *A. caudatus* and from 3.5 to 14% for *A. hypochondriacus*; the outcrossing rate was 31% for *A. cruentus.*[19] Using the red-green marker the author found that the outcrossing rates for landraces from different grain races varied from 7.6 to 41.1%; the cross-pollination rates were 7.6, 12.9, 15.6, and 41.4% for Aztec, Mercado, Nepal, and Mexican grain races, respectively.

Within a population, outcrossing rates of individual plants have varied from complete selfing to complete outcrossing. A bidirectional mass selection was initiated for high and low outcrossing lines to demonstrate that there is genetic variability for outcrossing rates in *A. cruentus*. In the S1 generation, outcrossing rates were significantly different between the high and low lines, the low lines exhibiting much higher male fertility than the high lines. Such response indicated quantitative control of sex expression and, consequently, of the potential for outcrossing in this species. Thus, the breeding system in amaranth can be modified to suit breeding procedures for either inbreeding or outbreeding species.[9,12]

VI. BREEDING FOR SPECIFIC CHARACTERS

Several characteristics have been mentioned to describe the desired model plant or ideotype of grain amaranth. This description will continue to evolve as more is learned about the physiology of the plant, as well as about the climatic niches where the amaranth demonstrates a comparative advantage. The ideotype might also vary with the different production areas and agricultural systems. In general terms the ideotype of grain amaranth may be described as a plant that has a height of 1.0 to 1.5 m, maturity to match climatic requirements in each region, seedling vigor, unbranched, tan seeded, flower above the leaf canopy, insect and disease tolerance, minimum shattering, synchronous dry down of plant at maturity, large seed size, and high yield.[7] For the improvement of the amaranth for forage or vegetable, total biomass production might be important.[44] Nutritional and functional components such as protein, oil, and starch fractions may be important to specific breeding programs especially for industrial applications of amaranth. In the following paragraphs, some ideas on the improvement for some characters will be discussed, as well as the sources of these desirable traits.

A. EARLY MATURITY

Amaranth varieties are usually labeled as early, middle, or late, but the length of growing period may vary so much from one region to another or among races that terms cannot be used properly without reference to environment. According to species, races, and region amaranth cultivars range from 70 to 240 days to maturity. The difference is not only varietal but also environmental, especially for factors of day length and temperature. In general terms, a conventional classification of length of growing period can be established for different regions where amaranth is grown (Table 3).

Early varieties are required for regions with a short growing season and for regions with dry farming systems to escape drought conditions. Diverse sources of early maturity for U.S. latitudes have been identified. Accessions from the African grain race when grown at 40° latitude mature the earliest; some of them mature in 70 days. Some accessions of the Prima grain race mature in 90 days, and the short landraces of the Spike grain race mature in 80 days. Early maturing segregates are found in the Mercado grain race.[7,28] In Mexico, the main sources of early maturing accessions are Spike, Prima, Mercado, and Nepal grain races. The races Nepal, Spike, and Mercado were the most suitable materials for avoiding the drought condition due to its early maturing.

In a study of bidirectional mass selection for time to first anthesis, the response was asymmetrical and the second cycle response was smaller than the first cycle. This suggests a large additive term in the total genetic variance for time to anthesis where early and late flowering selection lines diverged by 20.5 days. Earlier flowering may increase grain yield due to several interrelated factors including avoidance of cooler temperatures, increase of harvest index through reduction in plant height and number of leaves, lengthening of the reproductive period, decreased lodging due to decreased plant height, reduction in water use by shortening the growth cycle, and avoidance of the stress from pest infestation.[48] A major gene is involved in early flowering time, and this may be useful for developing very short season varieties.[25]

B. REDUCED PLANT HEIGHT

Plant height is influenced by both the environment and the genetic make-up of the plant. Landraces from Aztec and Mixteco grain races can reach 3 m in height under optimal growing conditions, while landraces Spike and Prima grain races can grow 0.5 m under limited soil moisture. Another factor influencing plant height is the stage of plant growth when the plant initiates flowering. Plants that flower early in the season tend to be shorter than the plants with long vegetative period. Besides the environment which influences plant height, several sources of germplasm with dwarfing genes have been identified.[7] Dwarfing lines were selected from African, Nepal, and South American grain races. Other sources of dwarfing lines are the *A. hypochondriacus* × *A. cruentus*, *A. hypochondriacus* × *A. hybridus*, and *A. hypochondriacus* × *A. quitensis crosses.*[7,41,48] In *A. caudatus,* a recessive gene determines reduced plant stature, and it is associated with several abnormal linked or pleiotropic effects on plant growth.[48] In a 6 × 6 diallel analysis both additive and nonadditive variances were important for plant height; the magnitude of specific combining ability variance was higher than the general combining ability variance. It suggests that recurrent selection should be adopted for rapid modification of this trait. After a couple of recurrent selection cycles, the selected lines should be subjected to multilocation tests for further evaluation. Reciprocal recurrent selection can also exploit additive as well as nonadditive types of genetic variances which are necessary for rapid improvement.[41,49]

Table 3 **Growth-period duration (days) for cultivated grain amaranth in different environments**

	Climate (temperature)		
Growth-Period	Warm (21–26 °C)	Medium (17–20 °C)	Cool (13–16 °C)
Early	60	80	110
Middle	100	120	135
Late	120	140	160

C. RESISTANCE TO LODGING

Lodging is a problem with amaranth, especially when it is grown in areas with abundant rainfall. The large inflorescence that develops on the amaranth plant is easily drooped by the wind. The stems or roots of many amaranth materials do not have the required strength to withstand the wind storms. Another cause of lodging is the high incidence of stem borer *Sciara* sp. which tunnel the stem, until it is weakened and lodged. Stem strength is an additional criterion that is required to improve the standability. Stem breakage is common on some germplasm during the rain and wind storms. The stem is particularly susceptible to breakage during periods of rapid growth when it is turgid.[2,4,7,51] Preliminary observations indicate that many *A. cruentus* accessions may develop a larger tap root than *A. hypochondriacus* accessions. In Mexico, some landraces from the Nepal grain race show good standability.[4,7]

D. INSECT AND DISEASE TOLERANCE

The primary insect of economic importance on grain amaranth is the *Lygus* bug. In Pennsylvania it has been observed that *A. hypochondriacus* is less damaged than *A. cruentus*. Another important pest on grain amaranth is the stem borer *Sciara* sp. that is present in all the production areas in Mexico. Landraces of Aztec grain race are less damaged by this insect pest.[2,14,52,55]

The most important disease on grain amaranth is the Damping Off. It is a disease that affects the seeds and seedlings and is caused by soil fungi (*Pythium* spp., *Rhizoctonia* spp., and *Fusarium* spp.). These fungi may kill both the germinating seed and seedlings. All accessions from *A. caudatus* expressed a high degree of susceptibility to this disease, but accessions of *A. cruentus* and *A. hypochondriacus* were observed to be more tolerant.[2,14] Several amaranth species show resistance to *Pythium* damping off.[56]

E. MALE STERILITY

Male sterility has potential application in plant breeding, both in facilitating genetic recombination and in the production of hybrid seed. The small flower size and the complex structure of the inflorescence, combined with a mixed mating system, make it desirable to develop an efficient control of pollination required in breeding work. Male sterility has been identified in landrace populations of *A. hypochondriacus*. An analysis of segregating progenies within landraces showed monogenic or digenic determination of male sterility. However, a gene-cytoplasmic mode of inheritance was also identified. The use of male sterility in grain amaranth may be useful for recombination in population improvement schemes or in the utilization of backcrosses in varietal improvement and also for production of commercial hybrids or synthetic varieties.[5,7,57,58]

F. SEED SIZE

The small seed size can make it difficult to establish amaranth in crusting soils and with sparse soil moisture.[59] The seed sizes of the genera range from 0.37 to 1.21 g per 1000 seed weight, the low values corresponding to wild and weedy species and the high values to cultivated grain species.[9,60,61] Perhaps the seed size in grain amaranth cannot be increased much; however, some sources of bold seeded material have been identified. *A. pumilus* has the biggest seed of the genera; however, it cannot be crossed with cultivated grain species. It is necessary to find methods of transfering genetic information from this species to grain amaranth.[10,62] Data on four grain species show that polyploidy may increase the grain size in amaranth significantly.[63] Seed from determinate plants was larger than seed from indeterminate plants, indicating that the use of determinate varieties might be used to increase seed size.[5,48] The grain races Aztec, Mercado, Nepal, Mexican, South American, and Edulis have medium to large seed.[28]

G. DROUGHT RESISTANCE

Crops growing under marginal lands face the high limitations caused by stress environments. Most of the world's arable land is classified as semiarid, and drought is the main limiting factor for crop production. Drought stress causes low economic yields per unit area and, consequently, low crop production in commercial fields. The options for solving this problem are limited. The only possible solutions are through cultural practices that increase the availability of stored soil moisture or by developing crop varieties or hybrids that can more efficiently avoid or tolerate the drought period, which said in other words, is through plant breeding. Because drought stress is a very complex phenomenon, the plant breeder requires a reasonable knowledge and understanding of the causes and consequences involved, particularly the mechanisms of drought resistance.[64-68]

In general there are two main responses: resistance and susceptibility. Stress resistance can be defined as the ability of the plant to survive the unfavorable factor and even to grow in its presence. By analogy, stress susceptibility can be defined as the inability of the plant to survive the unfavorable factor. Three major mechanisms of drought resistance have been defined: drought stress avoidance, drought stress tolerance, and drought stress escape.[69] A drought avoider must maintain a high water potential when exposed to an external water stress. There are two types of avoiders: (1) the water savers that avoid drought by conserving water and (2) the water spender which avoids drought by absorbing water rapidly enough to keep up with extremely rapid water loss. Both mechanisms maintain the plant turgid and with high water potential. In amaranth, several mechanisms permit it to conserve its supplies, like a water saver, under low water conditions. Some are stomatal closure, cuticular barriers to water loss, and decrease in transpiring surface. These mechanisms of water conservation tend to decrease growth as a result of low transpiration and photosynthetic rates caused by closure of stomata, wax accumulation in the cuticle, and reduction of leaf area. Leaf area reduction is due to hanging leaves when wilted under light water stress or by the falling of leaves during severe water stress.[70,71] The C4 metabolism present in amaranth plants shows high photosynthesis rates under high temperatures and light intensity, with low photorespiration and high efficiency in water use. On the other hand, amaranth as a water spender shows a high rate of water absorption. This is due to the main root of amaranth plant which penetrates into the soil up to 2 m with numerous lateral roots. This results in an especially surprising recovering capacity of the amaranth plant after a severe water stress.[18,70]

In opposition to the general protection conferred by the drought avoidance against all kinds of drought injury, drought tolerance must be highly specific in nearly all cases. Drought tolerance with low water potential is due to different mechanisms: maintenance of turgor and desiccation tolerance. The amaranth plant maintains the turgor due to osmotic adjustment.[70-72] Drought escapers are plants that complete their life cycles in arid habitats during the brief periods free of water stress. In amaranth, some experiences suggest that drought escape is due to rapid phenological development and phenotypic plasticity.[18,50,73] Some grain races such as Spike, Nepal, Prima, Mexican, and African can be sources of short cycle cultivars or earliness. Breeding for high yields under drought conditions has several limitations. A good knowledge and understanding of different factors involved in the effects of drought stress on plants, and especially the mechanisms of drought resistance, will enable the plant breeder to identify those traits with more possibilities for use in successful selection of desirable plants.

H. OTHER IMPORTANT TRAITS

Amaranth still is a partially domesticated crop and many cultural practices have yet to be developed to maximize production. A typical attribute of a less domesticated or wild primitive species is the seed shattering characteristic which is a survival and dispersal mechanism. This characteristic is a hazard in a cultivated crop because much grain can be lost at the seed maturity stage before the harvest. Shattering occurs when an abscission layer develops in the utricule. Variation for shatter resistance in amaranth has been reported. Indehiscent utricules are characteristic of some forms in *A. hybridus, A. powelli, A. bouchonii, A. australis, A. spinosus,* and *A. tuberculatus,* and these could be used as sources of shatter resistance because of their good crossing ability with cultivated grain species.[5,74,75] Another typical characteristic of wild species is branching, which can make mechanical harvesting difficult in grain amaranth. The Aztec grain race shows the lowest level of branching.[10,28]

The panicle structure can be important in developing varieties for mechanical harvesting. The drooping panicle orientation of many *A. caudatus* landraces is controlled by two major genes with other minor genes modifying the panicle orientation. The determinate panicle growth is controlled by one gene in *A. caudatus*. This trait has played a role in species description. In earlier reports, determinate plants were considered to be a separate species, *A. edulis* or *A. mantegazzianus*, but was later considered a conspicuous mutant of *A. caudatus*.[48] Recently, the author found determinant plants within a landrace from the Aztec grain race in *A. hypochondriacus*, and it seems to be controlled by a single recessive gene.

VII. ACHIEVEMENTS AND PROSPECTS

Progress has been made in collecting, characterizing, and identifying the amaranth germplasm adapted to different geographical regions in many countries. A scheme for classifying grain races that clearly emphasizes the concordance of some morphological variants and ecophysiological variation is already being developed. Classification and naming of some of these groups of landraces has helped in the understanding and knowledge of the natural genetic variability of amaranth; however, further studies are needed to clarify the clustering of these groups. Plant breeding programs were initiated to develop improved varieties for both mechanized production and traditional agricultural systems. The first improved lines grown by farmers were developed using mass and individual selection from the cataloged amaranth germplasm collections. There are now improved varieties of grain amaranth that have been released.

In Mexico, the variety "Revancha" developed from the *A. hypochondriacus* Mercado grain race has been released; it is earlier than the native varieties from the highlands and suitable for mechanical harvest.[76] In Peru, the varieties "Oscar Blanco" and "Noel Vietmeyer" from *A. caudatus* have been released.[31] In Ecuador, the varieties "INIAP Alegria" and "INIAP Ataco" from A. caudatus were released.[77] In Kenya, an improved line has been developed from the *A. hypochondriacus* Nepal grain race and was named "Jumla".[32] The first released varieties in the U.S. are "Amont" from the *A. cruentus* Mexican grain race and "Plainsman" originated from a cross between the *A. hypochondriacus* Mercado grain race and the *A. hybridus* Prima grain race.[1] In India, an improved variety from the *A. hypochondriacus* Nepal grain race named "Annapurna" has been released.[78]

The amount of breeding work conducted on amaranth has been minuscule compared with that done on staple crops. Even now, grain amaranth is considered an undeveloped crop because of its many problems. The primary limitations to amaranth production are in the seeding phase and in the harvest phase, difficulties that are related to the tiny seed, branching, plant height, lodging, seed shattering, and dry-down. These traits could be improved by using the genetic stocks described above and so obtain improved varieties without these problems.[79] Better commercial hybrids can now be developed by means of male sterility lines as female parents and predominantly male lines as male parents. This strategy can be useful in developing synthetic varieties. For the early breeding phase of a new crop, a variety of choices in the breeding methods is a valuable asset; recurrent selection and mass-pedigree selection should be explored, and heterosis among grain races should be sought.

Drought resistance is a feature of amaranth that make it attractive for productive areas deficient in water. Drought resistance and high yield are under genetic control; however, the expression of genes responsible for each mechanism is highly dependent on the environment. Therefore, it is of particular importance to choose the right environment for selection. On the basis of meteorological data of each region of interest, it is possible to determine which is the most adequate selection environment. A reasonable knowledge of the principles of plant breeding, particularly those aimed at improving crop yields under drought conditions, is necessary. Combining those factors with some experience and familiarity with genetic variability will result in better and specific varieties for different levels of drought. These varieties will have a different yield potential depending on the environment since the amaranth plant shows a high degree of phenotypic plasticity, which is the ability of an individual organism to alter its physiology and morphology in response to changes in environmental conditions.[80]

From the geneticist's point of view, some characteristics such as easy growing, short generations, large progeny per plant, control of matings, genetic diversity, developmental and morphological plasticity, and more experimental advantages make the amaranth a model organism for genetic studies. Now a broader concept of amaranth is needed; it will include the preferences and necessities of farmers, food industry, consumers, and scientists in order to enhance the possibilities of success in developing grain amaranth.

REFERENCES

1. **Kauffman, C.S.,** Realizing the potential of grain amaranth, *Food Rev. Int.,* 8(1), 5, 1992.
2. **Espitia, R.E.,** Amaranth germplasm development and agronomic studies in Mexico, *Food Rev. Int.,* 8(1), 71, 1992.

3. **Sauer, J.D.,** The grain amaranths: a survey of their history and classification, *Ann. Mo. Bot. Gard.*, 37, 561, 1950.
4. **Weber, E.L., W.W. Applegate, D.D. Baltensperger, M.D. Irwin, J.W. Lehman, and D.H. Putnam,** *Amaranth Grain Production Guide,* Rodale Press, Emmaus, PA, American Amaranth Institute, Bricelyn, MN.
5. **Kulakow, P.A. and S.K. Jain,** Grain-amaranth-crop species, evolution and genetics, in Proc. 4th Natl. Symp., Amaranth: Perspectives on Production, Processing and Marketing, Minneapolis, MN, August 23- 25, 1990, 105.
6. **Espitia-Rangel, E.,** Variabilidad genética e interrelaciones del rendimiento y sus componentes en alegría (*Amaranthus* spp.), master's thesis, Colegio de Postgraduados, Montecillo, México, 1991, 119.
7. **Weber, E.L. and C.S. Kauffman,** Plant breeding and seed production, in Proc. 4th Natl. Symp., Amaranth: Perspectives on Production, Processing and Marketing, Minneapolis, MN, 1990, 115.
8. **Joshi, B.D. and R.S. Rana,** Grain amaranths: the future food crop, NBPGR, New Delhi, India, 1991, 3.
9. **Sauer, J.D.,** The grain amaranths and their relatives: a revised taxonomic and geography survey, *Ann. Mo. Bot. Gard.,* 54(2), 103, 1967.
10. **Espitia-Rangel, E.,** Recursos genéticos de Amaranto (*Amaranthus* spp.), in Avances en el estudio de los recursos fitogenéticos de México, Ortega, P.R., H.G. Palomino, G.F. Castillo, H.V. González and M.M. Livera, Eds., SOMEFI, Chapingo, México, 1991, 197.
11. **Sauer, J.D.,** Identity of archeologic grain amaranths from the Valley of Tehuacan, Puebla, Mexico, *Am. Antiq.,* 34(1), 80, 1969.
12. **Sauer, J.D.,** The history of grain amaranths and their use and cultivation around the world, in Proc. 1st Amaranth Seminar, Maxatawny, PA, July 29, 1977, 9.
13. **Maldonado, A.U. and L.A. Estrada,** Amaranth genetic variation and utilization in Mexico, in Proc. 3rd Amaranth Conf., Washington, D.C., 1986, 120.
14. **Sanchez, E.M.C. R.E. Espitia, and K.S. Osada,** Etiología de la Mancha Negra del Tallo (*Macrophoma* sp.) en amaranto, in Primer Congreso Internacional del Amaranto, Oaxtepec Mor. México, Septiembre 22 to 27, 1991, 67.
15. **Wilson, R.L.,** Insect and disease pest amaranth, in Proc. 4th Natl. Symp., Amaranth: Perspectives on Production, Processing and Marketing, Minneapolis, MN, 1990, 163.
16. **Vazquez, C.G., R.E. Espitia and S.A. Marquez,** Potencial de reventado y calidad proteinica del amaranto, In Investigaciones recientes sobre amaranto, Reyna T.T., Ed., UNAM, México D. F., 1988, 79.
17. **Cavagnaro, J.B. and S.K. Jain,** Effect of water stress in four *Amaranthus* species, *Amaranth Newsletter,* September 1985.
18. **Espitia-Rangel, E.,** Efecto del déficit hídrico en dos etapas fenológicas en amaranto, in XIV Congreso Nacional de Fitogénetica, Tuxtla Gutiérrez Chis. México, Octubre 4 to 9, 1992, 56.
19. **Jain, S.K., H. Hauptli, and K.R. Vaidya,** Outcrossing rate in grain amaranths, *J. Hered.,* 73, 71, 1982.
20. **Hauptli, H. and S.K. Jain,** Genetic variation in outcrossing rate and correlated floral traits in a population of grain amaranth (*Amaranthus cruentus*), *Genetica,* 66, 21, 1985.
21. **Pal, M. and T.N. Khoshoo,** Grain amaranths, in *Diversity and Change in Indian Subcontinent,* J.B. Hutchinson, Ed., Cambridge University Press, 1974, 129.
22. **Coons, M.P.,** Relationships of *Amaranthus caudatus, Econ. Bot.,* 36(2), 129, 1982.
23. **Hunziker, A.T.,** Los pseudocereales de la agricultura indigena de América, Direc. Public. Universidad Nacional, Córdoba, Argentina.
24. **Pal, M.,** Evolution and improvement of cultivated amaranth. I. Breeding system and inflorescence structure, *Proc. Indian Natl. Sci. Acad.,* 38(1-2), 28, 1972.
25. **Kulakow, P.A. and S.K. Jain,** The inheritance of flowering in *Amaranthus* species, *J. Genet.,* 64, 85, 1985.
26. **Kigel, J. and B. Rubin,** *Amaranthus*, in *Handbook of Flowering,* Vol. 1, A.H. Halevy, Ed., CRC Press, Boca Raton, FL, 1985, 427.
27. **Kauffman, C.S. and C. Reider,** *Rodale Amaranth Germplasm Collection,* Rodale Pres, Emmaus, PA, 1984, 114.
28. **Kauffman, C.S.,** The status of grain amaranth for the 1990s, *Food Rev. Int.,* 8(1), 185, 1992.
29. **Espitia-Rangel, E.,** Razas mexicanas de amaranto, XIV Congreso Nacional de Fitogenetica, Tuxtla Gutiérrez, Chis. México, Octubre 4 to 9, 1992, 669.

30. **Espitia-Rangel, E.**, Caracterización y evaluación preliminar de germoplasma de Amaranthus, Thesis, Universidad Autónoma Agraria Antonio Narro, Saltillo Coah, México,1986,162.
31. **Sumar, K.L., N.J. Pacheco, R.C.A. Inca, H.G. Castelo, P.R. Aedo, Ch. Y. Callo, and J.E. Valdeiglesias,** Grain amaranth research in Peru, *Food Rev. Int.*, 8(1), 87, 1992.
32. **Gupta, V.K. and D. Thimba,** Grain amaranth: a promising crop for marginal areas of Kenya, *Food Rev. Int.*, 8(1), 51, 1992.
33. **Obilana, A.T. and A.R. Hallauer,** Estimation of variability of quantitative traits in BSSS by using unselected maize inbred lines, *Crop Sci.*, 14, 99, 1974.
34. **Ehdaie, B. and J.G. Waines,** Genetic variation, heritability and path-analysis in landraces of bread wheat from Southwestern Iran, *Euphytica,* 41, 183, 1989.
35. **Burton, G.W.,** Quantitative inheritance in pearl millet (*Pennisetum glaucum*), *Agron. J.*, 43, 409, 1951.
36. **Joshi, B.D.,** Genetic variability in grain amaranth, *Indian J. Agric. Sci.*, 56(8), 574, 1986.
37. **Robinson, P.,** Heritability: a second look, in *Statistical Genetics and Plant Breeding*, W.D. Hanson and H.F. Robinson, Eds., NAS-NRC, Publication 982, 1963, 609.
38. **Joshi, B.D., K.L. Mehra, and S.D. Sharma,** Cultivation of grain *Amaranthus* in the Northwestern hills India, *Indian Farming,* 32(12), 34, 1983.
39. **Vaidya, K.R. and S.K. Jain,** Response to mass selection for plant height and grain yield in amaranth *(Amaranthus), Plant Breed.*, 96, 61, 1987.
40. **Kulakow, P. and S. Jain,** Genetics of grain amaranths. IV. Variation and early generation response to selection in *Amaranthus cruentus* L., *Theor. Appl. Genet.*, 74, 113, 1987.
41. **Pandey, R.M.,** Genetics of agronomic traits in *Amaranthus, SABRAO J.*, 14(2), 121, 1982.
42. **Robinson, H.F., R.E. Comstock, and P.H. Harvey,** Genotypic and phenotypic correlations in corn and their implication in selection, *Agron. J.*, 43, 282, 1951.
43. **Goldenberg, J.B.,** El empleo de la correlación en mejoramiento de plantas, Fitotec. Latinoam., 5(2), 1, 1958.
44. **Pandey, R.M.,** Genetic associations in *Amaranthus, Indian J. Genet. Plant Breed.*, 41(1), 78, 1981.
45. **Espitia-Rangel, E.,** Caracterización y evaluación preliminar de germoplasma de Amaranto, in *Primer Coloquio Nacional del Amaranto*, Querétaro, Qro. México, Agosto 13 to 15, 1987, 113.
46. **Lehman, J.W., R.L. Clark, and K.F. Frey,** Biomass heterosis and combining ability in interspecific and intraspecific matings of grain amaranths, *Crop Sci.*, 31, 1111, 1991.
47. **Kauffman, C.S. and P.W. Hass,** Grain amaranth: an overview of research and production methods, Rodale Research Report N-C-84-6, Rodale Press, Emmaus, PA, 1984, 13.
48. **Kulakow, P.A.,** Genetics of grain amaranth. II. The inheritance of determinance, panicle orientation, dwarfism, and embryo color in *Amaranthus caudatus, J. Hered.*, 78, 293, 1987.
49. **Comstock, R.E., H.F. Robinson, and P.H. Harvey,** A breeding procedure designed to use of both general and specific combining hability, *Agron. J.*, 4, 360, 1949.
50. **Espitia-Rangel, E.,** Evaluación de germoplasma de amaranto en regiones de temporal deficiente en La Mesa Central, XIV Congreso Nacional de Fitogenética, Tuxtla Gutiérrez, Chis. México, Octubre 4 to 9, 1992, 475.
51. **Espitia-Rangel, E.,** Situacion actual y problemática del amaranto en México, in *Primer Congreso Nacional del Amaranto*, Chapingo, México, Octubre 25 to 27, 1984, 1986, 101.
52. **Espitia-Rangel, E.,** Guía para cultivar amaranto en los Valles Altos de La Mesa Central, Folleto para Productores 18, INIFAP-SARH, Chapingo, México, 1989, 12.
53. **Wilson, R.L.,** Entomological research on amaranth at the North Central Plant Introduction Station, in Proc. 3rd Amaranth Conf., Kutztown, PA, September 11 to 13, 1984, 192.
54. **Wilson, R.L.,** Studies of insects feeding on grain amaranth in the Midwest, *J. Kans. Entomo. Soc.*, 62(4), 440, 1989.
55. **Espitia-Rangel, E.,** Plagas y enfermedades del amaranto en México, in Primer Congreso Nacional del Amaranto, Octubre 25 to 27, 1984, 1986, 233.
56. **Sealy, R.L., C.M. Kenerley, and E.L. McWilliams,** Evaluation of *Amaranthus* accessions for resistance to Dampping Off by *Pythium myriotylum, Plant Dis.*, 72, 985, 1988.
57. **Peters, I. and S.K. Jain,** Genetics of grain amaranths. III. Gene-cytoplasmic male sterility, *J. Hered.*, 78, 251, 1987.
58. **Gudu, S. and V.K. Gupta,** Male sterility in the grain amaranth (*Amaranthus hypochondriacus* ex-Nepal) variety Jumla, *Euphytica,* 37, 23, 1988.

59. **Lehman, J.W.,** The potential of grain amaranths in the 1990s and beyond, in Proc. 4th Amaranth Symp., Amaranth: Perspectives on Production, Processing and Marketing, Minneapolis MN, 1990, 1.
60. **Azurdia-Pérez, C.A.,** Aspectos relativos a la botánica del bledo (*Amaranthus* spp.), *Tikalia,* 7(1), 23, 1989.
61. **Joshi, B.D.,** Exploration for amaranth germplasm, *Plant Genetic Resources Newsletter* No. 48, IBPGR-FAO, Rome, Italy, 1981, 41.
62. **Brenner, D.,** The grain amaranth gene pools, In Proc. 4th Amaranth Symp., Amaranth: Perspectives on Production, Processing and Marketing, Minneapolis, MN, 1990, 193.
63. **Misra, P.S., M. Pal, C.R. Mitra, and T.N. Khoshoo,** Chemuric studies on some diploid and tetraploid grain amaranths, in Proc. Indian Acad. Sci. LXXIV (3), 155, 1971.
64. **Lewis, C.F. and M.N. Christhiansen,** Breeding plants for stress environments, in *Plant Breeding II,* K.J. Frey, Ed., Iowa State University, Ames, 1981, 151.
65. **Quinzenberry, J.E.,** Breeding for drought resistance and plant water use efficiency, in *Breeding for Less Favorable Environments,* M.N. Christhiansen, and C.F. Lewis, Eds., John Wiley & Sons, New York, 1981, 7.
66. **Sulivan, C.Y. and J.D. Eastin,** Plant physiological responses to water stress, *Agric. Meteorol.,* 14, 113, 1974.
67. **Hsiao, T.C. and E. Acevedo,** Plant responses to water deficits, water-use efficiency, and drought resistance, *Agric. Meteorol.,* 14, 59, 1974.
68. **Seropian, C. and C. Planchon,** Physiological responses of six bread wheat and durum wheat genotypes to water stress, *Euphytica,* 33, 757, 1984.
69. **Levitt, J.,** Responses of plants to environmental stress, Water radiation, salt and other stresses, Vol. II 2nd ed., Academic Press, New York, 1980, 25.
70. **Villasana G.F.A.,** *Características morfológicas y fisiológicas que confieren resistencia a la sequa al* amaranto (*A. Hypochondriacus*), Thesis, Universidad Autónoma Chapingo, Méx. México, 1989, 119.
71. **Cavagnaro, J.B.,** Effect of Water Stress and Dry Matter Partitioning in Four Species of *Amaranthus,* M.S. thesis, University of California, Davis, 1985.
72. **Trejo C., R., L., M.C. López, and O.J.L. Rodriguez,** Ajuste osmótico en células de amaranto *Amaranthus hypochondriacus* en medio salino (NaCl), In Primer Congreso Internacional del Amaranto, Oaxtepec, Mor. México, Septiembre 22 to 27, 1991, 81.
73. **Espitia-Rangel, E.,** Estabilidad del rendimiento en amaranto, in Primer Congreso Internacional del Amaranto, Oaxtepec, Mor. México, Septiembre 22 to 27, 1991, 65.
74. **Scaeffer, J.R.,** Considerations and constraints to the marketing of amaranth grain as a new comodity, in Proc. 4th Amaranth Symp., Amaranth: Perspectives on Production, Processing and Marketing, Minneapolis MN, 1990, 95.
75. **Brenner, D. and H. Hauptli,** Seed shattering control with indehiscent utricules in grain amaranths, *Legacy,* 3, 2.
76. **Espitia-Rangel, E.,** Revancha: Variedad mejorada de amaranto para los Valles Altos de México, in Primer Congreso Internacional del Amaranto, Oaxtepec, Mor. México, Septiembre 22 to 27, 1991, 64.
77. **Nieto C.C., N.C. Visos and M.M. Rivera,** INIAP alegra e INIAP ataco primeras variedades de amaranto (*Amaranthus caudatus*) para Ecuador, in Primer Congreso Internacional de Amaranto, Oaxtepec, Mor. México, Septiembre 22 to 27, 1991, 134.
78. **Joshi, B.D.,** Annapurna, a new variety of grain amaranth, *Indian Farming,* November 29, 1985.
79. **Putnam, D.H.,** Agronomic practices for grain amaranth, in Proc. 4th Amaranth Symp., Amaranth: Perspectives on Production, Processing and Marketing, Minneapolis MN, 1990, 151.
80. **Schlichting, C.D.,** The evolution of phenotypic plasticity in plants, *Annu. Rev. Ecol. Syst.,* 17, 667, 1986.

Chapter 4

Development and Ecophysiology of Amaranths

Jaime Kigel

CONTENTS

0-8493-5374-2/94/$0.00+$.50

I. INTRODUCTION

Amaranthus (Amaranthaceae) is a genus of tropical origin but widely distributed all over the world, including temperate regions.[1,2] About 50 species are native to the Americas and another 15 can be found in Europe, Asia, Africa, and Australia. Most amaranth species are pioneer nitrophilous annuals of open habitats and produce many small dormant seeds. This extensive seed production is associated with prolonged seed-dormancy and with prompt germination after soil disturbance and seed exposure to light. This assemblage of reproductive traits enables amaranths to survive by continuous colonization of new disturbed sites with full sunlight and little competition from other plants. With such strategy of reproduction it is not surprising that several species of amaranth were preadapted for invasion of habitats drastically modified by human activities. Thus, today, amaranths are best known as opportunistic weeds, associated with soil cultivation (e.g., *A. hybridus* L., *A. powellii* S. Watt., *A. retroflexus* L., *A. spinosus* L., etc.). However, other amaranth species have been domesticated in tropical and subtropical regions and are cultivated for their grain (*A. hypochondriacus* L., *A. cruentus* L., *A. caudatus* L. = *A. edulis* Spegazzini) or for leaf consumption (*A. tricolor* L. = *A. gangeticus* L., *A. viridis* L. = *A. gracilis* Desf., *A. blitum* L. = *A. lividus* L.).*

Grain amaranths originated in the New World, e.g., *A. hypochondriacus* and *A. cruentus* in Mexico and Central America; *A. caudatus* in the northern Andean region of South America. Several pot-herb species, such as *A. tricolor, A. melancholicus,* and *A. blitum* originated in the Old World (Asia and Europe).[1,3]

Taxonomically, the genus amaranth is divided into two sections, recognized mainly by the following characteristics:[1,2,3]

	Section	
	Amaranthus	***Blitopsis***
Number of tepals	5	3
Utricule dehiscence	Circumscissile	Irregular dehiscence or indehiscent
Inflorescence	Terminal and very prominent, usually branched	Mostly axillary, small when terminal

Grain amaranths belong to the section *Amaranthus*, while many other amaranth species that belong to the section *Blitopsis* have small terminal inflorescences and are grown for leaf consumption.

* Taxonomy in amaranths is controversial and synonymy is a common problem that contributes to taxonomic confusion. In this chapter, species identification is as in the cited references.

Most of the research on the physiology and development of amaranth has been carried out with the weedy species. Nevertheless, the data acquired can be applied to cultivated amaranths because of the taxonomic proximity between the cultivated and some of the weedy species as well as the similarities in their growth patterns and responses to environmental factors. Based on this approach and due to insufficient basic research on cultivated amaranths, the information in this review relies mainly on research pertaining to noncultivated amaranths.

II. PLANT MORPHOLOGY AND ANATOMY

Amaranths are herbaceous annual plants, with prostrate or erect stems, that may reach 2 m height. Different morphological types of grain amaranth have been described according to their branching pattern, height, inflorescence size and form, days to maturity, seed size, and color, etc.[4]

A. LEAF STRUCTURE AND FUNCTION

Amaranth leaves are alternate, entire, and petiolate. In *A. powellii* stomatal density of the adaxial leaf epidermis is about 70% of the abaxial epidermis density.[5] Amaranth leaves possess the Krantz-type leaf anatomy, characteristic to plants with the C_4-pathway of photosynthetic CO_2 assimilation. Each minor vein is surrounded by a chlorenchymatic bundle-sheath, which, in turn, is encircled by a layer of radially arranged mesophyll cells.[6] Mesophyll and bundle-sheath cells differ in their photosynthetic activities.[7,8] The initial fixation of CO_2 into C_4 acids is located in the mesophyll cells and due to the phosphoenolpyruvate carboxylase (PEP carboxylase) activity. The C_4 acids produced in the mesophyll cells are transported to the bundle-sheath cells where CO_2 is released for refixation by ribulose-1,5, biphosphate carboxylase/oxygenase (RuBP carboxylase/oxygenase, Rubisco). Thus, the C_4 pathway concentrates CO_2 near the Rubisco, reducing the photorespiration losses due to the oxygenase activity of the enzyme. In consequence, at high temperatures (>25°C) and high light intensities, the photosynthetic fixation of CO_2 by C_4 plants is about twofold higher than by C_3 plants, in which Rubisco is used directly for CO_2 fixation.[7,8] Amaranths belong to the NAD-malic enzyme type of C_4 plants, in which aspartate is produced from oxaloacetate in the mesophyll cells after CO_2 fixation (aspartate formers) and moves into the bundle-sheath cells, where it is converted back to oxaloacetate by transamination. Later, the oxaloacetate is reduced to malate, which is oxidatively decarboxylated by the NAD-malic enzyme, and the released CO_2 is refixed by the Rubisco present in the bundle-sheath cells. As in other aspartate formers,[7] the bundle-sheath cells in amaranth lack a suberin lamella and their chloroplasts and mitochondria are concentrated on the cell side adjacent to the vascular bundle.[9] Most of the mesophyll cells are in direct contact with the bundle-sheath cells. Chloroplasts of the spongy parenchyma cells have well developed grana, while those of the palisade parenchyma are poorly developed ones.[9] The large number and size of mitochondria in the bundle-sheath cells are characteristic to NAD-malic enzyme type species and play an essential role in CO_2 fixation.

Expression of genes encoding C_4 photosynthetic enzymes during early leaf development has been examined in *A. hypochondriacus.*[10] Genes encoding bundle-sheath or mesophyll specific cell proteins show distinct patterns of expression during leaf development. PEP-carboxylase is specifically localized in the mesophyll cells during early and late stages of leaf growth. However, in very young leaves (<10 mm) the chloroplast-encoded large subunit and the nuclear-encoded small subunits of Rubisco are present in both cell types. Rubisco subunits become more specific to the bundle-sheath cells as the leaf expands. The large and small subunits of Rubisco are not synthesized in the absence of light in *A. hypochondriacus* cotyledons.[11] In *A. cruentus* a chloroplast genomic library was constructed, and the gene for the Rubisco large subunit has been cloned.[12] Restriction analysis of the clone showed that the structure of this gene is similar to the one found in corn.

B. STEM ANATOMY

Amaranthaceae plants are characterized by anomalous secondary thickening of the stem, in which phloem is present within the secondary xylem (included phloem).[13,14,15] Secondary growth in the stem is due to an extra fascicular cambial meristem that is derived from parenchymatous cells on the periphery of the primary vascular bundles. The function of this meristem is abnormal. It forms internal derivatives which are added to the bulk of the secondary tissue, but externally the meristem only produces enough derivatives to perpetuate itself. The internal derivatives differentiate into secondary xylem, composed of fibers, parenchyma cells, and scattered clusters of vessels. Externally to these vessel clusters, the cambial cells produce small groups of phloem cells. Cells sandwiched between the vessels and the phloem mature into parenchyma. Externally to the phloem, the cambium continues its activity and produces more secondary xylem. This anomalous cambium is, therefore, a unidirectional cambium from which all the secondary vascular tissue is derived and lies on the internal side of the meristem. The outer cells of the cambium retain the capacity to divide, so that the layers of initials move continuously outward, as the secondary tissue increases in diameter. Cork-cambium originates in the subepidermal layers of the stem.[16]

C. ROOTS

Information on root morphology and spatial distribution in the soil, as well as its responses to nutrient and water availability, are of primordial importance for crop management and yield improvement. However, roots have been barely investigated in amaranths. Secondary thickening of the root is also anomalous, as described for the stem.[16] In *A. retroflexus* the root system of plants, grown without competition, develops rapidly, reaching nearly its maximum extent (2.4 m depth, 1.8 m spread) after only 10 weeks of growth.[17] This fast development of the root system may contribute to the competitive ability of amaranths. In *A. palmeri,* water extraction was smaller, compared to other weeds with similar root profiles.[18]

D. FLOWERS AND INFLORESCENCES

Species of amaranth are monoecious or dioecious.[1,2,19,20] In both cases, the basic unit of the inflorescence are little dichasial cymes (the glomerulus) producing staminate (male), pistillate (female), or both types of flowers. Structure of the cymose glomeruli is basically the same in both sections of the genus in monoecious and dioecious species. In monoecious species two kinds of flower arrangement are found:

1. The first flower of each glomerulus is staminate, while the following flowers in the dichasium are all pistillate. Thus, each glomerulus has only one staminate flower. This arrangement is typical to grain amaranth species.
2. Flowers of the upper glomeruli in the inflorescenes are all staminate, while in the lower glomeruli they are all pistillate (e.g., *A. spinosus*).

Flowers are small and inconspicuous, with a trimerous (section *Blitopsis*) or pentamerous (section *Amaranthus*) organization. Each flower is subtended by a bract and has 3 or 5 scarious tepals. Staminate flowers produce 3 or 5 stamens and, in mixed glomeruli, these flowers abscise shortly after pollen shedding.[3] Pollination is anemophilous. Pollen grains are trinucleate when released from the anther.[21] Pollen size ranges from 10 to 35 μm in different amaranth species.[21,22] They possess numerous germination pores, mostly more than 6 (i.e., 15 to 20 in *A. viridis*[21]), that are sunken and covered by flecks of sexine.[20] In the pistillate flowers, the ovary has only one ovule and a short solid style with a glandular stigma, branched 2 or 3 times.[21] In each glomerulus, pistillate flowers continue to be formed until the onset of seed-set.[23] When fertilization is prevented, many pistillate flowers are successively produced in the glomerulus.

E. FRUIT

The fruit is a one seed membranous utricule, with circumscissile dehiscence (a pyxide) in section *Amaranthus*, or with irregular dehiscence or indehiscent (an achene) in section *Blitopsis*.[2,3] In species with indehiscent utricules, the utricule itself abscises, enabling seed dispersal. Since seed shattering is one of the main problems of grain amaranth production, species or varieties with indehiscent and non-abscising utricules are of interest for breeders. Possible sources for these characteristics are indehiscent species with some cross-compatibility with grain amaranth, or available semi-indehiscent accessions. The genetic control of fruit dehiscence and abscission has not yet been worked out. Nevertheless, theoretically, it may be possible to combine both the indehiscent and the non-abscising characteristics of the utricules for production of new cultivars with reduced seed shattering.[24]

F. SEEDS

1. Development and Structure

Amaranths have small, lenticular seeds, with a curved embryo that surrounds a starchy perisperm.[25] As in other families of the Centrospermales, the main storage tissue in Amaranthaceae seeds is the perisperm. Thus, their storage tissue develops from the diploid nucelus and not from the triploid endosperm or embryonic cotyledons, as in most Angiosperms. Ovule and seed development have been described in detail in *A. caudatus*[26,27] and in *A. viridis*.[21] The one-locule ovary contains a single basal campylotropic ovule, with a very short funiculus. The ovule has two integuments (bitegemic), each containing two layers of cells. The micropyle is formed by the inner integument. Embryo sac is of the Polygonum type. In *A. caudatus,* the three antipodal cells degenerate at anthesis and collapse by the time of fertilization.[27] The pollen tube enters through the micropyle and after it reaches the embryo sac, syngamy and triple fusion occur simultaneously. The primary endosperm nucleus divides earlier than the zygote. As embryo development proceeds and reaches the heart stage, the nuclear endosperm that fills the embryo sac becomes cellular, starting at the micropylar end and extending towards the chalaza. Embryo development corresponds to the Chenopodial type.[21,27,28] During its development the embryo digests the endosperm. In the mature seed the endosperm is reduced to a thin layer, 1 to 2 cells in thickness, surrounding the entire embryo in *A. viridis* (except the side in contact with the perisperm),[21] or appears as a cap covering the radicle and hypocotyl in *A. caudatus.*[26] Endosperm cells have thick cell walls and contain many small, undefined spherical bodies imbibed in a spongy matrix, as viewed by scanning electron microscopy.[29] The membranous endosperm may play a regulatory role in seed germination (see Section III.B.2). Perisperm cells of mature seeds are densely packed with unusually small, e.g., 1 to 3 μm, starch grains that have a polyhedral form.[29] Seeds of grain amaranth may have opaque (waxy) or translucent (nonwaxy) perisperms. In *A. hypochondriacus,* starch grains from the perisperm of translucent seeds are nonglutinous and contain amylase and amylopectin, whereas those from opaque seeds are glutinous and contain almost entirely amylopectin.[30] This characteristic of the starch is controlled by a single major gene with nonglutinous starch dominant.[31] Amaranths were the first reported dicots containing both glutinous and nonglutinous starch, a feature found in cereal grains.

The testa of the mature seed is composed of three layers of cells that develop from the two integuments.[26] The external layer of the outer integument produces tall cells with a dark cell content and thickened outer tangential wall containing dark staining stalactiti.[32,33] In the inner integument, the outer cell layer collapses during seed development, while in the persistent inner layer the internal tangential wall becomes thickened and cutinized.[26,32]

The color of the seed coat in amaranth varies between black, brown, yellow, and cream-white. The chemical basis of the color differences is unclear. Tannins are present in the seed coat of brown and black seed of *A. hypochondriacus,* but are absent in white seeds.[33] When present, tannins occur in the stalactiti structures of the exotesta. Thus, seed color may be associated with the presence of anti-nutritive factors in the seed coat. In *A. caudatus,*[34,45] *A. hypochondriacus,*[34] and *A. cruentus*[36] crosses between black and white, or yellow and white, seeds show single gene inheritance, in which black and yellow seeds dominate. However, according to Kulakow et al.,[37] inheritance of seed color in *A. caudatus* and *A. hybridus* is controlled by two loci. Apparently, more than one gene system may be involved in controlling seed color.

2. Seed Dispersal

Seeds of amaranth lack special devices to increase dispersal in undisturbed sites and most of them remain near the parent plants.[38] Nevertheless, their small size allows some dispersion by wind when they drop from the inflorescence of tall amaranth species. Moreover, their round form and polished surface facilitates dispersion by running water (rain runoff, floods).[38] Amaranth seeds are also dispersed by birds and animals, either externally (mixed with mud on their feet) or internally (after being eaten — endozoochory). Herbivorous mammals swallow the seeds with the foliage. Some of the seeds withstand digestion, and their germination may even increase due to seed coat scarification.[38] Human agricultural activities greatly contribute to amaranth dispersal and today, several species are between the most noxious and persistent weeds all over the world.[39]

3. Seed Size

Seed size of amaranth species varies roughly between 1 to 2 mm in length and 0.2 to 1.1 mg in weight (Table 1). Grain amaranths are at the upper range of the seed weight. Seed size is a trait of ecological and agronomic importance, since it may affect germination behavior, seedling size, and its competitive ability. Even small differences in seed size may be crucial for successful emergence and establishment, particularly in small seed species. Available data indicate great intraspecific variability in seed size of weedy[40,41,42] and grain amaranths.[3] Differences of up to 20% from population seed-weight average were found in *A. dubius* even between individuals of the same population.[43]

Interpopulational variation in seed size has been reported for amaranth species occurring along geographic ranges. An elevational clina occurs in California for seed weight of *A. retroflexus,* with lighter seed produced by populations at higher elevations.[44] However, in field seed collections, variation in seed size might be induced phenotypically by local environmental conditions or due to genetic differences

Table 1 **Seed weight (mg/100 seeds) in Amaranthus spp.**

Weedy Amaranths		Grain amaranths	
A. albus	34[66]	A. caudatus	112[3]
A. blitopsis	54[66]	A. cruentus	99[3];71[188]
A. hybridus	36[66]	A. Edulis	101[3]
A. lividus	64[66]	A. hypochondriacus	103[3];54[188]
A. powellii	51[45];41[188]		
A. retroflexus	56[66];32–47[46];36[188];37[45]		
A. spinosus	15–24[187]		
A. tricolor	80[66]		

between populations adapted to local conditions. Interpopulation phenotypic variation can be reduced by growing plants from the different populations in the same environment, where seeds are produced under similar conditions. Using this approach, it has been found that, in populations of *A. retroflexus* and *A. hybridus* inhabiting a transect from central Canada to the southern U.S. (52° to 35° latitude), seed weight is positively correlated with latitude, e.g., larger seeds were produced by the northern populations.[39,45] Furthermore, while germination at 35°C was high and similar in all populations of both species, germination at 20°C (near the minimum germination temperature) was higher in the northern populations of *A. retroflexus*.[40] Larger seeds and the ability to germinate at lower temperatures may be of advantage in northern regions with shorter growing seasons, ensuring earlier germination and faster establishment. However, in *A. hybridus* no correlation was found between latitude and germination at 20°C.

Based on the results obtained by McWilliams et al.[40] additional populations of *A. retroflexus* were studied by Schimpft,[42] who found that seed weight correlates better with average rainfall than with latitude or length of growing season. Larger seeds were associated with dryer environments, probably due to their capacity to establish seedlings from deeper, and therefore wetter, soil layers. These studies clearly indicate genetic differences of seed size in populations of *A. retroflexus* adapted to a wide range of environmental conditions.

4. Environmental Effect on Seed Size

Evidence for environmental phenotypic effects on seed size can be obtained from studies under controlled conditions. In *A. retroflexus*, plants from different populations produced seeds heavier by about 20% under short days (SD) at 20/15°C than those produced at 27°C.[40] In another study[46] (Table 2), seed weight tends to increase when parent plants are grown at lower temperature or in shorter days. A reduction of 70% in solar radiation did not affect seed weight under (SD), but decreased it slightly under long days (LD). Water stress reduces plant growth and seed production in *A. retroflexus*, but bigger seeds are produced by the stressed plants.[47] Furthermore, a linear positive relationship between seed size and germinability was found in seeds produced under different irrigation treatments.

In grain amaranths, variation in mean seed weight occurs between different years at the same location.[3] However, the effects of individual environmental factors on seed weight of grain amaranth has not been investigated so far.

III. SEED DORMANCY AND GERMINATION

A. SEED LONGEVITY

Changes in seed germinability during dry storage have been studied in grain amaranth species.[3] *A. edulis* maintained high seed viability during 12 years of storage, while *A. caudatus* and *A. hypochondriacus* had a faster loss of viability (Figure 1).

Longevity of seeds buried in the soil is much shorter than under dry storage conditions.[48] After incorporation into the soil some seeds may be consumed or destroyed by different herbivores and pathogens, while others gradually senesce and decay. However, *in situ* germination appears to be the main fate of buried seeds. As in other domesticated crops, seeds of amaranth cultivars lack prolonged primary seed dormancy. Therefore, most of them germinate after sowing and do not persist in the soil. In contrast, weedy species of amaranth usually have prolonged seed dormancy and may persist for several years in the soil, e.g, seeds of *A. blitoides*, *A. graecizans*, and *A. retroflexus* germinated after 6 years of burial.[49] Seed longevity in the soil has been extensively studied in *A. retroflexus*. In Beal's[50] and Duvel's[51] long-

Table 2 **Environmental effects on seed weight (mg/100 seeds) of *A. retroflexus***

Effects of Photo- and Thermoperiod

Thermoperiod[2]	Photoperiod[1] LD	SD	LD →F SD
32/17°C	32.2[c]	44.4[a]	—
27/22°C	38.3[b]	43.6[a]	41.1[b]
22/17°C	44.7[a]	—	46.6[a]

Effects of Solar Radiation Level (at 27/22°C)

Solar radiation[2]	LD	SD	LD →F SD
100%	40.4[b]	41.1[ab]	41.1[ab]
30%	37.8[c]	43.2[a]	—

Based on Kigel, J., Ofir, M., and Koller, D., *J. Exp. Bot.*, 28, 1125, 1977 and unpublished data.

[1]LD = 16 h; DS = 8 h; LD →F SD: plants transferred from LD to SD at onset of flowering; [2]Solar radiation reduced by screens.

Note: Means with different letters are significantly different ($p = 0.05$) by Duncan's test. DW was determined after 24 h drying at 75°C.

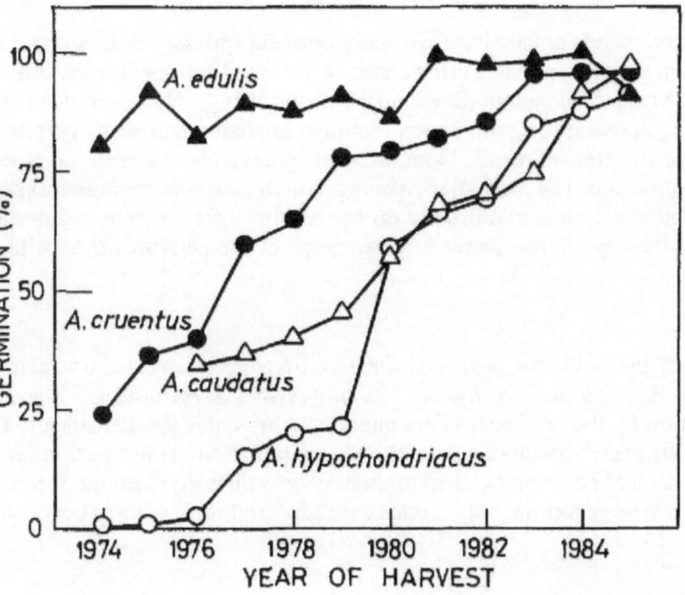

Figure 1 Changes in germinability of grain amaranths during storage. (Adapted from Josi, B.D. and Rana, R.S., *Grain Amaranths: The Future Food Crop*, National Bureau of Genetic Research, Shimla Science Monograph No. 3, India, 1991.)

term burial experiments, a few seeds still germinated after 40 and 10 years of burial, respectively, in undisturbed sites. Under controlled conditions (20°C in darkness) seed viability was maintained during 6 years of wet storage.[52] On the other hand, factors favoring germination reduce seed longevity and persistence in the soil. Soil disturbance by tillage that results in increased soil aeration and exposure to light induces germination of light, responsive-seeds. In experimental tilled plots, the soil population of *A. retroflexus* seeds decreased exponentially and was practically nil after 3 years of burial.[53] Seed viability decreased from 94% after harvest to less than 10% after 1.5 years of burial, regardless of the depth of burial (between 8 and 38 cm). Very few seeds germinated after 2.5 years of burial, even though there was no decrease in viability during this period in dry-stored seeds.[54] In other experiments, about 20% of the seeds buried 5 cm deep at the end of the summer remained dormant at the end of the following spring, and only 1 to 2% of the original population was still viable, but deeply dormant, after 2 years of burial.[50] In these experiments a few seeds always remained viable in the soil after 5 years of burial, producing between 1 and 5 emerged seedling per square meter. Thus, seed longevity is not the only trait in the reproductive biology of *A. retroflexus* that contributes to its persistence as a weed. Its persistence is due to a combination of great seed output per plant and the survival of a few seeds each year. These seeds will be enough to produce a few plants that, due to their great seed production, will ensure reinfestation of the field, even if the seed life span in the soil is shortened by frequent tillage.

Seed dormancy and, therefore, prolonged longevity can be imposed by light filtering through the canopy. In *A. patulus*[55a] a most of the seeds buried at 10 cm depth germinated shortly after burial, about 20% of them were still dormant the next year, but no viable seeds remained 3 years after burial. In contrast, most (>80%) of the seeds buried at soil surface under a closed canopy did not germinate during the first year, and some of them were still viable 3 years later. Dormancy imposed by canopy filtered light does not endure, since seeds readily germinate after reburial at 10 cm depth.

B. SEED DORMANCY

Amaranths are annual plants that establish every year from seeds. Hence, the timing of germination is of crucial importance for successful establishment of the young seedling and for the ability of the mature plant to set seed before the end of the growing season. As in other species, the seed bank of wild and weedy amaranths is composed of seeds differing in their genetic and physiological characteristics. This results in a heterogeneous seed population, in which seeds differ in dormancy level and responsiveness to germination stimuli. Possible sources for this physiological heterogeneity in the seed-bank are endogenous correlative factors (seed position, plant age, etc.) or environmental factors (temperature, light conditions, water availability, etc.) that act during seed development. After dispersal, seed behavior is also modulated by conditions experienced during burial (temperature, water availability, gas composition, etc.).

At dispersal time, seeds of wild and weedy amaranths are usually dormant (primary dormancy).[56] As in many other species,[57] these dormant seeds germinate at first under very limited and specific conditions, e.g., light and high temperature in the case of amaranth (Section III.B.3 and II.B.4). However, due to the after-ripening processes, the light requirement for germination diminish gradually and seeds germinate under an increasingly broader range of temperatures.[58] Domesticated species of amaranth have been selected for lack of primary dormancy and/or fast seed after-ripening, which causes germination shortly after harvest. Therefore, seeds of cultivated amaranth usually do not require light for germination (are dark germinators) and, when fresh, they germinate under a wider range of temperatures than wild or weedy amaranths.

1. Seed Coat Effects

Dormancy in amaranth is imposed by the seed coat. Seed scarification by concentrated sulfuric acid or sand paper enhances germination of *A. caudatus*[59], *A. lividus*[60], *A. quitensis*[61], *A. retroflexus*[58,62] and *A. graecizans*.[63] Inhibition of germination by the seed coat is not due to lack of water uptake, since in the presence of water seeds imbibe even under conditions that prevent germination. Thus, restriction of germination is probably due to mechanical constraints, or to limited oxygen diffusion into the seeds. In *A. caudatus* seed coat scarification increases germination in otherwise low inhibitory oxygen concentrations.[54]

2. Stages of Germination

Three morphological stages have been described in the germination of *A. retroflexus*:[64]

Stage 1. Embryo expands due to water uptake and seed coat splits around the seed.
Stage 2. Radicle protrudes through the splitted seed coat, but is still enveloped by the endosperm.
Stage 3. Radicle penetrates through the endosperm and normal root growth follows.

Similar stages have been described for *Chenopodium album*.[65] Stages 1 and 2 are considered as incomplete germination. The fact that germination of seeds with a split seed coat is arrested indicates that the endosperm layer by itself is an obstacle for germination. Germination is arrested at stages 1 or 2 by

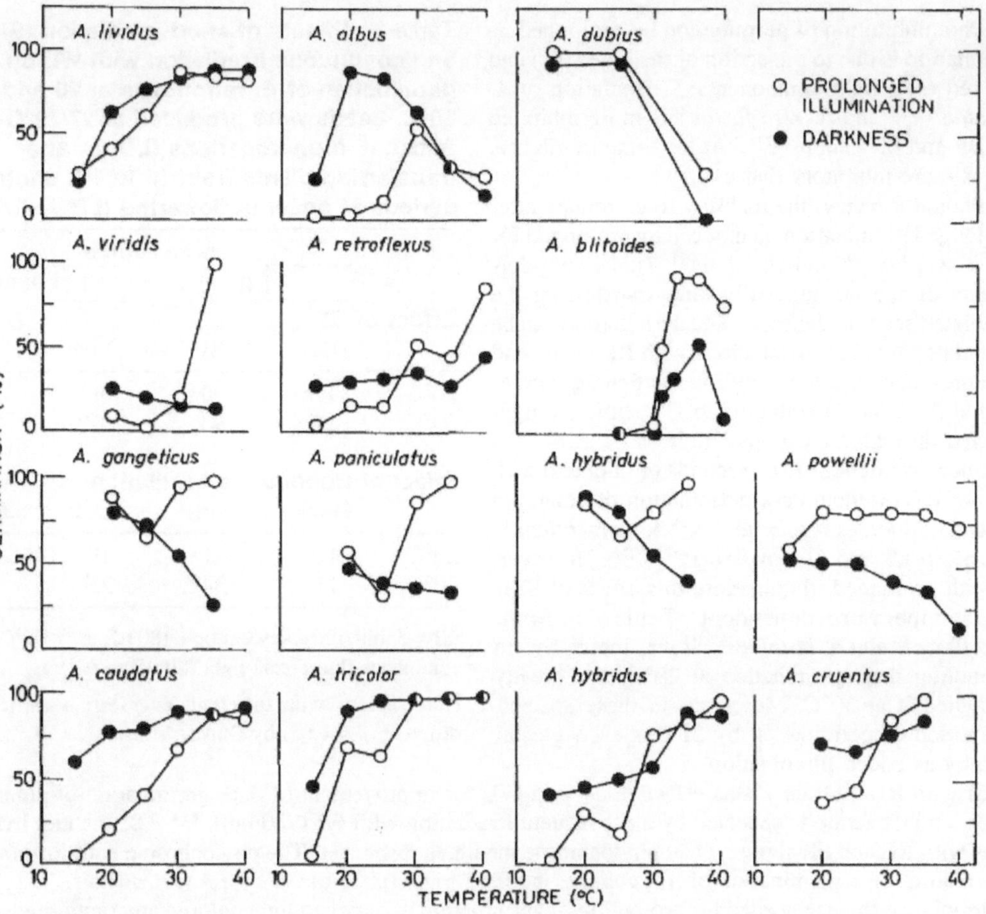

Figure 2 Seeds germination of *Amaranthus* spp. in darkness (●) or prolonged illumination (○) as affected by temperature. (Data is based on References 43, 66, 67, 69, and 100.)

low water potential or very low CO_2 levels. Light, ethylene, and higher CO_2 levels promote radicle penetration of the endosperm in seeds arrested by incubation at low water potential.[64]

In *Chenopodium* spp. incompletely germinated seeds remain viable.[65] Thus, arrested stage has been considered as an adaptation to environmental stresses, which permits a fast response of the seeds to improved conditions for germination and plant growth. However, in *A. retroflexus* many buried seeds that began germination in the soil and were arrested after seed coat splitting, for undetermined causes, do not complete germination after 20 months of burial and their embryos deteriorate.[55]

3. Light Effects

Most examined amaranth species produce photoblastic (light responsive) seeds, and their germination is under phytochrome control. However, the direction in which germination is affected by light (inhibition *vs.* promotion), as well as the extent of the seeds' response to light, are temperature dependent[43,59,66-69] (Figure 2). Prolonged illumination (continuous light or daily photoperiods, e.g., 12/12 or 16/8 h light/dark) with white light (WL) usually inhibits germination at temperatures below 25 to 30°C. At higher temperatures, prolonged illumination promotes germination in species with low dark germination and does not inhibit it in species showing high dark germination. Thus, extended illumination with WL has a dual effect on amaranth germination, e.g., it may promote or inhibit germination according to the temperature at which seeds imbibe.

A few exceptions to this general pattern of response to light × temperature interaction have been reported, e.g., lack of photoinhibition at lower temperatures in *A. retroflexus,*[69] or of photopromotion at higher temperatures in *A. albus,*[66] as well as lack of response to light in *A. dubius*[43] (Figure 2).

Photoinhibition of germination by prolonged illumination is due to the action of the blue (BL) and far red (FR) light components. Germination of *A. caudatus*[59,68] and *A. retroflexus*[70] is more inhibited by BL and FR than by WL. At the same irradiance, FR is more inhibitory than BL.[68]

Photodormancy (the inability to germinate after prolonged illumination) is induced by prolonged FR in *A. caudatus*[69] and *A. albus*.[71] This photo-dormancy disappears gradually after transferring the irradiated seeds to darkness, and germination can be stimulated by short irradiations with RL at the end of a prolonged illumination.[68] In the field, germination of *A. patulus* is inhibited by FR enriched light filtered through the canopy.[55a]

Short irradiations (SI) (seconds or minutes) with WL or RL promote dark germination of amaranth seeds (*A. blitoides*,[67] *A. cruentus*,[69] *A. gangenticus*,[69] *A. cruentus*,[69] and *A. retroflexus*[46,72,73,74]). However, as with prolonged illumination, this effect of SI is also temperature dependent (Table 3). In *A. retroflexus*[46] and *A. blitoides*[67] SI was ineffective in promoting dark germination at 20°C, but greatly promoted it at 30°C. Moreover, in these species, promotion of germination by SI was even greater than by extended illumination.

Table 3 Effects of short irradiation (SI) and continuous irradiation with WL on the germination of A. retroflexus at 20 and 30°C. Seeds were produced at 27/22°C under 16 h photoperiods (LD) or after transferring plants from 16 to 8 h photoperiods at onset of flowering (LD →SD)

	Seed source			
	LD		LD →SD	
Effect of SI	Dark	SI[1]	Dark	SI
20°C	17[c]	0[d]	36[b]	0[d]
30°C	14[c]	98[a]	37[b]	97[a]
Effect of Continuous Irradiation	Dark	Cont[2]	Dark	Cont
20°C	10[c]	1[d]	19[b]	2[d]
30°C	11[c]	34[a]	25[b]	49[a]

[1]SI = 5 min of fluorescent light (80 $\mu E\ m^{-2}s^{-1}$); [2]Cont = continuous fluorescent light (80 $\mu E\ m^{-2}s^{-1}$).

Note: Means with different letters are significantly different ($p = 0.05$) by Duncan's test.

SI with RL (660 nm) is as effective as with WL for improvement of dark germination of amaranth seeds, and its action is canceled by a subsequent irradiation with FR (730 nm).[68,69,70] Sequential irradiations with RL and FR show typical phytochrome mediated responses. The phytochrome equilibrium (P_{fr}/P_{tot}) required for germination of *A. caudatus* is >0.02[68] and 0.001 to 0.02 for *A. retroflexus*.[72]

Irradiance fluence needed for promotion of germination of dark and light colored amaranth seeds may differ due to variation in the light transmission characteristics of their seed coat. In *A. retroflexus* the dark seed coat reduces by *ca* 70% and 40% the fluence rate that reaches the embryo at 660 and 730 nm, respectively.[72] Thus, seed coats may act as light filters, and different phytochrome photoequilibria may be established under the same light conditions in seeds differing in their seed coat color.

Responsiveness to light appears gradually during imbibition. In *A. retroflexus*, hydration of 19% water content is reached 3 to 4 h after onset of imbibition and makes the seeds fully receptive to RL.[75] In imbibed seeds of *A. caudatus*, 25% of the phytochrome appears as P_{fr}; and phytochrome synthesis begins after 8 h of imbibition at 25°C.[68] In *A. retroflexus*, effectiveness of SI with WL or RL increases with imbibition time and is probably due to the gradual accumulation of P_r and the ensuing of larger amounts of P_{fr} obtained after the SI.[72]

In *A. albus*, SI with RL is followed by an increase in the relative amount of unsaturated polar lipids (mainly linoleic acid), while in seeds continuously imbibed in darkness, an opposite trend was observed.[76] These responses suggest the involvement of membrane lipids in the transition from the dormant to the nondormant state in these seeds.

Comparative analysis of published data on dark germination of amaranth species within the range of critical temperatures (10 to 45°C) reveals three main patterns of germination responses to increasing temperature in darkness (Figure 2):

1. Promotion of dark germination by higher temperature (*A. lividus*,[66] *A. hybridus*,[66] *A. blitoides*[67])
2. Inhibition of dark germination by higher temperature (*A. gangenticus*,[69] *A. hybridus*,[69] *A. dubius*,[43] *A. albus*[66])
3. Lack or small effect of temperature in species with low (*A. retroflexus*,[46,66] *A. viridis*[69]) or high (*A. caudatus*,[69] *A. tricolor*,[69] *A. cruentus*,[69] *A. paniculatus*[69]) dark germination

Amaranth species with high dark germination (mainly amaranth cultivars) probably accumulate a higher level of P_{fr} in their seeds during seed maturation, or have a lower P_{fr} requirement for germination, which enables dark germination. On the other hand, accelerated thermal reversion of P_{fr} to P_r in darkness may explain the inhibitory effect of increasing temperatures on dark germination of some amaranth species, e.g., at high temperature P_{fr} cannot be maintained long enough to promote germination. Finally, it is conceivable that in amaranth species in which high temperature promotes germination, the processes involved in breaking dormancy are accelerated by the high temperature and are accomplished before P_{fr} falls below a critical level.

The photoinhibition of germination in amaranth usually observed under lower temperatures is probably due to high irradiance reaction (HIR) of photomorphogenesis.[77] In *A. hypochondriacus* (a dark germinator), continuous WL at 25°C totally inhibits germination. At least 15 hours of dark incubation prior to the light are needed to escape the inhibitory effect of prolonged illumination.[78] In the case of inhibition by the HIR, germination processes initiated by P_{fr} can be arrested at later stages (incipient radicle emergence) by a prolonged period of FR (or WL enriched with FR) that acts before radicle elongation. It is assumed that the prolonged FR effect is not due simply to a reversal of P_{fr}, but involves a more complex operation of phytochrome in which the inhibition requires a high cycling rate of the $P_{fr} \rightleftarrows P_r$ photoconversion.[75]

4. Temperature Effects

Critical minimum and maximum temperatures for amaranth germination are approximately 10 and 45°C, respectively.[68,79] Freshly produced seeds need high temperatures for germination. The range of temperatures at which germination occurs broadens gradually during after-ripening in dry storage or burial in the soil.[80,81] Seed dormancy in amaranth is alleviated and dark germination improved by exposing imbibed seeds to low temperatures (chilling), brief shifts from high to low temperature, or daily temperature alternations.

a. Chilling

Exposure of imbibed seeds to 5 to 15°C in darkness for several days increases dark germination of *A. retroflexus*[46,82] and *A. quitensis*[61] after they are transferred to higher temperatures. Chilling also increases responsiveness to SI with WL or RL.[82] Apparently P_{fr} appears in the seeds during imbibition. At low temperatures (<15°C) high levels of P_{fr} are maintained, while at higher temperatures it is lost due to the fast thermal reversion to P_r, resulting in lower dark germination. To break dormancy by chilling, P_{fr} must be present during the chilling period. FR irradiation during chilling cancels its promotive effect. Although P_{fr} breaks dormancy even at low temperature, germination occurs only after temperature rises over a threshold minimal temperature.

b. Temperature Shifts

Seeds of *A. retroflexus* are stimulated to germinate in darkness by abrupt, brief (*ca* 2 h) shifts from 30 to 40°C to a lower temperature if P_{fr} is generated in the seeds before the shift.[83] The critical temperature below which dormancy is broken (P_{fr} action is initiated) is about 32°C. This critical dependence of phytochrome action upon temperature has been interpreted in terms of a cell membrane transition which is needed for P_{fr} action.[84] At high temperatures, the state of the membrane is apparently modified, preventing its interaction with P_{fr}.

c. Daily Temperature Alternations (Thermoperiods)

Temperature alternations and light are the recommended optimal conditions for amaranth germination.[85] The interactive effects between thermoperiod and light (12/12 hour photoperiods *vs.* darkness) on the germination of cultivated amaranths have been studied by Olaridan and Mumford.[69] In *A. viridis* dark germination was promoted by 30/20°C or 30/25°C, and light germination by 25/20°C and 30/25°C thermoperiods, compared to controls at constant 20, 25, and 30°C. However, in *A. cruentus, A. gangeticus, A. hybridus,* and *A. paniculatus* even 35/20°C thermoperiods (15°C day/night differential) did not affect germination, compared to light and dark controls at constant temperatures.

Temperature shifts are implicit in daily temperature alternations. Thus, lack of germination promotion by them appears to contradict the promotive effects of temperature shifts. However, when seeds are incubated under daily temperature alternations, temperature shifts are less abrupt, and seeds are kept for a longer time at each of the alternating temperatures. This may render the daily change in temperature ineffective, due to thermal decay of P_{fr}.

5. CO_2 and O_2 Effects

Reduction of CO_2 below its level in the air allows embryo expansion and seed coat splitting during imbibition of *A. retroflexus* seeds, but prevents radicle protrusion (incomplete germination).[64] Low levels of CO_2 also inhibit germination of seeds recovered after burial in the soil.[55] Germination of dormant seeds of *A. retroflexus* is not improved by raising CO_2 level above its concentration in the air.[55,64]

Lowering oxygen concentration below 70% results in a gradual linear reduction in the germination of *A. caudatus* seeds.[59] Nevertheless, limited germination (approximately 10%) was observed even at 1% O_2. Photoinhibition by continuous WL is alleviated by increasing O_2 above air level.[59]

6. Osmotic and Salinity Effects

Lowering water potential during seed imbibition by increasing concentration of PEG-6000 inhibits the germination of nondormant seeds of *A. hybridus*,[86] *A. caudatus*,[87,88] and *A. retroflexus*.[89] In *A. caudatus* practically no germination occurs below -0.35 to -0.40 MPa. However, in *A. retroflexus*, after-ripening increases the germination capacity at lower water potentials, and older seed stocks germinate even at -0.8 MPa.[89]

Osmotic inhibition is counteracted by GA_3,[90] as well as by ethylene,[64] by Ethephon[88] (an ethylene releasing compound), and by the ethylene precursor ACC.[87,88] The cytokinin benzyladenine (0.05 to 50 ppm) is ineffective in removing osmotic inhibition of germination in *A. caudatus*.[90]

In *A. retroflexus* incomplete germination occurs in a large proportion of seeds incubated under –0.4 MPa.[64] Although seeds imbibe and the embryo splits the seed coat, radicle penetration through the endosperm is inhibited by osmotic stress. This inhibition is relieved by ethylene and, to a smaller extent, by light or increased CO_2 level.[64]

Germination tolerance to salinity differs in different amaranth cultivars (Figure 3).[91] *A. tricolor* is much more tolerant to salinity than *A. blitum* and *A. cruentus* (e.g., 50% inhibition at 160 to 230 *vs.* 90 to 100 mM NaCl, respectively). In *A. retroflexus*, populations from alkali soils can germinate in substrates saturated with –1.25 MPa NaCl solutions.[45]

7. Hormonal Effects

Amaranth seed germination is promoted by GAs and ethylene, but inhibited by ABA.[88,92] The cytokinin benzyladenine is ineffective.[92]

a. Ethylene

Ethylene occurs naturally in the soil atmosphere[93] and is also produced by imbibed seeds.[87] Thus, changes in ethylene concentration in the soil or in ethylene metabolism in the seed are important for the regulation of seed dormancy and germination. Endogenous ethylene is required during germination of nondormant seeds of *A. caudatus* (dark germinator).[87] Their germination is inhibited by 2,5 norboradiene, a competitive inhibitor of ethylene action. This inhibition is counteracted by ethylene, by ethylene releasing Ethephon, and the ethylene precursor ACC, while it is enforced by AVG (an inhibitor of ethylene synthesis).

Exogenous ethylene induces germination of dormant seeds of *A. albus*,[94] *A. spinosus*,[94] *A. retroflexus*[94,95] in light and darkness, of osmotically inhibited seeds of *A. caudatus*,[87] and of light inhibited seeds of *A. hypochondriacus*.[78] Saturation response is usually in the range of 1 to 10 µl/l.[94,95] Exogenous ethylene apparently promotes two different consecutive processes during germination of *A. retroflexus* seeds.[64] The primary action stimulates embryo expansion that results in seed coat splitting. The secondary action is observed under stress conditions that inhibit radicle elongation and penetration through the endosperm (low water potential, CO_2 removal). These processes differ in their dose-response relationships. The first process is less sensitive to ethylene (threshold *ca* 0.5 µl/l, saturation at 50 µl/l) compared to the later process involved in radicle growth (threshold $<$0.02 µl/l, saturation *ca* 0.5 µl/l). In *A. retroflexus*, ethylene can stimulate germination of seeds with primary dormancy, or after secondary dormancy has been induced by preincubation at high temperatures (≥25°C) in darkness.[93] In the later case, responsiveness to ethylene increases with longer preincubation before exposure to ethylene,[95] or by seed burial overwinter in the soil.[55,96] Ethylene responsiveness of buried seeds is increased by higher nitrate levels in the soil.[55] Thus, seeds of *A. retroflexus* maintain their responsiveness to ethylene in the soil, which even increases during burial. Injections of ethylene into the soil stimulates the germination of *A. retroflexus*.[97] Thus, the stimulatory effects of ethylene on the germination of amaranth seeds has been proposed as a methodology for amaranth control in the field.[98]

b. Gibberellins

Exogenous GAs (usually GA_3) promote germination of dormant seeds of amaranth (*A. hybridus*,[99] *A. lividus*,[60] *A. powellii*,[100] *A. quitensis*,[61] *A. retroflexus*[73,74,92,100]). Promoting effects were observed even in darkness.[60,92] Effective concentrations of GA are 10^{-4} to 10^{-3}M and higher. These results suggest that endogenous GA is needed for germination processes or radicle emergence in amaranth seeds. This possibility is supported by the fact that in non-dormant (dark germinators) seeds of *A. caudatus*[101,102] and

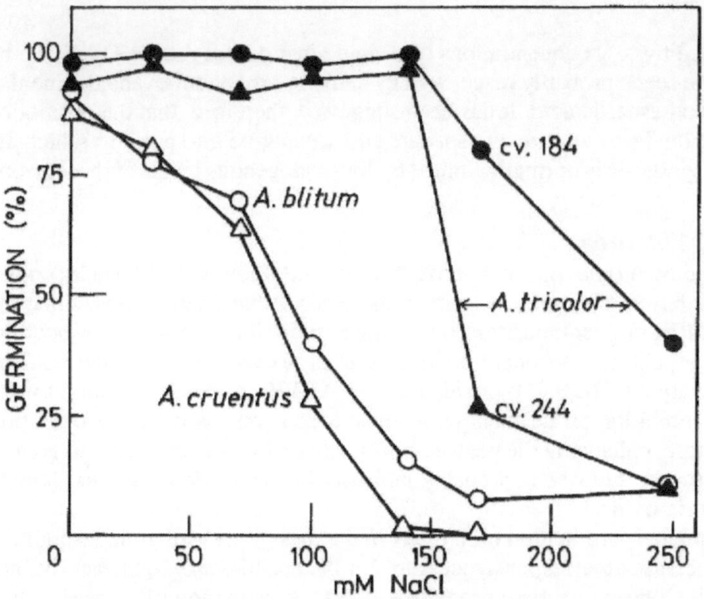

Figure 3 Seed germination of cultivated amaranths as affected by salinity. Germination was studied at 27°C, under prolonged illumination. (Adapted from Macler, et al., *Proc. 4th Amaranth Symp.*, Minneapolis, MN, 1990, 129.)

A. paniculatus,[103] inhibitors of GA synthesis (e.g., AMO-1618, ancymidol, paclobutrazol, tetcyclasis) greatly reduced dark germination. This reduction was reversed by exogenous GA_3[101,102,103] or GA_{4+7}.[88] Imbibition of *A. caudatus* seeds in water for more than 12 h results in escape from ancymidol inhibition.[102] Thus, synthesis of GA is needed during the early stages of amaranth germination.

Inhibition of germination in *A. caudatus* and *A. paniculatus* by GA biosynthesis inhibitors is also reversed by Ethephon and ACC.[88,101,102,103] Furthermore, in *A. caudatus*, simultaneous application of AVG increases the inhibitory effect of tetcyclasis.[88] Since AVG reduces ethylene synthesis by seeds of *A. caudatus*[87] it has been proposed that the promotive action of GA on amaranth seed germination depends on ethylene synthesis and that ethylene acts on the same or similar processes as GA. A phase-sequence model for seed germination responses of *A. retroflexus* to ethylene and other stimuli has been proposed by Schoenbeck and Egley.[104]

c. ABA

ABA (10^{-5} to 10^{-4} M) inhibits germination of *A. retroflexus*[92] and *A. caudatus.*[105] Endogenous ethylene production is probably involved in ABA inhibition, since Ethephon and ACC counteract the inhibition, while AVG increases the inhibitory effect of ABA.[105]

8. Effects of Herbicides and Other Chemicals

a. Herbicides

Spraying *A. retroflexus* plants with 2,4-D in the field did not affect seed weight. However, seed dormancy was lowered and seed germination was enhanced at 25°C under daily photoperiods.[62] Imbibition of *A. retroflexus* seeds in 5 to 50 ppm 2,4-D[62] or glyphosate[106] for a few hours improved their germination. Higher concentrations or longer imbibition periods had an inhibitory activity.

b. Fatty Acids

Behenic acid, a long chain fatty acid, delays the germination of *A. retroflexus* seeds at 24 or 35°C, in light and darkness, and totally inhibits it at 15°C.[107]

c. Anesthetics

Compounds with anesthetic-like properties in animals (e.g., diethyl ether, chloroform, n-alcohols, benzylalcohol, etc.) stimulate germination of dormant seeds of *A. albus*[71] and *A. retroflexus.*[108] However, in order to increase germination by anesthetics, the presence of the active form of phytochrome (P_{fr}) is required. In *A. retroflexus* the correlation between the membrane/buffer partition coefficient of the active substances and their relative promotion of germination is very low.[108] This fact does not support a connection between anesthetic activity and seed membrane lipophilic components.

d. Cyanide

Germination of *A. albus* is promoted by low concentrations of cyanide that do not reduce O_2 uptake by the seeds.[109] Cyanide supplied to the seeds probably reacts with cysteine to produce ß-cyanoalanine that is further converted to asparagine and aspartic acid. It has been suggested, therefore, that the promotion of germination by cyanide is due to the incorporation of aspartate and asparagine into proteins which are required for germination, and their synthesis is normally limited by low endogenous levels of these amino acids.

e. NO_3^-, NO_2^-, Hydroxylamine, Thiourea

Germination of *A. albus* is enhanced by nitrate, nitrite, hydroxylamine, and thiourea.[110,111] Their promotive effect is correlated with the inhibition of catalase activity in the seeds. It has been proposed that the metabolically derived H_2O_2, spared from decomposition by catalase inhibition, is used by a putative peroxidase which is coupled to the pentose-phosphate pathway that is needed for germination. This pathway requires O_2 for the regeneration of NADP by oxidation of $NADPH_2$, a reaction limited by the low levels of O_2 in the seed. Thus, inhibition of the catalase by the above chemicals increases oxidation of $NADPH_2$ by the putative peroxidase, enhancing the pentose-phosphate pathway and leading to greater germination. The effect of nitrate on the catalase is probably indirect, due to nitrite or hydroxylamine formed upon its reduction or by nitric oxide.

Nitrate (as $NaNo_3$ or KNO_3) promotes germination of *A. albus* in darkness[110] as well as under light.[112] Moreover, in *A. hybridus* KNO_3 increases the responsiveness to 2 h of incandescent light, but did not promote germination in darkness.[113] Contrary to these promotive effects, germination of *A. graecizans, A. hybridus, A. palmeri, A. retroflexus,* and *A. spinosus* under light at 30/22°C was unaffected by KNO_3 (0.125 to 0.25% solutions), and higher concentrations inhibited germination.[112] A similar lack of response has been reported for seeds of *A. retroflexus* imbibed in darkness[55] or buried in the soil.[98]

These contradictory effects of nitrate on amaranth germination arise probably from interspecific genetic differences, varying levels of seed dormancy, or testing conditions (e.g., seed O_2 levels during imbibition).

9. Modulation of Dormancy During Seed Development

a. Effects of Mother Plant

Variability in the dormancy of seeds produced by inflorescences located on different branches of the plant or produced at different developmental stages of the mother plant have been examined in *A. retroflexus*. Dormancy of seeds from lateral inflorescences is greater than in seeds produced by the main shoot inflorescence in plants grown in the field, but their germinability can be increased by removal of the main inflorescence.[114] An opposite trend was found in plants grown under controlled conditions, at 20°C LD. In these plants, seeds from lateral inflorescences were less dormant than those from the main shoot inflorescence, even though they started to flower at the same time.[115]

Seeds collected from senescing plants (yellow leaves, red stems) are smaller[41] and more dormant[46,47,74] than seeds from mature green plants. Similar trends were observed for plants in the field,[46] or under controlled conditions.[73,74]

Dormancy is also affected by the age (or size) of the plant when flowering is induced by SD. Older plants which are induced to flower by a few SD[84] produce seeds which are more dormant than those formed on younger plants under subsequent LD.[116]

b. Environmental and Seasonal Effects

Germination responses of amaranth seeds are modulated by the photothermal conditions experienced by the parent plants and the seeds during their development. In *A. caudatus* and *A. retroflexus*[46,74,115] seeds produced under LD were more dormant than those produced under SD. In *A. caudatus* these differences disappear shortly after harvest, while in *A. retroflexus* they persist during dry storage[46,74] or after they were buried in the soil over winter.[74] Seeds from SD have a higher dark germination, as well as greater responsiveness to chilling, SI, and GA_3. Germinability and responsiveness increase by transferring parent plants at the onset of flowering from LD to SD and decrease when plants are transferred at the same stage from SD to LD.[46] Dormancy of *A. retroflexus* seeds is increased by lowering the temperature at which parent plants are grown under SD[46] and LD.[74,115] Lowering the level of solar irradiation at which parent plants develop has opposite effects under LD or SD, e.g., it reduces dormancy at SD but increases it at LD.[46] These differences in dormancy induced by the parental photothermal environment were not correlated with differences in seed coat thickness or seed dry weight.[46]

Water availability during parent plant development affected seed dormancy and seed size in *A. retroflexus*.[47] Increasing water stress reduced plant size and seed production, but increased seed weight, as well as their dark germination and responsiveness to GA_3, or to SI with RL. Characteristics of seeds produced by plants transferred between irrigation treatments are determined by water availability during seed development, and to a lesser extent, by conditions previously experienced by the parent plant. A linear positive relationship was found between seed size achieved under the different irrigation treatments and germinability, e.g., bigger seeds obtained in the dryer treatments were the less dormant.

Seasonal changes in environmental conditions occurring during the development of the mother plant and their seeds affect seed dormancy. Germination behavior of seeds collected in the same site from populations of *A. hybridus* and *A. retroflexus* changed gradually during the year (Figure 4). In both species, seeds collected during the summer were more dormant than seeds collected during winter, as measured by dark germination and responsiveness to chilling, or to SI with WL. A similar seasonal trend was observed in *A. retroflexus* for seeds produced by plants sown between March and July in the field.[73] For each sowing date, seeds produced later in the season (from summer to autumn) by the same plants were less dormant than earlier produced seeds. Thus, conditions during seed maturation are more important than plant age or size in determining seed dormancy. The lower dormancy of seeds produced during the autumn and winter months is probably due to shorter days, lower light intensity, and lower temperatures. These environmental factors are known to reduce dormancy in *A. retroflexus*.[46,74,115]

10. Seedling Emergence in the Field

a. *Seasonality of Emergence, Dormancy Cycles*

Amaranth seeds require relatively high temperature for germination. Therefore, germination and seedling emergence in the field occurs during the warmer season (mid to late spring and summer). In *A. hypochondriacus,* best emergence occurs in the range of 24 to 34°C, although seeds are able to germinate above and below these temperatures.[79] In *A. palmeri,* emergence occurs above 18°C soil temperature and increases linearly with higher temperatures.[125] Minimal soil temperature for germination of grain amaranths in the field is about 18 to 20°C[117,118] and 25°C for *A. retroflexus*.[55] The time of the year during which these temperatures are reached differs in different regions and determines the length of the growth season until the autumn frosts. In weedy amaranths, emergence begins during mid or late spring[49,56,80,99,119], in response to increasing temperatures. Rate of emergence increases with higher temperatures in *A. hybridus, A. powellii,* and *A. retroflexus*.[112,120] After rising temperatures have stimulated the spring flush of emergence, rainfall and irrigation are the most significant factors influencing time-distribution of seedling emergence.[119,120,121] Seeds of *A. retroflexus* that do not germinate when temperatures first become favorable for germination do not enter secondary dormancy and may germinate later during the summer.[80] However, when summer progresses, nongerminated seeds enter secondary dormancy and show low germination during autumn and early winter, even when transferred to higher temperatures.[55] Seeds come out of dormancy again during the winter, but their germination in the field is prevented until soil temperature rises above a threshold level. This cycling of dormancy in the seed-bank is also characteristic of other summer annuals.[123] Order of emergence in fields infested with different amaranth species has been correlated to the extent of seed dormancy at dispersal time. In southern Spain, *A. blitoides* and *A. retroflexus* are less dormant and emerge before *A. hybridus* and *A. cruentus*.[99] In Ontario, Canada, seeds of *A. powellii* require lower temperature than *A. retroflexus* and *A. hybridus* for germination after overwinter burial in the soil.[45] Indeed, under controlled conditions, *A. powellii* germinates faster than *A. retroflexus* at low temperatures under both dark and daily photoperiods.[81]

b. *Depth of Emergence*

Amaranths have small seeds and limited food reserves to support germination and seedling emergence. This limitation restricts the depth from which amaranth seedlings can emerge. Practically, no emergence occurs in seeds of *A. hypochondriacus,*[79] *A. quitensis,*[61] *A. retroflexus,*[120] *A. caudatus,*[124] and *A. palmeri*[125] buried at 50 mm or deeper in the soil. Best emergence is achieved from 10 to 15 mm depth. Amaranth emergence is better in coarse textured soils than in fine textured soils, in which emergence is reduced by soil crusting under dryer conditions.[120] Sowing deeper than 15 mm may be needed to obtain adequate soil moisture for amaranth germination in dry land areas. This may be practical in friable soils if seeding rates compensate for reduced emergence due to the deeper sowing.[120]

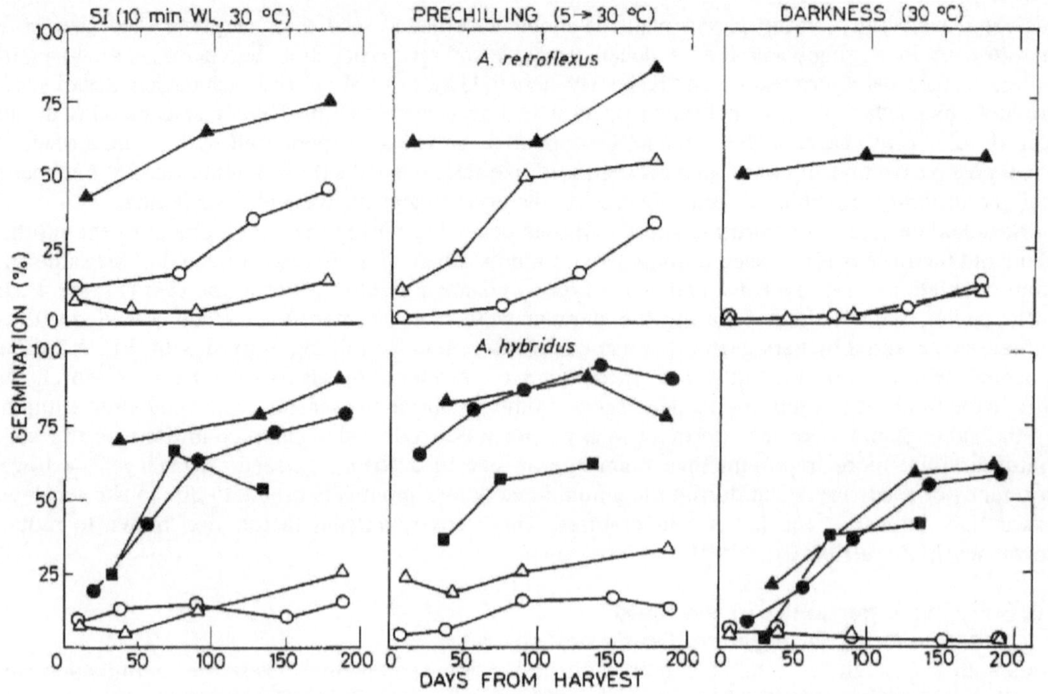

Figure 4 Seasonal effects on dark germination at 30°C of *A. hybridus* and *A. retroflexus* seeds, and on their responsiveness to a short irradiation (SI) with WL, or to 3 days prechilling at 5°C. Seeds were collected during 1986 to 1987: (●) December 12; (o) June 20; (△) - Aug. 10; (▲) January 29; (■) February 19 (Kigel, J., unpublished.)

In the tropics, high temperatures of 40 to 50°C at soil surface are quite common and may occur in freshly sown seed-beds of amaranth. A heat stress of 45°C for 10 h on sowing day strongly reduces seedling emergence of *A. caudatus*, but is less effective if it occurs 2 d after sowing.[122] Deeper sowing may reduce the chances for heat stress during the early stages of germination.

IV. VEGETATIVE DEVELOPMENT: ENVIRONMENTAL EFFECTS

A. LIGHT AND TEMPERATURE

Amaranths are more abundant in relatively warm regions and thrive under high solar irradiation. Since they are adapted to these conditions, their growth is greatly diminished by decreasing light intensity and low temperature.[126-130] As in most other C_4 species, the optimal temperature for amaranth biomass accumulation is relatively high and ranges between 30 and 40°C at high light intensity. Ontogenetic and interspecific variation in biomass allocation to different parts of the plant and environmental effects on these patterns of allocation have been little studied in amaranths. In *A. hybridus,* the relative allocation to root growth (as expressed by the root/shoot ratio) is not influenced by temperature[128] and light intensity,[129] or is slightly reduced by shade.[130] On the other hand, partitioning to stem growth is increased by higher temperature,[128] while partitioning to leaf growth is either unaffected[129] or reduced[128] by it. In *A. retroflexus* higher temperatures increase plant weight and leaf area, but the leaf area produced by unit of plant weight (leaf area ratio, LAR) is reduced.[127]

LAR is usually larger in C_4 than in C_3 plants, a trend which is also observed when amaranths are contrasted with comparable herbaceous C_3 species.[126,131] Reduction of light intensity increases relative biomass allocation to leaf growth and LAR of amaranths. However, the effects of irradiance on the relative growth rate (RGR) of *A. hybridus*[129a] and *A. retroflexus*[126] are relatively small compared to its effects on photosynthetic CO_2 assimilation (see Section VI.B). This small overall growth response is due to the compensatory opposite responses of the net assimilation rate (NAR, change in mass per unit area per day) and LAR, e.g., lower irradiance reduces NAR but increases LAR.

B. NITROGEN AVAILABILITY

1. Effects on Plant Growth

Amaranths are nitrophilous plants with strong positive growth responses to increasing nitrogen (N) availability. In cultivated amaranth, increased N application results in higher leaf yields.[132,136] However, plant parts are differentially affected by N availability. In *A. retroflexus,*[137,138] *A. spinosus,*[139] and *A. viridis,*[139] root growth is less influenced by N supply than shoot growth, so higher root/shoot ratios are observed in plants grown with low N supply. The stronger reduction in shoot growth by low N is mainly due to less branching and fewer and smaller leaves.[5,138] In *A. powellii,* nitrate application enlarges leaf area of mature leaves due to greater cell division and cell expansion.[5] Increasing levels of nitrate did not affect leaf water potential but resulted in higher cell solute concentration and a concomitant increase in turgor pressure. This may partly explain the larger cell expansion observed with increased nitrate application.

N effects on biomass accumulation and plant morphology were studied by growth analysis techniques in *A. powellii*[5] and *A. retroflexus.*[138] RGR, NAR, and LAR decreased with lower N levels. In *A. retroflexus,* the decrease in RGR with decreasing N supply was due mainly to a decline in LAR that was relatively greater than the decline in NAR. The lower availability of N reduced biomass allocation to the leaves, but did not affect the leaf area produced by unit leaf weight (specific leaf area, SLA). However, at a very low N level, leaf necrosis occurs before normal leaf senescence. These response patterns of RGR, NAR, and LAR to N deficiency have been reported for several other crops.[140] However, in *A. powellii* opposite trends were found.[5] While RGR and NAR decrease with N deficiency, leaf expansion is less affected than biomass accumulation, and SLA is even increased at low N. As a result, LAR increases with N deficiency. Thus, *A. powellii* is able to generate a relatively large leaf area even under limiting N availability and improve its competitive ability compared to other herbaceous species grown under similar conditions. Amaranths, like other C_4 species, usually have lower leaf N content than C_3 species.[5,138,141,142] The smaller amount of N required to produce leaf area and its photosynthetic machinery in C_4 species (see Section VII) probably allows proportionally more N to be diverted to leaf production, thus increasing amaranth capacity for light interception compared to C_3 plants.[138]

Interspecific differences in growth response of amaranths to N application have been found.[132] In *A. spinosus,* growth promotion by N was greater than in *A. viridis,* but *A. viridis* grows better at low levels of N. Increasing levels of P do not affect the growth rate of *A. viridis,* but are increasingly toxic for *A. spinosus.*

2. Effects on Leaf Composition

Leaves of *A. retroflexus* have a higher N content and nitrate reductase activity (NRA) than other parts of the plant, while nitrate content is lowest in the leaves and highest in the stem tissues.[137] Increased supply of N results in higher organic N, crude protein, and nitrate content in leaves of cultivated amaranths.[137,138,143,144,145] Attempts to increase maximum organic N content in leaves of *A. retroflexus* by increasing N supply resulted in nitrate accumulation, without further increase in organic N.[138]

Crude fiber content is reduced by higher supply of N in *A. tristis*[143] and *A. tricolor.*[144] In *A. tristis,* N fertilization increases carotene,[143] oxalic acid,[145] and hydrocyanic acid[146] content in leaves, but did not affect Ca content[147] or reduced the levels of Fe and Mg.[147] Ascorbic acid content declined with increased supply of N in *A. tristis*[147] and other amaranths cultivated for their leaves.[145]

3. Effects on Yield

According to Weber *et al.,*[148] amaranth grain yield of 1350 kg/ha removes about 41 kg of N, 8 kg of P, and 7 kg of K per hectare. Thus, adequate levels of soil fertilizers must be applied to ensure high yields. Research in Minnesota has shown that soil N is a main limiting factor and amaranth grain yield is increased by application of 100 to 200 kg N per hectare.[149] However, high rates of fertilizers also increase lodging, which may limit N application, especially in wetter regions. Since the effectiveness of N application and its economic feasibility depends on many factors, such as soil characteristics, weather conditions, activity of soil microorganisms, etc., more comprehensive studies are required for a better understanding of the relationships between grain amaranth yield and the availability of N in the soil.

C. SALINITY

1. Sodium Requirement

Na was shown to be an essential micronutrient in C_4 plants.[150] In *A. tricolor,* Na deficiency reduced growth and chlorophyll contents, and these responses are specifically prevented by NaCl at concentration >0.1 mM/l.[150] Increased CO_2 concentrations (1500 µl/l and higher) alleviate, but did not completely overcome, the symptoms of Na-deficiency.[151,152] This response to CO_2 enrichment suggests that mesophyll cells, the site of initial CO_2 fixation by PEP carboxylase, are affected by Na deficiency. Indeed, *A. tricolor* Na-deficient mesophyll chloroplasts have less thylakoids stacked in their grana than normal plants.[153] On the other hand, no marked differences in grana structure were observed in the bundle-sheath chloroplasts.

The impairment of thylakoids in mesophyll chloroplasts of Na-deficient plants may reduce photophosphorylation and ATP production. This, in turn, may limit the transport of pyruvate into mesophyll chloroplasts and its subsequent conversion to the PEP needed for CO_2 fixation.[154] Nevertheless, even under Na-deficiency, there is no shift in the balance between C_3 and C_4 photosynthesis as shown by the lack of change in the carbon isotope discrimination ratio, CO_2 compensation point, and in the labeling pattern of carbon incorporation to intermediates of the C_4 pathway in Na-deficient plants.[155]

Application of a small amount of NaCl (*ca* 0.5 mg/l) to Na-deficient *A. tricolor* plants promotes plant growth,[156] as well as light-dependent O_2 evolution, but it also increases extractable NRA activity.[157] This suggests that the Na-sensitive metabolic process is not confined to photosynthesis. It has been proposed that Na promotes nitrate uptake by roots[158] that, in turn, induces higher levels NRA.[159] Stimulation of nitrate uptake by the roots preceeds the enhancement of NRA, and the stimulatory threshold of Na concentration in the leaves is very low (*ca* 0.1 to 0.2 mM). The greater NRA promotes nitrate reduction, its subsequent incorporation into protein, and probably results in the enhanced growth observed after Na application.[159]

2. Salinity Effects

Even though Na is required for normal development of amaranths, higher levels are toxic. Growth is gradually reduced by increasing salinity[160,161,162] and is practically inhibited at NaCl >70 mM.[162] In amaranths cultivated for leaf consumption, salinity reduces leaf area and leaf pigment content.[161] Leaf pigments are affected differentially, e.g., leaf chlorophyll a/b ratio is increased, while chlorophyll/carotenoid ratio decreases with higher salinity.[161] In *A. cruentus* leaf protein, starch content, and photosynthesis are reduced by increased salinity, but dark respiration is unaffected.[162] Amaranth seedlings do not develop salinity tolerance when salinity is gradually increased,[162] but osmotic responses to salinity have not been studied yet. Productivity in relation to soil salinity does not differ in *A. hypochondriacus* and *A. retroflexus* (C_4 species) compared to *Beta vulgaris* and *Chenopodium album* (C_3 species).[160] Nevertheless, considerable variability for salinity tolerance has been found in amaranth species (*A. tricolor* > *A. cruentus* = *A. hybridus* > *A. blitum*).[162] Yield tolerance of grain amaranths to salinity has not been examined in field experiments.

D. ACID SOIL, AL AND MN TOXICITY

In tropical and subtropical regions amaranths are frequently grown in marginal acids soils. In mineral soils, low pH increases Al and Mn solubility, which may reach toxic levels.[163] Al is also antagonistic to soil P, a nutrient required in amaranth in moderate amounts.[164] Thus, Al and Mn toxicity are between the most important limiting factors in amaranth grown under subsistence conditions in regions with acid soils where liming may not be economically feasible.

Growth of *A. retroflexus* is severely reduced by soils with a pH lower than 5.2.[166] Reducing soil pH from 8.4 to 4.5 adversely affects leaf production by *A. tricolor* and seed yield in *A. leucocarpus.*[167] Of the two species, *A. leucocarups* showed greater tolerance to pH variation. However, according to Makus,[164] pH reduction from 7.3 to 4.4 increased plant biomass, leaf area, and chlorophyll content of *A. tricolor*. Addition of Al at all pH levels reduced leaf chlorophyll, N, and P content, but did not affect plant weight. The major negative effect of Al in plant metabolism is apparently due to conformational changes on the membrane-bound calmodulin.[165] Nevertheless, *A. tricolor* leaves are able to accumulate up to 2000 mg Al/kg.[164] Under certain soil conditions this species can be considered as an Al accumulator.[168] The ability of *A. tricolor* to accumulate Al in the leaves without detrimental effects on their growth may be due to the high total oxalate levels (up to 13%) in their leaves[169] that may chelate excess Al.

Differential tolerance to Al and Mn has been found between amaranth species.[170] Based on the above-ground biomass accumulation, tolerance to Al was in the order *A. tricolor* >> *A. hypochondriacus* = *A. caudatus* > *A. cruentus*. Also, *A. retroflexus* is very sensitive to Al and Mn.[171] Species and strains more tolerant to Al-toxicity were also more tolerant to Mn-toxicity, but interspecific differences in Mn-toxicity were relatively smaller.[170] A relatively great variability has been found between cultivated amaranths in their Al tolerance[170] that can be increased by selection.[172] On the other hand, it has been stated that cultivation of amaranth in soils with pH <5.0 is not economically feasible due to its high intolerance to Al.[170]

V. FLOWERING AND SEED PRODUCTION

Flowering time is closely associated with plant size, extent of branching, and inflorescence development and is a most important trait for the determination of yield in grain amaranth. In general, days to flowering is negatively correlated with yield and plant height.[173] Flowering physiology of amaranths has been extensively reviewed by Kigel and Rubin,[174] and anatomical changes in the shoot apex associated to transition to flowering have been described by Nougarede *et al.*[175] All examined amaranth species behaved as obligatory or facultative SD plants.[176-180] No vernalization or low temperature requirements for flowering have been reported. Morphologically, the greater inductiveness of SD is expressed by a reduction in the number of leaves produced by the main shoot apex prior to its transition to the reproductive stage and also by the ensuing reduction in stem length due to fewer and shorter nodes.[174] As in other SD plants, brief red irradiations in the middle of the inductive dark period (night breaks) prevent flowering, and their effect is reversed by far FR irradiations. Reported critical day-lengths are between 12 to 14 h for *A. retroflexus*[174] and 15 to 16 h for *A. caudatus.*[176]

Responsiveness to SD induction appears immediately after seedling emergence in *A. retroflexus*[181] (lacks a juvenility phase), while in *A. caudatus*[176] responsiveness is attained about 4 weeks after germination. Once the responsive stage is reached, sensitivity to SD induction increases gradually with plant age,[176,181] as expressed by shorter periods between the actual SD induction and macroscopic flowering, or by reduced leaf production before flowering. Since the transition of the apex to the reproductive stage in *A. retroflexus* occurs only after the apex reaches a certain size (*ca.* 120 μm long) under both inductive and non-inductive conditions, the shortening of the period between SD induction and flowering observed in older plants is partly due to progressive increase in shoot apex size that occurs under noninductive conditions.[177]

Inductive conditions (shorter days) reduce inflorescence length and branching,[177,181] while abnormal inflorescence development is frequently observed under noninductive photoperiods.[174] Solar irradiation levels do not affect onset of flowering under SD, but delay it under LD conditions.[174]

Gibberellins (GA_3 and GA_{4+7}) enhance stem elongation but are ineffective for flower initiation in *A. caudatus* and *A. tricolor.*[178] However, GA_{13} which reduces the number of days to flower initiation and shortened the critical dark period does not promote stem elongation.

Genetics of flowering has hardly been studied in amaranths. One major gene for earliness (Ea) dominant over late flowering has been identified under 8 h daylength.[179] Based on segregation patterns within individual progenies and transgressive segregation, a model of three genes has been proposed to explain the ratios obtained in crosses between *A. retroflexus* and *A. cruentus*. Intraspecific variability in flowering time occurs in weedy and cultivated amaranths. A cline in date of flowering was found in plants of *A. retroflexus, A. hybridus*, and *A. powellii* collected in North America and grown in the same location, with earlier flowering in plants from the more northern latitudes.[182] All three species flowered eventually under 16 h day-length, but flowering was much earlier under 8 h.[45] Under 16 h day-length, order of flowering was *A. retroflexus* → *A. powellii* → *A. hybridus*. A significant correlation was found between date of flowering and latitude of collection for *A. hybridus* and *A. retroflexus.*[182] The first fall frost is probably the selection factor acting upon the genetic variability of flowering date. As a result, the population maintained is a compromise between large, late-flowering plants, capable of producing many seeds and smaller plants which almost always produce some seeds before frost.

Variability in flowering data between species and strains of grain amaranths has been frequently observed in field trials.[183] In trials sown between April and August (13 → 12 h day-length) in Thailand[184] differences of 40 to 50 d in time to flowering were found between the first and last sowing date for *A. caudatus*, compared to 15 to 18 d for *A. cruentus* African type and *A. hybridus* Prima type. Order of flowering in trials carried out in Mexico[185] and in two sites in Thailand was *A. hybridus* (Prima type) → *A.hypochondricuas* (Mercado and Nepal types), *A. cruentus* (Mexican type) → *A. hypochondriacus* (Aztec type) → *A. caudatus*. In these trials, better growth and higher grain yields were reached in the earlier plantings, in which plants had a longer vegetative growth period (photoperiod conditions were less inductive). However, later plantings had the advantage of shorter growing period and dryer conditions at harvest. But when sowing was further delayed and plants developed under inductive shorter days, plant size and yield were severely reduced due to the very early initiation of flowering. Thus, if amaranth is grown at a time when the photoperiod is inadequate for the selected species or strains, yield will be reduced even if soil and air temperatures are adequate.[183]

Amaranth species, especially those of section *Amaranthus* with their well-developed terminal inflorescences, are characterized by huge seed production, in the range of 10^{-4} to 10^{-6} seed/plant under optimal conditions.[187,188] This great reproductive output stands behind amaranth potential as a grain crop, as well as a successful weed.

VI. PHOTOSYNTHESIS, TRANSPIRATION, AND WATER USE EFFICIENCY (WUE)

To a great extent, the presence of the C_4-photosynthesis pathway and its CO_2-concentration mechanism determines the gas exchange response of amaranths to temperature, light intensity, and water availability.

Studies in the field, as well as under controlled conditions, show that when grown under similar conditions most C_4 plants have much lower rates of water loss and greater WUE (fixed CO_2/transpired H_2O) than C_3 plants.[189] The higher WUE results from the presence of the CO_2-concentration mechanism in C_4 plants. Due to its action, lower levels of internal CO_2 concentration are reached in the leaf and, therefore, stepper gradients of CO_2 are produced between the leaf and the external atmosphere. These steeper gradients enable C_4 plants to maintain a greater CO_2 flux for any given stomatal conductance compared to C_3 plants, where the CO_2 gradients are smaller. Thus, photosynthetic CO_2 assimilation in C_4 plants, including amaranths, is less limited by stomata resistance than in C_3 plants. Even at low stomata conductances that severely reduce transpiration, C_4 plants are able to maintain a higher CO_2 concentration at the site of CO_2 fixation and, hence, relatively greater rates of photosynthesis are still reached. Thus, the CO_2-concentration mechanism allows C_4 plants to partially close their stomata during water stress periods with relatively little resulting loss in net CO_2 assimilation.

Under high irradiance and optimal temperatures, amaranth leaves adjust stomatal conductance and assimilatory capacity, so that their internal CO_2 concentration (C_i) is close to the level at which photosynthesis just becomes saturated by CO_2 (Figure 5).[189,190] Net photosynthesis in most amaranth species is nearly saturated at C_i values of 120 to 150 μl/l.[189,191-196] This range of saturating C_i is found also in plants growing under normal conditions. Thus, under these conditions, stomata conductance is at the minimum value, below which CO_2 fixation is limited by C_i. The range of C_i found in amaranths overlaps values reported for other C_4 plants[197] and is lower than in C_3 species that usually operate above 200 μl/l.[189] Only in *A. tricolor*, a cultivated amaranth adapted to shade, a high C_i of 200 μl/l was reported.[191]

Quantum yield of amaranth plants grown under optimal conditions varies between 0.050 and 0.053 mol CO_2/mol photons.[195,198] This yield is similar to the quantum yield of C_3 plants, but lower than the yield found in dicot C_4 species possessing other decarboxylation pathways and in C_4 grasses with the NAD-malic enzyme pathway.[198] C_4 dicots lack the suberized lamella that is found in the walls separating mesophyll and bundle sheath cells in the leaves of all C_4 monocots. This lamella apparently acts as a barrier that limits CO_2 diffusion into or out of the bundle-sheath cells. Lack of this suberized lamella in C_4 plants leads to reduced quantum yields since additional ATP is required to refix the CO_2 lost due to the greater leakiness of the bundle-sheath cells.[189]

A. TEMPERATURE EFFECTS

Temperature dependency of net CO_2-photosynthetic assimilation has been studied in cultivated, as well as in wild and weedy, amaranths.[5,191,193,195,199-201a] All the studied species showed a similar response pattern to temperature, e.g., net photosynthesis increases almost linearly between 10 and 30°C, then a relatively narrow optimum usually occurs at a high temperature of about 35°C, and finally the rate declines steeply at temperatures higher than the optimum (Figures 5, 7). This pattern of response to temperature is typical to most C_4 species.[189,190] Compared to C_3 species, C_4 species (amaranth included) have higher temperature optima, but slower rates of net photosynthesis at low temperatures. Furthermore, while their photosynthesis is strongly enhanced by rising temperature in the range of 20 to 40°C, photosynthesis in C_3 plants is often hardly affected in the range of 20 to 35°C. Thus, C_3 species usually show a much broader temperature optima for photosynthesis than C_4 species. The smaller response of net photosynthesis to temperature in C_3 plants is partly due to: (1) the increasing restriction of photosynthesis by the progressively lower C_i resulting from the faster CO_2 fixation at higher temperature and (2) to the differential effects of temperature on the oxygenase/carboxylase activities of Rubisco. According to Edwards *et al.*,[190] Rubisco oxygenation activity increases more than the carboxylation activity under increasing temperature because the solubility of O_2 in the aqueous phase is less reduced than the solubility of CO_2, and the relative increase in the K_m (CO_2) of Rubisco is greater than for the K_m (O_2).

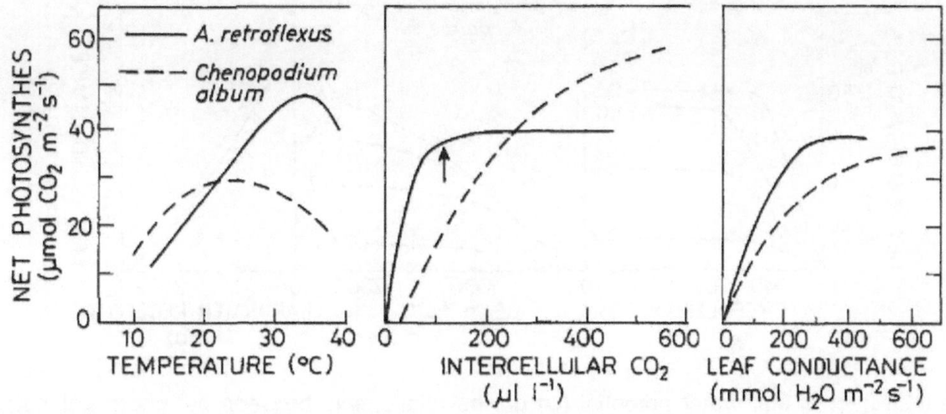

Figure 5 Relationships between net photosynthesis (CO_2 fixation) and temperature, leaf intercellular CO_2 concentratioon (C_i) and leaf conductance in *A. retroflexus* (C_4 plant) *vs. Chenopodium album* (C_3 plant). Arrow indicates C_i of *A. retroflexus* at normal CO_2 concentration in the air (330 µl l^{-1}). (Adapted from Chu *et al.*[127] *Crop Sci., 18, 308, 1978;* Pearcy, R. W. and Ehleringer, J., *Plant Cell Environ.*, 7, 1, 1984.)

In C_4 plants, the oxygenation activity of Rubisco is prevented and internal CO_2 limitations are largely reduced, due to the action of their CO_2-concentration mechanism. Nevertheless, photorespiration activity is present in amaranths. In *A. graecizans*[202] and *A. retroflexus*[189] net photosynthesis and quantum yield are increased by reducing oxygen levels below 21%. However, under normal atmospheric conditions the oxygenation activity of Rubisco is too low to have significant effects on net photosynthesis of C_4 plants at different temperatures.

At temperatures higher than the optimum, an irreversible time-dependent reduction in rate of photosynthesis occurs in several grain amaranths.[195] This reduction is presumably due to heat injury of the chloroplast membranes due to changes in properties of its lipid components, rather than enzyme denaturation that usually occurs at higher temperatures.[203]

It is important to note that properties inherent to the C_4 pathway do not necessarily render the photosynthetic apparatus of C_4 plants more resistant to damage by heat in comparison to C_3 plants. Variation in heat tolerance occurs even between amaranth species. *A. palmeri,*[193] a Sonoran desert summer annual, has a much higher temperature optima for photosynthesis than the shade adapted *A. tricolor,*[200] e.g., 40 *vs.* 30°C. Thus, other factors, such as thermostability of thylakoid membranes, the electron transport system, or the enzymes related to CO_2 fixation, are probably also involved in the determination of heat tolerance. Acclimation to heat stress has not been studied in amaranths.

B. LIGHT INTENSITY

In C_4 plants, rates of photosynthesis usually continue to increase up to full sunlight, while in C_3 plants photosynthesis saturates at one half or one third of full sunlight at normal CO_2 levels. Photosynthesis of amaranth plants, grown under high irradiance is not light-saturated, even at quantum fluxes above that of full sunlight (Figures 6 and 7),[5,191,193,195,199,201] due to their CO_2-concentration mechanism. Stomata response in *A. hypochondriacus* to light intensity is mediated by C_i effects on stomata conductance and not by direct effects of light on stomata aperture.[192]

The light intensity required for light-saturation of photosynthesis gradually diminishes as temperature decreases. This reduction has been attributed to the fact that low temperature is more limiting for CO_2-fixation and metabolite transport processes than for the photochemical reactions. In *A. edulis* inhibition of sucrose translocation from the leaf by low temperature results in negative feedback inhibition of photosynthesis due to cessation or slower translocation of sucrose.[196]

Adaptation of the photosynthetic mechanism to light intensity (sun- *vs.* shade-adapted leaves) has been barely studied in amaranths. In *A. retroflexus,* quantum yield is not affected by irradiance levels during plant growth.[198] However, plants grown under decreasing sunlight levels saturate at lower light intensity and show a concomitant decrease in Rubisco activity.[204] In *A. gangeticus,* a reduction of up to 90% in sunlight intensity during plant growth results in 20 to 40% increase in the photosynthetic unit size, while the ratio cyt f/total chlorophyll decreases progressively in the shade.[205] Chlorophyll fluorescence

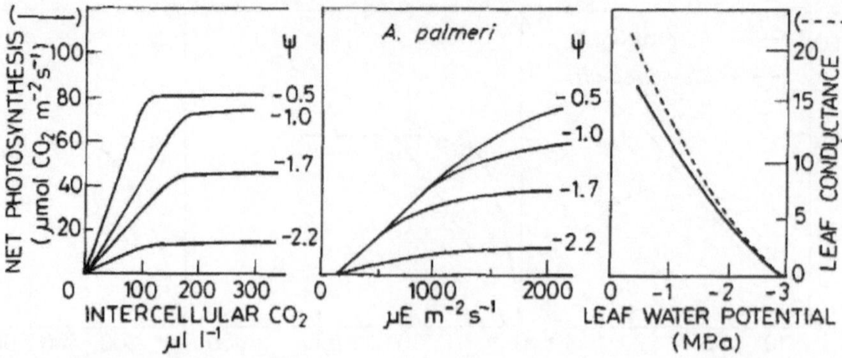

Figure 6 Effect of leaf water potential (ψ) on the relationship between net photosynthesis (CO_2 fixation), leaf intercellular CO_2 concentration (C_i) and irradiance in *A. palmeri.* (Adapted from Ehleringer, J., *Oecologia,* 57, 107, 1983. With permission.)

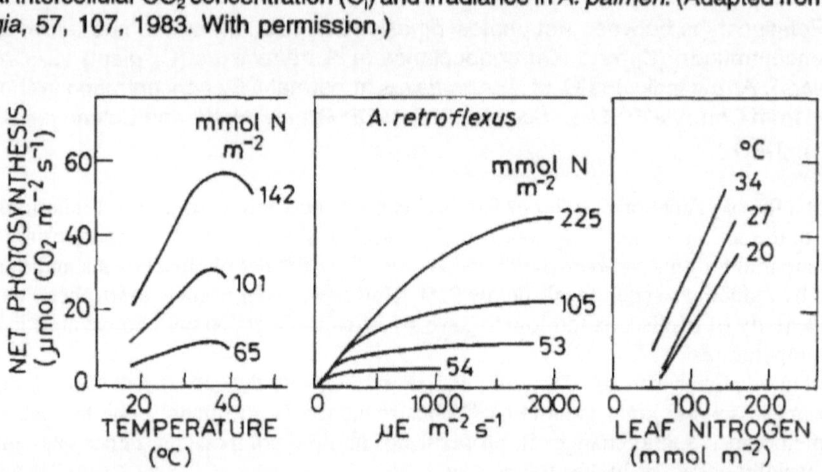

Figure 7 Effects of leaf N content (mmol N/m²) on the relationships between net photosynthesis (CO_2 fixation), temperature and irradiance in *A. retroflexus.* (Adapted from Sage, R. F. and Pearcy, R. W., *Plant Physiol.,* 84, 959, 1987.)

measurements of dark adapted leaves also indicate an impairment in electron transport between photosystem I and II in the shaded plants.

Leaf solar tracking (diaheliotropic leaf movements) have been reported only for the desert-adapted *A. palmeri.*[193]

C. WATER STRESS

Closure of stomata in response to water stress is the major physiological control for the reduction of transpiratory water loss and prevents deleterious dehydration of leaf tissues. Since, in amaranths, stomata conductance under optimal conditions is maintained at values ensuring that C_i is just sufficient for photosynthesis saturation, reduction in conductance due to decreasing leaf water potential or air humidity results in lower C_i and a concomitant reduction in net photosynthesis. Stomata conductance in amaranths is gradually reduced by decreasing leaf water potential (Figure 6).[191,193,206-209] Stomata start to close at –0.5 MPa in *A. palmeri*[193] and –0.8 MPa in *A. hybridus.*[208] Minimal leaf water potential observed in these species was –1.8 and –2.9 MPa, respectively. In *A. palmeri,* net photosynthesis approaches 0 at –2.9 MPa.[193] In the field, leaf water potentials found in *A. hybridus* were similar to those found in coexisting C_3 weeds.[208]

Under mild water stress, the ensuing reduction in net photosynthesis is fully accounted by the changes in stomata conductance (Figure 6). However, under prolonged drought, stomata and nonstomatal inhibition of photosynthesis occur in *A. palmeri*[193] and *A. hypochondriacus.*[206,207,210] In these species, decreased leaf water potential does not affect C_i, but reduces the rate of photosynthesis. This fact indicates that nonstomatal inhibition of photosynthesis occurs, otherwise C_i would have been reduced. In *A. palmeri*[193] photosynthesis inhibition is accompanied by an increment in quantum yield and a decrease in the carboxylation capacity that, in turn, was correlated with a decrease in leaf protein content. This indicates an increased protein catabolism and N remobilization as water stress develops. In *A. hypochondriacus*, similar relative changes in net photosynthesis and stomata conductance were found under water stress or after ABA application. Thus, it has been suggested[211] that both the reduction in photosynthetic capacity and in stomata conductance by water stress is mediated by ABA. However, in *A. powellii* ABA does not affect the photosynthetic capacity, but induces stomata closure and greater stomata sensitivity to CO_2.[212]

Amaranth stomata are sensitive not only to changes in leaf water potential, but also to air humidity, e.g., stomata conductance decreases when the leaf to air vapor pressure deficit (VPD) increases.[191,193,206,207,209] In *A. tricolor*, Lin and Ehleringer[191] observed that in addition to the effects of humidity on net photosynthesis that are mediated through stomata conductance, a direct inhibition of photosynthetic capacity occurs under low humidity. However, the nature of this nonstomatal inhibition was not elucidated.

In many species, stomata sensitivity to leaf water potential or VPD depends on previous growth conditions, mainly exposure to water stress. However, this adaptive response has not been studied in amaranths.

D. NITROGEN EFFECTS

Effects of nitrate application on leaf nitrogen content and photosynthetic activities were comprehensively studied in *A. powellii*[213] and *A. retroflexus*.[201,214] Increasing nitrate applications in the range of 1 to 45 mM augments organic N content of the leaves per unit leaf weight or area, e.g., 40 to 240 mM N m^{-2} for *A. powellii* and 50 to 160 mM N m^{-2} for *A. retroflexus*. Attempts to raise organic N content above the observed maxima by increasing nitrate application to *A. retroflexus* were unsuccessful and resulted in nitrate accumulation in the leaves.[201]

A positive linear relationship was found in both species between N/leaf area on one hand, and CO_2-fixation at light saturation (Figure 7), chlorophyll a+b content, and stomata conductance, on the other.[201,204,213] Concomitantly, activities of Rubisco and PEP increased in leaves with higher organic N content, and the increase was relatively higher in PEP activity.[214] Thus, at low N availability, a larger proportion of organic N is invested in Rubisco than in PEP, compared to plants at high N availability. In *A. retroflexus*, the fraction of organic N allocated to Rubisco ranges between 5 to 9%, compared to 2 to 5% for PEP.[214] These ranges are consistent with reports for other C_4 species.

In *A. powellii*[213] and *A. retroflexus*,[214] decreasing N/leaf area results in lower stomata conductance, but at the same time C_i is unaffected or even increases. Thus, the reduction in CO_2 assimilation at lower leaf N content is not due to limited stomata conductance but to biochemical constrains ensuing from the lower enzymatic activities.

WUE is also affected by leaf N content. Since higher N/leaf area enhances the photosynthetic capacity, it also reduces C_i and, thus, for a given VPD, it also increases the WUE. In *A. retroflexus*, WUE is raised from 3 to 12 µmol CO_2/mmol H_2O by increasing N/leaf area.[201]

E. CO_2 ENRICHMENT

Photosynthetic CO_2-fixation is enhanced in C_3 plants by increasing CO_2 above atmospheric levels, *ca* 300 µl/l. However, in C_4 plants, photosynthesis is nearly saturated at these levels due to their CO_2-concentration mechanism.[215] Thus, CO_2 enrichment has little or no effect on amaranth photosynthesis[210] or biomass accumulation.[216,217] Furthermore, LAR of *A. retroflexus* was unaffected by elevated CO_2.[217] On the other hand, one of the major effects of increased CO_2 is a reduction in stomata conductance. Stomata sensitivity to CO_2 is usually larger in C_4 plants.[215] In *A. retroflexus* transpiration is halved by increasing CO_2 from 300 to 1200 µl/l.[215] As a result, its WUE is also increased by elevated CO_2.

VII. NITROGEN USE EFFICIENCY (NUE)

In C_4 plants, smaller amounts of N are needed to generate leaf area and its photosynthetic machinery than in C_3 plants.[189,190] Under normal growth conditions, maxima ranges of N/leaf area are 120 to 180 *vs.* 200 to 260 mmole N/m^2, for C_3 and C_4 species, respectively.[201] On the other hand, the photosynthetic capacity of both C_3 and C_4 plants is closely correlated with N leaf content, since it also reflects the allocation to photosynthetic enzymes.[142,189] Allocation of N to Rubisco is especially large. In C_3 plants, it accounts for 30 to 60% of the soluble proteins in the leaves, but only for 5 to 20% in C_4 plants.[214] Since in C_4 plants Rubisco operates close to CO_2 saturation due to their CO_2-concentration mechanism, high rates of photosynthesis can be maintained with less Rubisco than in C_3 plants.[189,190] Thus, the photosynthetic rate per unit of N leaf content (photosynthetic NUE) should be higher in C_4 than in C_3 plants, due to their lower investment in Rubisco, the major soluble leaf protein. Indeed, most C_4 species, amaranths included, have greater photosynthetic NUE than comparable herbaceous C_3 species.[142]

The smaller amount of N needed for leaf area production, together with their higher photosynthetic NUE, suggests a higher NUE at the whole plant level in amaranths as well as in other C_4 plants. However, the C_4 pathway and the higher photosynthetic NUE do not automatically invest C_4 plants with a greater photosynthetic capacity compared to C_3 plants. This was shown when *A. retroflexus* was compared to the herbaceous C_3 annual *Chenopodium album* that coexist in similar habitats. At identical N/leaf area, the photosynthetic capacity of *A. retroflexus* is greater than that of *C. album*, and the differences between the species become larger as N/leaf area increases.[201] However, for a given N soil content, *C. album* has a greater N/leaf area than *A. retroflexus*,[138] so that the actual light-saturated CO_2 assimilation capacities are quite similar.[138,201] In this case, differences between species in the amount of N allocated to individual leaves compensate for the advantages resulting from higher intrinsic photosynthetic rates or photosynthetic NUE. Nevertheless, the greater N cost of photosynthesis in C_3 plants may limit allocation of N to other plant organs (e.g., root and leaf production), compared to C_4 plants. Thus, amaranths, as other C_4 plants, should be able to invest relatively more N into new root and leaf production than other C_3 crops and, therefore, have a greater whole plant CO_2 assimilation and mineral uptake, particularly when grown under high N availability.

VIII. ADAPTATION TO WATER STRESS

Even though the high intrinsic WUE of C_4 plants may represent an important trait for survival in dry habitats and greater biomass production under limited water supply, there is no evidence that the C_4 pathway *per se* confers a greater drought tolerance or a greater ability to withstand tissue dehydration. In fact, both C_3 and C_4 plants exhibit wide variation in their tolerance to low water potentials, with complete overlap in the range of maximal tolerance.[189,218] Plant adaptation to water stress depends on several other traits, such as root development, leaf anatomy, osmotic adjustment, desiccation tolerance, leaf movements, etc., that are not necessarily related to a specific photosynthetic pathway. However, these adaptive traits were hardly studied in amaranths.

Osmotic adjustment has been reported for *A. palmeri*[193] and *A. hypochondriacus*.[210,219] In the latter, accumulation of glycinebetaines and proline in the leaves occurs at relative water contents (RWC) lower than 70%, the threshold below which leaf water potential starts to decrease rapidly. Lowest reported leaf water potentials in amaranths are relatively high, e.g., –3.0, –2.9, and –1.9 MPa for *A. hypochondriacus*,[219] *A. palmeri*,[193] and *A. hybridus*,[208] respectively. These values are well above the lethal water potentials reported for C_3 and C_4 plants, with high tolerance to water stress and leaf dehydration.[218] Moreover, in amaranths, stomata start to close at relatively high leaf water potentials (–0.5 in *A. palmeri*[193] and –0.8 in *A. hybridus*[208]), and conductance decreases linearly with declining leaf water potential. The possibility of stomata adjustment (e.g., reduced sensitivity to water stress) has not been examined in amaranths. Thus, at this stage it is difficult to decide to what extent amaranths avoid or tolerate leaf dehydration.

IX. ALLELOPATHY

Presence of inhibitory and promotive allelochemicals in amaranths has been shown in field studies, as well as in laboratory growth and germination bioassays, in which aqueous and organic extracts of amaranth residues were tested. Residues of *A. retroflexus*,[220] *A. palmeri*,[221] and *A. dubius*[222] incorporated into the soil inhibit the germination and growth of several crops and weeds. Residues of *A. retroflexus* were the most phytotoxic among several other plant residues.[223] Water-soluble germination inhibitors are present in *A. retroflexus*[224,225] and *A. palmeri*,[225,226] with highest concentration in the roots and inflorescence tissues that remain after seed dispersal.[225] These inhibitors appear to be anionic and soluble in methanol. Concentrations of the phenolic acids were too low to account for the observed inhibitory effects.[225] Crude organic extracts from *A. palmeri* also have allelopathic effects.[226,227] From the compounds present in these extracts, vanillin and 2,6-dimethoxybenzoquinone slightly promote germination of sorghum, oats, and rye grass. Chondrillosterol greatly promotes germination of *A. palmeri* and *A. retroflexus*, but has no effect on *A. caudatus* and *A. tricolor*.[226] Other sterols with allelopathic activity extracted from amaranths are: ß-sistosterol and stigmasterol (*A. spinosus*[228]), amosterol (*A. viridis*[229]). Volatile compounds are the most active chemical produced by soil-containing residues of *A. palmeri*, or by dried and partially rehydrated tissue of this species.[230] They inhibit the germination of onion, carrot, tomato, and *A. palmeri* seeds. These phytotoxic effects may adversely affect vegetable stands, particularly under suboptimal environmental conditions.

Methyl-ketones, aliphatic alcohols, and aldehydes were identified in volatile mixtures produced by aerial parts of *A. retroflexus, A. hypochondriacus,* and *A. palmeri.*[231,232] The activities of some of these compounds appears to be additive and depends on their relative volatility and hydrophilicity.

Volatile compounds produced by amaranth tissues also act as insect attractans. Spraying *A. retroflexus* extracts on soybean plants resulted in increased parasitization of *Heliothis zea* eggs by the parasitic wasp *Trichogramma* spp.[234] Presence of oviposition-deterring chemicals has been proposed to explain the protective action of sprays with extracts of *A. hybridus* against *Spodoptera exigua.*[235]

While many studies indicate the presence of allelochemicals in amaranth tissues and their effects on crops, the effects of crop and weed residues on amaranth germination and growth have been examined only in a few cases. For example, seedling growth of *A. retroflexus* is strongly inhibited by aqueous extracts of rye (most active compounds are ß-phenyl acetic acid and ß-hydroxybutyric acid[236]), as well as by volatile compounds produced by the leaves of *Ambrosia psilostachys.*[237] Allelopathic interactions between grain amaranths and other crops grown in succession in the same fields are probably of importance in the determination of yield and require further research.

X. COMPETITIVE INTERACTIONS

Actual crop yield is strongly influenced by competitive interactions in the field. However, the interactions between grain amaranths and competing weeds have not been studied yet. The greater WUE and higher temperature optima for photosynthesis and growth of grain amaranths suggest a possible better performance, compared to weedy C_3 species in mixed populations growing under high irradiation, high temperature, and limited water supply. Under these conditions, *A. retroflexus* outperforms C_3 annuals when grown at high N availability, due to their greater plant NUE.[138,213] Similar trends will probably occur with grain amaranth, but this prediction has not been tested in the field. Detailed data are available only for the competitive ability of *A. retroflexus* grown in mixed culture with other herbaceous C_3 species. Its relative competitiveness (as measured by relative biomass accumulation) depends on the environmental conditions under which the mixed population is grown. For example, temperature changes the relative competitive ability of *A. retroflexus* and *Chenopodium album,* and these changes are correlated with the temperature effects on their photosynthetic performance, e.g., *A. retroflexus* has an advantage at high temperature, and *C. album* at low temperature.[127,238] The competitive outcome is determined mainly by differences in RGR prior to canopy closure. At each temperature the species with the highest photosynthetic rate and resulting faster growth overtops and shades the other species at the time of canopy closure. Thus, the outcome is determined by temperature even before actual extreme competition occurs. Similar trends were observed in mixtures of *A. hybridus* and soybean or *Xanthium pennsylvanicum,*[128] but no differences were found in the competitive ability of *A. dubius* grown in mixtures with *Sorghum bicolor* or soybean under 31/24° and 25/18°C.[239] In the field, order of competitive ability was *A. powelli* >> *A. retroflexus* > *A. hybridus* when grown in mixed stands at constant density.[240]

Increased concentration of CO_2 enhances the competitiveness of C_3 plants more than of C_4 plants.[241] In mixed cultures of *A. retroflexus* and three other C_3 annuals, *A. retroflexus* had a competitive advantage at ambient CO_2 and drier soil. However, this advantage was partially cancelled by increased CO_2 that improves photosynthesis and biomass accumulation of C_3 species. These differential effects of temperature and CO_2 concentration on the competitiveness of amaranths in relation to C_3 species are of interest when the effects of predicted global long term increases in temperature and atmospheric CO_2 concentration are taken into consideration for future crop yield.

XI. CONCLUSIONS

Prospects for future expansion of grain amaranth cultivation are very encouraging. In addition to its nutritional advantages, grain amaranth is a promising C_4 crop for semiarid and marginal lands due to its high WUE, high photosynthetic rates, and high NUE. However, amaranth cultivation and breeding in semiarid regions will require a deeper understanding of the mechanisms involved in its putative greater heat and drought tolerance, as well as of the effects of these limiting factors on plant morphology, assimilate partitioning, and harvest index. Therefore, information on amaranth capacity for osmotic adjustment, dehydration tolerance, root and stomata responses to decreasing soil water availability, heat acclimation, etc. is urgently needed for germplasm selection and development of improved plant types for semiarid regions.

ACKNOWLEDGMENT

This review was prepared with the support of the German-Israel Fund for Research and International Development (GIFRID).

REFERENCES

1. **Sauer, J.D.,** The grain amaranth and their relatives: a revised taxonomic and geographic survey, *Ann. Mo. Bot. Gard.*, 54, 103, 1967.
2. **Aellen, P.,** Die Amaranthaceen Mittel Europas, In *Illustratierte Flora Von Mittles Europa*, Vol. 3, 2nd ed., Hegi, G., Ed., 1961, 461.
3. **Joshi, B.D. and R.S. Rana,** *Grain Amaranths: The Future Food Crop*, National Bureau of Genetic Research Shimla Science Monograph, No. 3, India, 1991.
4. **Kauffmnan, C.S.,** The status of grain amaranths for the 1990s, *Food Rev. Int.*, 8, 165, 1992.
5. **Hunt, E.R., J.A. Weber and D.M. Gates,** Effects of nitrate application on *Amaranthus powellii* Wats. I. Changes in photosynthesis, growth rates and leaf area, *Plant Physiol.*, 79, 609, 1985.
6. **Fisher, D.G. and R.F. Evert,** Studies on the leaf of *Amaranthus retroflexus* L. Morphology and anatomy, *Am. J. Bot.*, 69, 1133, 1982.
7. **Edwards, G.E. and S.C. Huber,** The C_4 pathway, in *The Biochemistry of Plants*, Vol. 8, Hatch, M.D. and Boardman, N.K., Eds., Academic Press, 1981, 237.
8. **Edwards, G.E. and M.S.B. Ku,** Regulation of the C_4 pathway of photosynthesis, in *Perspectives in Biochemical and Genetic Regulation of Photosynthesis*, Zelitch, E., Ed., Alan R. Liss, New York, 1990, 175.
9. **Fisher D.G. and R.F. Evert,** Studies on the leaf of *Amaranthus retroflexus* (Amaranthaceae): chloroplast polymorphism, *Bot. Gaz.*, 143, 146, 1982.
10. **Wang, J.L., D.F. Klessig and J.O. Berry,** Regulation of C_4 gene expression in developing amaranth leaves, *Plant Cell*, 4, 173, 1992.
11. **Berry, J.O., D.E. Breiding and D.F. Kiessig,** Light mediated control of translational initiation of ribulose-1,5-biphosphate carboxylase/oxygenase in amaranth cotyledons, *Plant Cell*, 2, 995, 1990.
12. **Gan, L., X.B. Li, N.B. Hu, S.X. Yue and L.H. Zhu,** Construction of chloroplast genomic library of *Amaranthus cruentus* R104 and cloning of the gene for the large subunit of ribulose-1,5-biphosphate carboxylase/oxygenase, *Acta Bot. Sinica*, 31, 245, 1989.
13. **Wilson, C.L.,** Medullary bundles in relation to the primary vascular system in Chenopodiaceae and Amaranthaceae, *Bot. Gaz.*, 78, 175, 1924.
14. **Balfour, E.,** Anomalous secondary thickening in Chenopodiacea, Nyctaginaceae and Amaranthaceae, *Phytomorphology*, 8, 243, 1958.
15. **Philipson, W.R., J.M. Ward and B.G. Butterfield,** *The Vascular Cambium*, Chapman and Hall, London, 1971, chap. 7.
16. **Metcalfe, C.R. and L. Chalk,** *Anatomy of the Dicotyledons*, Vol. 2, Clarendon Press, 1950.
17. **Davis, R.G., W.C. Johnson and F.O. Wood,** Weed root profiles, *Agron. J.*, 59, 555, 1967.
18. **Davis, R., A. Wiese and J. Pafford,** Root moisture extraction profiles of various weeds, *Weeds*, 13, 98, 1965.
19. **Brenan, J.P.M.,** *Amaranthus* in Britain, *Watsonia*, 4, 261, 1961.
20. **Eliasson, U.H.,** Floral morphology and taxonomic relations among the genera of Amaranthaceae in the New World and the Hawaiian Island, *Bot. J. Linn. Soc.*, 96, 235, 1988.
21. **Kajale, L.B.,** A contribution to the embryology of the Amaranthaceae, *Proc. Nat. Inst. Sci. India, 6,* 597, 1940.
22. **Baquar, S.R. and O.O. Olusi,** Cytomorphological and phylogenetic studies in the genus *Amaranthus* from Nigeria, *Kromosomo* (Tokyo), 2, 1665, 1988.
23. **Murray, M.J.,** The genetics of sex determination in the family Amaranthaceae, *Genetics,* 25, 409, 1940.

24. **Brenner, D. and H. Hauptli,** Seed shattering control with indehiscent utricules in grain amaranths, *Legacy* (Amaranth Inst.), 3, 2, 1990.
25. **Corner, E.J.H.,** *The Seeds of Dicotyledons*, Cambridge Univsity Press, London, 1976.
26. **Woodcock, E.F.,** Seed development in *Amaranthus caudatus* L., Michigan Academy Science (Arts Lett. Papers), 15, 173, 1931.
27. **Naidu, K.C.,** Embryology of two species of *Amaranthus, Beitr. Biol. Pflanz.*, 62, 179, 1987.
28. **Soueges, R.,** Embryogenie des Amarantacees. Developpement de l'embryon chez l'*Amaranthus retroflexus* L., *C.R. Acad. Sci.*, Paris, 204, 892, 1937.
29. **Irving, D.W., A.A. Betchart and R.M. Saunders,** Morphological studies on *Amaranthus cruentus*, *J. Food. Sci.*, 46, 1170, 1981.
30. **Okuno, K. and S. Sakaguchi,** Glutinous and non-glutinous starches in perisperm of grain amaranths, *Cereal Res. Commun.*, 9, 305, 1981.
31. **Okuno, K. and S. Sakaguchi,** Inheritance of starch characteristics in perisperm of *Amaranthus hypochondriacus, J. Hered.*, 73, 467, 1982.
32. **Kowal, T.,** Morphological and anatomical features of the seeds of the genus *Amaranthus* and keys for their identification, *Monogr. Bot.*, 2, 162, 1954.
33. **Ramos, S. and F. Engleman,** Tannin deposits in the testa of *A. hypochondriacus, Agrociencia*, 42, 35, 1980.
34. **Gupta, V.K. and S. Gudu,** Inheritance of some morphological traits in grain amaranthus, *Euphytica*, 46, 79, 1990.
35. **Coons, M.P.,** Relationships of *A. caudatus, Econ. Bot.*, 36, 129, 1982.
36. **Hauptli, H. and S. Jain,** Genetic variation in outcrossing rate and correlated floral traits in a population of grain amaranth (*Amaranthus cruentus* L.), *Genetica*, 66, 21, 1985.
37. **Kulakow, P.A., H. Hauptli and S.K. Jain,** Genetics of grain amaranth. I. The inheritance genetics of six color characteristics, *J. Hered.*, 76, 27, 1985.
38. **Ridley, N.H.,** *The Dispersal of Plants Throughout the World*, L. Reeve & Co., Kent, 1930.
39. **Holm, L.G., D.L. Pluknett, L. Pancho and J.P. Herberger,** *The World's Worst Weeds*, University Press Hawaii, Honolulu, 1977.
40. **McWilliams, E.L., R.Q. Landers and J.P. Mahlstede,** Variation in seed weight and germination in populations of *Amaranthus retroflexus* L., *Ecology*, 49, 290, 1968.
41. **McWilliams, E.D.,** Ecotypic differentiation within *Amaranthus retroflexus* L., *A. hybridus* L. and *A. powellii* Wats, *Diss. Abstr.*, 27, 11, 1967.
42. **Schimpf, D.J.,** Seed weight of *Amaranthus retroflexus* in relation to moisture and length of growing season, *Ecology*, 58, 450, 1977.
43. **Wulff, R.D.,** Intraspecific variation in germination requirements and growth of *Amaranthus dubius*, *Am. J. Bot.*, 75, 1307, 1988.
44. **Baker, H.G.,** Seed weight in relation to environmental conditions in California, *Ecology*, 53, 997, 1972.
45. **Weaver, S.E. and E.L. McWilliams,** The biology of Canadian weeds. 44. *Amaranthus retroflexus* L., *A. Powellii* S. Wats. and *A. hybridus* L., *Can J. Plant Sci.*, 60, 1215, 1980.
46. **Kigel, J., M. Ofir and D. Koller,** Control of the germination responses of *Amaranthus retroflexus* L. by their parental photothermal environment, *J. Exp. Bot.*, 28, 1125, 1977.
47. **Chadoeuf-Hannel, R. and G. Barralis,** Influence de differents regimes hydriques sur la croisssance vegetative, le poids et la germination des grains d'une mauvaise herbe cultivee en serre: *Amaranthus retroflexus* L., *Agronomie*, 2, 835, 1982.
48. **Roberts, E.H.,** Dormancy: a factor affecting seed survival in the soil, in *Viability of Seeds*, Roberts, E.H., Ed., Chapman and Hall, London, 1972.
49. **Chapil, W.S.,** Germination of weed seeds, *Sci. Agric.*, 26, 307, 1946.
50. **Kivilaan, A. and R.S. Bandurski,** The one hundred-year period for Dr. Beel's seed viability experiment, *Am. J. Bot.*, 68, 1290, 1981.
51. **Toole, E.H. and E. Brown,** Final results of the Duvel buried sed experiment, *J. Agric. Res.*, 72, 201, 1946.
52. **Crocker, W. and L.V. Barton,** *Physiology of Seeds*, Chronica Botanica, Waltham, MA, 1953.
53. **Egley, G.H. and P.D. Williams,** Decline of weed seeds and seedling emergence over five years as affected by soil disturbance, *Weed Sci.*, 38, 504, 1990.
54. **Egley, G.H. and T.M. Chandler,** Germination and viability of weed seeds after 2.5 years in a 50-year buried seed study, *Weed Sci.*, 26, 230, 1978.

55. **Egley, G.H.,** Some effects of nitrate-treated soil upon the sensitivity of buried redroot pig weed (*Amaranthus retroflexus* L.) seeds to ethylene, temperature, light and carbon dioxide, *Plant Cell Environ.*, 12, 581, 1989.

55a. **Washitani, I.,** Field fate of *Amaranthus patulus* seeds subjected to leaf-canopy inhibition of germination, *Oecologia*, 66, 338, 1985.

56. **Taylorson, R.B.,** Changes in dormancy and viability of weed seed in soils, *Weed Sci.*, 18, 265, 1970.

57. **Vegis, A.,** Climatic control of germination, bud break and dormancy, in *Environmental Control of Plant Growth*, Evans, L.T., Ed., Academic Press, New York, 1963.

58. **Evans, C.R.,** Effect of temperature on germination of *Amaranthus retroflexus*, *Bot. Gaz.*, 73, 213, 1922.

59. **Gutterman, Y., F. Corbineau and D. Come,** Interrelated effects of temperature, light and oxygen on *Amaranthus caudatus* L. seed germination, *Weed Res.*, 32, 111, 1992.

60. **Teitz, A.Y., S.F. Gorskiand and M.B. McDonald,** The dormancy of livid amaranth (*Amaranthus lividus* L.) seeds, *Seed Sci. Technol.*, 18, 781, 1990.

61. **Faccini, D. and E. Barat,** A study of the germination behaviour of *Amaranthus quitensis* H.B.K., *Malezas*, 17, 53, 1989.

62. **Rojas-Garciduenas, M. and T. Kommedahl,** The effects of 2,4-D on germination of pigweed seeds, *Weeds*, 8, 1, 1960.

63. **Martin, J.N.,** Germination studies of the seeds of some common weeds, *Iowa Acad. Sci. Proc.*, 50, 221, 1943.

64. **Schonbeck, M.W. and G.H. Egley,** Effect of ethylene and some other environmental factors on different stages of germination in redroot pigweed (*Amaranthus retroflexus* L.) seeds, *Plant Cell Environ.*, 11, 189, 1988.

65. **Cumming, B.S.,** The dependence of germination on photoperiod, light quality and temperature in *Chenopodium* spp., *Can. J. Bot.*, 41, 1211, 1963.

66. **Pita, J.M. and J.M. Duran,** Germination in the genus *Amaranthus* L. I. Light and temperature, *ITEA*, 15, 56, 1984.

67. **Kadman-Zehavi, A.,** The effect of light and temperature on the germination of *Amaranthus blitoides* seeds, *Bull. Res. Counc. Isr.*, 4, 370, 1955.

68. **Kendrick, R.E. and B. Frankland,** Photocontrol of germination in *Amaranthus caudatus*, *Planta*, 85, 326, 1969.

69. **Oladiran, J.A. and P.M. Mumford,** The stimulation of seeds germination by temperature and light in agronomic *Amaranthus* species, *Biochem. Physiol. Pflanzen.*, 180, 45, 1985.

70. **Kadman-Zehavi, A.,** Effects of short and continuous illumination on the germination of *Amaranthus retroflexus* seeds, *Bull. Res. Counc. Isr.*, *9D, 1*, 1960.

71. **Chadoeuf-Hannel, R. and R.B. Taylorson,** Anaesthetic stimulation of *Amaranthus albus* seed germination: interaction with phytochrome, *Physiol. Plant.*, 65, 451, 1985.

72. **Taylorson, R.B. and S.B. Hendricks,** Changes in phytochrome expressed by germination of *Amaranthus retroflexus* L. seeds, *Plant Physiol.*, 47, 619, 1971.

73. **Chadoeuf-Hannel, R. and G. Barralis,** Comportment germinatif des grains d'*Amaranthus retroflexus* L. recoltées dans les conditions naturelles, *Weed Res.*, 22, 361, 1982.

74. **Chadouef-Hannel, R. and G. Barralis,** Evolution de l'aptitude a germer des grains d'*Amaranthus retroflexus* L. recoltées dans differentes conditions, au cours de leur conservation, *Weed Sci.*, 23, 109, 1983.

75. **Taylorson, R.B. and S.B. Hendricks,** Rehydration of phytochrome in imbibing seeds of *Amaranthus retroflexus* L., *Plant Physiol.*, 49, 663, 1972.

76. **Chadouef-Hannel, R. and R.B. Taylorson,** Changes in lipid fatty acids associated with dormancy breaking in *Amaranthus albus* seeds, *J. Seed Technol.*, 11, 15, 1987.

77. **Bewley, J.D. and M. Black,** *Seeds: Physiology of Development and Germination*, Vol. 2, Plenum Press, New York, 1978.

78. **Kang, H. and B.H. Kwack,** Effects of certain environments and ethephon on light-inhibited germination of *Amaranthus hypochondriacus* seeds, *J. Korean Soc. Hortic. Sci.*, 30, 311, 1989.

79. **Webb, D.M., C.W. Smith and T. Schulz-Schaeffer,** Amaranth seedling emergence as affected by seeding date and temperature on a thermogradient plate, *Agron. J.*, 79, 23, 1987.

80. **Baskin, J.M. and C.C. Baskin,** Role of temperature in the germination ecology of three summer annual weeds, *Oecologia*, 30, 377, 1977.

81. **Weaver, S.E.,** Differential growth and competitive ability of *Amaranthus retroflexus, A. powellii* and *A. hybridus*, *Can. J. Plant Sci.*, 64, 715, 1984.

82. **Taylorson, R.B. and S.B. Hendricks,** Action of phytochrome during prechilling of *Amaranthus retroflexus* L. seeds, *Plant Physiol.*, 44, 821, 1969.
83. **Hendricks, S.B. and R.B. Taylorson**, Dependence of phytochrome action in seeds on membrane organization, *Plant Physiol.*, 61, 17, 1978.
84. **Hendricks, S.B. and R.B. Taylorson,** Dependence of thermal responses of seeds on membrane transitions, *Proc. Natl. Acad. Sci., U.S.A.*, 76, 778, 1979.
85. **Grubben, G.J.H.,** The utilization of Amaranth as a tropical leaf vegetable with special reference to South Dahomey, Communication 67, Department Agricultural Research Royal, Tropical Institute, Amsterdam, 1976.
86. **Adebona, A.C. and B.E. Ayisire**, Effect of polyethylene glycol induced moisture stress on the germination of some tropical seeds, *Turrialba*, 29, 318, 1979.
87. **Kepczynski, J. and C.M. Karsen,** Requirement for the action of endogenous ethylene during germination and non-dormant seeds of *Amaranthus caudatus*, *Physiol. Plant.*, 63, 49, 1985.
88. **Kepczynski, J.,** Ethylene-dependent action of gibbberellin in seed germination of *Amaranthus caudatus* L., *Physiol. Plant.*, 67, 584, 1986.
89. **Schonbeck, D.W. and J.H. Egley,** Redroot pigweed (*Amaranthus retroflexus*) seed germination responses to after ripenning, temperature, ethylene and some other environmental factors, *Weed Sci.*, 28, 543, 1980.
90. **Kepcynski, J.,** The effect of plant growth regulators on *Amaranthus caudatus* seed germination under water stress, *Proc. Inst. Sadow.*, 8, 85, 1983.
91. **Macler, B., E. Bamberg, E. Moffet, H. Bui, E. Minor and L. Nishioka,** Effects of salinity and nitrogen on growth, productivity and food value of *Amaranthus* in controlled culture, in *Proc. 4th Amaranth Symp.*, Minneapolis, MN, 1990, 129.
92. **Holm, P.E. and M.R. Miller,** Weed seed germination responses to chemical and physical treatments, *Weed Sci.*, 20, 150, 1972.
93. **Smith, A.M.,** Ethylene in soil biology, *Annu. Rev. Phytopathol.*, 14, 53, 1976.
94. **Taylorson, R.B.,** Response of weed seeds to ethytlene and related hydrocarbons, *Weed Sci.*, 27, 7, 1978.
95. **Schonbeck, M.W. and G.H. Egley,** Changes in sensitivity of *Amaranthus retroflexus* L. seeds to ethylene during preincubation. I. Constant temperatures, *Plant Cell Environ.*, 4, 229, 1981.
96. **Schonbeck, M.W. and G.H. Egley,** Changes in sensitivity of *Amaranthus retroflexus* L. seeeds to ethylene during preincubation. II. Effects of alternating temperature and burial in soil, *Plant Cell Environ.*, 4, 237, 1981.
97. **Egley, H.H.,** Stimulation of common cocklebur (*Xanthium pensylvanicum*) and redroot pigweed (*Amaranthus retroflexus*) seed germination by injections of ethylene into soil, *Weed Sci.*, 28, 510, 1980.
98. **Egley, G.H.,** Ethephon reduction of redroot pigweed (*Amaranthus retroflexus*) seed populations, *Weed Technol.*, 4, 808, 1990.
99. **Bedin, P., M.T. Ochoa and C. Zaragoza,** Contribucion al estudio de la germinacion en algunas especies del genero *Amaranthus* encontradas en la provincia de Zaragoza (Espana), *An. Inst. Nac. Invest. Agr.* (Ser. Agric. Madrid), 16, 81, 1981.
100. **Weaver, S.E. and A.G. Thomas,** Germination responses to temperaure of atrazine-resistant and susceptible biotypes of two pigweeds (*Amaranthus*) species, *Weed Sci.*, 34, 865, 1986.
101. **Kepcynski, J., E. Kepcynska and J.S. Knypl,** Effects of gibberellic acid, ethephon and 1-aminocyclopropane-1-carboxylic acid on germination of *Amaranthus caudatus* seeds inhibited by paclobutrazol, *J. Plant Growth Regul.*, 7, 59, 1988.
102. **Kepcynski, J. and J.S. Knypl,** Inhibition by growth rretardants AMO-1618 and ancymidol, and its reversal by GA_3 in *Amaranthus caudatus* seeds, *J. Plant Physiol.*, 132, 584, 1988.
103. **Kepczynski, J. and E. Kepczynska,** Reversing the inhibitory effect of paclobutrazol on seed germination of *Amaranthus paniculatus* by GA_3, ethephon or ACC, *Plant Growth Regul.*, 7, 47, 1988.
104. **Schoenbeck, M.W. and G.H. Egley,** Phase-sequence of redroot pigweed seed germination responses to ethylene and other stimuli, *Plant Physiol.*, 68, 175, 1981.
105. **Kepczynski, J.,** Inhibition of *Amaranthus caudatus* seed germination by polyethylene glycol-6000 and abscisic acid and its reversal by ethephon or 1-aminocyclopropane-1-carboxylic acid, *Physiol. Plant.*, 67, 588, 1986.
106. **Egley, G.H. and R.D. Williams,** Glyphosate and paraquat effects on weed seed germination and seedling emergence, *Weed Sci.*, 26, 249, 1978.

107. **Wanner, H., Santosa and H. Stocker**, Different effects of long chain fatty acids on seed germination, *Biochem. Physiol. Pflanzen*, 171S, 391, 1977.
108. **Taylorson, R.B.**, Responses of redroot pigweed (*Amaranthus retroflexus*) and witch grass (*Panicum capillare*) seeds to anaesthetics, *Weed Sci.*, 37, 93, 1989.
109. **Taylorson, R.B. and S.B. Hendricks,** Promotion of seed germination by cyanide, *Plant Physiol.*, 52, 23, 1973.
110. **Hendricks, S.B. and R.B. Taylorson,** Promotion of seed germination by nitrate, hydroxylamine and ammonium salts, *Plant Physiol.*, 54, 304, 1974.
111. **Hendricks, S.B. and R.B. Taylorson,** Breaking of seed dormancy by catalase inhibition, *Proc. Natl. Acad. Sci., U.S.A.*, 72, 306, 1975.
112. **Santelman, P.W. and L. Evetts,** Germination and herbicide susceptibility of six pigweed species, *Weed Sci.*, 19, 51, 1971.
113. **Engelhardt, M., M. Vicente and K. Silberschmidt,** The stimulating effect of light and KNO_3 on the germination of *A. hybridus* L., *Rev. Brasileira Biol., 22, 1*, 1962.
114. **Solymosi, P.,** Comparative study on the number and germination of *A. retroflexus* seeds originated from lateral and terminal flowers, *Novenyvedelem*, 19, 252, 1983.
115. **Chadoeuf-Hannel, R. and J. Monin**, Recherche des causes possibles de l'heterogeneite germinative des grains d'*Amaranthus retroflexus* L. recoltees dans les conditions naturelles, *Bull. Soc. Ecophysiol.*, 5, 149, 1980.
116. **Kigel, J., A. Gibly and M. Negbi,** Seed germination in *Amaranthus retroflexus* L. as affected by the photoperiod and age during flower induction of the parent plants, *J. Exp. Bot.*, 30, 997, 1979.
117. **Kauffmann, C.S., N.N. Bailey, B.T. Volek, L.E. Weber and N.R. Volok,** *Amaranth Grain Production Guide,* Rodale Research Report, Rodale Press, PA, 1983.
118. **Kaufman, C.S.,** Grain amaranth research: an approach to the development of a new crop, in *Proc. 2nd Amaranth Conf., Ro*dale Press, Emmaus, PA, 1979, 81.
119. **Ogg, A.G. and J.H. Dawson,** Time of emergence of light weed species, *Weed Sci.*, 32, 327, 1984.
120. **Wiese, A.F. and R..G. Davis,** Weed emergence from two soils at various moistures, temperature and depths, *Weeds*, 15, 118, 1967.
121. **Roberts, H.A. and M.E. Potter,** Emergence patterns of weed seedlings in relation to cultivation and rainfall, *Weed Res.*, 20, 377, 1980.
122. **Stoller, E.W. and L.M. Wax,** Periodicity of germination and emergence of some annual weeds, *Weed Sci.*, 21, 574, 1973.
123. **Baskin, J.M. and C.C. Baskin,** The annual dormancy cycle in buried seeds: a continuum, *Bioscience,* 25, 492, 1985.
124. **Onwveme, I.C. and S.A. Adegoroye,** Emergence of seedlings from different depths following heat stress, *J. Agric. Sci.*, 84, 525, 1975.
125. **Keeley, P.E., C.H. Carter and R.J. Thullen,** Influence of planting date on growth of Palmer Amaranth (*Amaranthus palmeri*), *Weed Sci.*, 35, 199, 1987.
126. **Patterson, D.T., C.R. Meyer and P.C. Quimby,** Effect of irradiance on relative growth rates, net assimilation rates and leaf area partitioning in cotton and three associated weeds, *Plant Physiol.*, 62, 14, 1978.
127. **Chu, C., P.M. Ludford, J.L. Ozbun and P.D. Sweet,** Effect of temperature and competition on the establishment and growth of redroot pigweed and common lambsquarters, *Crop Sci.*, 18, 308, 1978.
128. **Flint, E.P. and D.T. Patterson**, Interference and temperature effects on growth in soybean (*Glycine max*) and associated C_3 and C_4 weeds, *Weed Sci.*, 311, 193, 1983.
129. **Fawusi, M.D.A., D.P. Amrod and A.M. Eastham,** Influence of temperature on the growth of *Solanum nigrum* and *Amaranthus hybridus*, *Sci. Hortic.*, 18, 305, 1982.
129a. **Fawusi, M.D.A., D.P. Amrod and A.M. Eastham,** Effects of levels of photosynthetically active radiation on the growth of *Amaranthus hybridus* and *Celosia argentea, J. Hortic. Sci.*, 58, 555, 1983.
130. **Eze, J.M.,** Growth of *Amaranthus hybridus* (African spinach) under different daylight intensities in the dry season in Southern Nigeria, *Exp. Agric.*, 23, 193, 1987.
131. **Potter, J.R. and J.W. Jones,** Leaf area partitioning as an important factor in growth, *Plant Physiol.*, 59, 10, 1977.
132. **Ramachandra, H.A. and K.R. Thimmaraju,** Effect of different levels of nitrogen and phosphorous on growth components and yield of amaranthus (*Amaranthus gangeticus* L.), *Mysore J. Agric. Sci.*, 17, 158, 1983.

133. **Nawawi, M., W.I. Santoso and S. Notodimejo,** The effect of shading and nitrogen pertilizer application on the yield of spinach (*Amaranthus tricolor* L.), *Agrivita*, 8/9, 17, 1985.
134. **Keskar, B.G., D.P. Bore and S.R. Masleker,** A note on the effects of nitrogen on yield of Chauli (*Amaranthus blitum* L.), *Haryana J. Hortic. Sci.*, 12, 241, 1983.
135. **Hilman, Y. and Z. Abidin,** The effect of nitrogen fertilization on the growth and yield of four amaranth cultivars, *Bull. Penelitian Hortic.*, 15, 22, 1987.
136. **Singh, U.C., S. Sundarajan and D. Veeraragavathatham,** Effect of split application of nitrogen on growth and yield of *Amaranthus tristis* L., *South Indian Hortic.*, 33, 100, 1985.
137. **Gebauer, G., M.I. Schuhmacher, B. Krstic, H. Rehder and H. Ziegler,** Biomass production and nitrate metabolism of *Atriplex hortensis* L. (C_3 plant) and *Amaranthus retroflexus* L. (C_4 plant) in cultures at different levels of nitrogen supply, *Oecologia*, 2, 303, 1987.
138. **Sage, R.F. and R.W. Pearcy,** The nitrogen use efficiency of C_3 and C_4 plants. I. Leaf nitrogen, growth and biomass partitioning in *Chenopodium album* L. and *Amaranthus retroflexus* L., *Plant Physiol.*, 84, 954, 1987.
139. **Ramakrishnan, P.S.,** Comparative biology of two closely related nitrophilous species of *Amaranthus* living in the same area, *Trop. Ecol.*, 17, 100, 1976.
140. **Novoa, R. and R.S. Loomis,** Nitrogen and plant production, *Plant Sci.*, 58, 177, 1981.
141. **Brown, R.H.,** A difference in N use efficiency in C_3 and C_4 plants and its implications in adaptation and evolution, *Crop Sci.*, 18, 93, 1978.
142. **Field, C. and H.A. Mooney,** The photosynthetic-nitrogen relationship in wild plants, in *On the Economy of Plant Form and Function*, Givnish, T., Ed., Cambridge University Press, 1986, 25.
143. **Singh, U.C., S. Sundararagan and D. Veerargavathathan,** Effect of split application of nitrogen on crude protein, carotene and ascorbic acid content of *Amaranthus tristis* L., *South Indian Hortic.*, 34, 150, 1986.
144. **Walters, R.D., D.C. Coffey and C.E. Sams,** Fiber, nitrate and protein content of Amaranth accessions affected by soil nitrogen application and harvest date, *Hortic. Science*, 23, 338, 1988.
145. **Makus, D.J.,** Vegetable amaranth response to nitrogen fertility under moderately low and very low residual soil phosphorous levels, *Hort Sci.*, *21*, 697, 1986.
146. **Singh, U.C., S. Sundararagan and D. Veeraragavathathan,** Effect of split application of nitrogen on certain qualitative characters of *Amaranthus tristis* L., *South Indian Hortic.*, 33, 230, 1985.
147. **Singh U.C., S. Sundararagan and D. Veeraragavathathan,** Effect of split application of nitrogen on certain minimal nutrient content of clipping amaranth (*Amaranthus tristis* L.), *South Indian Hortic.*, 34, 341, 1986.
148. **Weber, L.E., W.W. Applegate, D.D. Baltensperger, M.D. Irwin, J.W. Lehmann and D.H. Putman,** *Amaranth Grain Production Guide*, Rodale Press, Amaranth Institute, Bricelyn, MN, 1990.
149. **Putman, D.H.,** Agronomic practices for grain amaranth, in *Proc. 4th Natl. Amaranth Symp.*, Minneapolis, MN, 1990, 151.
150. **Brownell, P.F. and C.J. Crossland,** The requirement for Na as a microelement by species having the C_4 dicarboxylic photosynthesis pathway, *Plant Physiol.*, 49, 794, 1972.
151. **Johnston, M., C.P.L. Grof and P.F. Brownell,** Responses to ambient CO_2 concentrations by sodium deficient C_4 plants, *Aust. J. Plant Physiol.*, 11, 137, 1984.
152. **Johnston, M., C.P.L. Grof and P.F. Brownell,** Sodium deficiency in the C_4 species *Amaranthus tricolor* L. is not completely alleviated by high CO_2 concentrations, *Photosynthetica*, 20, 476, 1986.
153. **Grof, C.P.L., M. Johnston and P.F. Brownell,** Effects of Na nutrition on the ultrastructure of chloroplasts of C_4 plants, *Plant Physiol.*, 89, 539, 1989.
154. **Johnston, M., C.P.L. Grof and P.F. Brownell,** The effect of Na nutrition on the pool sizes of intermediates of the C_4 photosynthetic pathway, *Aust. J. Plant Physiol.*, 15, 749, 1988.
155. **Boag, S. and P.E. Brownell,** C_4 photosynthesis in sodium-deficient plants, *Aust. J. Plant Physiol.*, 6, 431, 1979.
156. **Matoh, T., D. Ohta and E. Takahashi,** Effect of sodium application on growth of *Amaranthus tricolor* L., *Plant Cell Physiol.*, 27, 187, 1986.
157. **Ohta, D., T. Matoh and E. Takahashi,** Early responses of sodium-deficient *Amaranthus tricolor* L. plants to sodium application, *Plant Physiol.*, 84, 112, 1987.
158. **Ohta, D., T. Matoh and E. Takahashi,** Sodium stimulated NO_3^- uptake in *Amaranthus tricolor* L. plants, *Plant Physiol.*, 87, 223, 1988.
159. **Ohta, D., S. Yasuoka, T. Metoh and E. Takahashi,** Sodium stimulates growth of *Amaranthus tricolor* L. plants through enhanced nitrate assimilation, *Plant Physiol.*, 89, 1102, 1989.

160. **Miller, T.E., J.S. Wing and A.R. Huete,** The agricultural potential of selected C_4 plants in arid environments, 7, 275, 1984.
161. **Makus, D.J.,** Effect of soil salinity and nitrogen fertility on vegetable amaranth, *Hortic. Sci.*, 24, 757, 1989.
162. **Macler, B., E. Bamberg, E. Moffet, H. Bui, E. Minor and L. Nishioka,** Effects of salinity and nitrogen on growth, productivity and food value of *Amaranthus* in controlled culture, in *Proc. 4th Natl. Amaranth Symp.*, Minneapolis, MN, 1990, 129.
163. **Foy, C.D., R.L. Chaney and M.C. White,** The physiology of metal toxicity in plants, *Annu. Rev. Plant Physiol.*, 29, 511, 1978.
164. **Makus, D.J.,** Aluminum accumulation in vegetable amaranth grown in a soil with adjusted pH values, *Hortic. Sci.*, 24, 460, 1989.
165. **Hang, A.,** Molecular aspects of aluminum toxicity, *CRC Crit. Rev. Plant Sci.*, 1, 245, 1983.
166. **Buchanan, G.A., C.S. Howland and M.C. Harris,** Response of weeds to soil pH, *Weed Sci.*, 23, 473, 1975.
167. **Udayarajan, V. and G. Oblisani,** Effects of soil pH on the growth and yield of *Amaranthus* species, *South Indian Hortic.*, 35, 294, 1987.
168. **Chenery, E.M. and K.R. Sporne,** A note on the evolutionary status of aluminum accumulation among dicotyledons, *New Phytol.*, 76, 551, 1976.
169. **Srivastava, S.K. and P.S. Krishman,** Oxalate content of plant tissues, *J. Sci. Ind. Res. India*, 18C, 146, 1959.
170. **Foy, C.D. and T.A. Campbell,** Differential tolerance of *Amaranthus* strains to high levels of aluminum and manganase in acid soils, *J. Plant Nutr.*, 7, 1365, 1984.
171. **Bilski, J.J. and C.D. Foy,** Differential tolerance of weed species to aluminum, manganase and salinity, *J. Plant Nutr.*, 11, 93, 1988.
172. **Campbell, T.A. and C.D. Foy,** Selection of grain Amaranthus species for tolerance to excess aluminum in an acid soil, *J. Plant Nutr.*, 10, 249, 1987.
173. **Hauptli, H. and S.K. Jain,** Genetic structure of landrace populations of the New World grain amaranths, *Euphytica*, 33, 875, 1984.
174. **Kigel, J. and B. Rubin,** *Amaranthus*, in *Handbook of Flowering*, Vol. 1, Halevy, A.H. Ed., CRC Press, Boca Raton, FL, 1988, 427.
175. **Nougarede, A., E.M. Gifford and P. Rondet,** Cytohistological studies of the apical meristem of *Amaranthus retroflexus* under various photoperiodic regimes, *Bot. Gaz.*, 126, 281, 1965.
176. **Zabka, J.J.,** Photoperiodism in *Amaranthus caudatus*, *Am. J. Bot.*, 48, 21, 1961.
177. **Koller, D., J. Kigel and S. Ovadia,** A kinetic analysis of the facultative photoperiodic response in *Amaranthus retroflexus* L., *Planta*, 136, 13, 1977.
178. **Kohli, R.K. and S. Sawhney,** Promotory effect of GA_{13} on flowering of *Amaranthus* — a short day plant, *Biol. Plant.*, 21, 206, 1979.
179. **Kulakow, P.A. and S.K. Jain,** The inheritance of flowering in *Amaranthus* species, *J. Genet.*, 64, 85, 1985.
180. **Fuller, H.J.,** Photoperiodic responses of *Chenopodium quinoa* Willd. and *Amaranthus caudatus* L., *Am. J. Bot.*, 36, 175, 1949.
181. **Koller, D., J. Kigel and S. Ovadia,** The facultative photoperiodic response in the reproductive development of *Amaranthus retroflexus* L. Changes in the dose response during ontogeny, *Planta*, 137, 133, 1977.
182. **McWilliams, E.L., P.R. Landers and J.P. Mahlstede,** Ecotypic differentiation in response to photoperiodism in several species of *Amaranthus*, *Iowa Acad. Sci. Proc.*, 73, 44, 1966.
183. **Kauffman, C.S.,** The status of grain amaranth for the 1990's, *Food Rev. Int.*, 8, 165, 1992.
184. **Senthong, C., S. Julsrigival, D.T. Tiyawalee and P. Wivutvongveana,** Germplasm screening of grain amaranth in Chiang Mai, Thailand, *Food Rev. Int.*, 8, 143, 1992.
185. **Espitia, E.,** Amaranth germplasm development and agronomic studies in Mexico, *Food Rev. Int.*, 8, 71, 1992.
186. **Duriyaprajan, S., P. Buranasilpin, S. Tanpanich, S. Chitnawasarn and J. Watanakul,** Preliminary evaluation and screening of introduced amaranth in Thailand, *Food Rev. Int.*, 8, 125, 1992.
187. **Gopal, B.,** Autoecological study of *Amaranthus spinosus*, *Ann. Arid Zone*, 13, 187, 1974.
188. **Hauptli, H. and S.K. Jain,** Biosynthetics and agronomic potential of some weedy and cultivated amaranths, *Theor. Appl. Genet.*, 52, 177, 1978.

189. **Pearcy, R.W. and J. Ehleringer,** Comparative ecophysiology of C_3 and C_4 plants, *Plant Cell Environ.*, *7*, 1, 1984.
190. **Edwards, G.E., M.S.B. Ku and R.K. Monson,** C_4 photosynthesis and its regulation, in *Photosynthetic Mechanisms and the Environment*, Barber, J. and N.B. Baker, Eds., Elsevier, 1985, 289.
191. **Lin, Z.F. and J.E. Ehleringer,** Photosynthetic characterization of *Amaranthus tricolor* a C_4 tropical leafy vegetable, *Photosynthesis Res.*, 4, 171, 1983.
192. **Ramos, C. and A.E. Hall,** Effects of photon fluence rate and intercellular CO_2 uptake rate of *Capsicum* and *Amaranthus*, *Photosynthetica*, 17, 34, 1983.
193. **Ehleringer, J.,** Ecophysiology of *Amaranthus palmeri*, a Sonoran desert annual, *Oecologia*, 57, 107, 1983.
194. **Breen, P.J., J.D. Hesketh and D.B. Peters,** Field measurements of leaf photosynthesis of C_3 and C_4 species under high irradiance and enriched CO_2, *Photosynthetica*, 20, 281, 1986.
195. **Harley, P.C. and J. Ehrleringer,** Gas exchange characteristics of leaves of four species of grain amaranth, *Field Crops Res.*, 17, 141, 1987.
196. **Blechschmidt-Schneider, S., P. Ferrar and C.B. Osmond,** Control of photosynthesis by the carbohydrate level in leaves of the C_4 plant *Amaranthus edulis* L., *Planta*, 177, 515, 1989.
197. **Korner, C.H., J.A. Scheel and H. Bauer,** Maximum leaf diffusive conductance in vascular plants, *Photosynthetica*, 13, 45, 1979.
198. **Ehleringer, J. and R.W. Pearcy,** Variation in quantum yield for CO_2 uptake among C_3 and C_4 plants, *Plant Physiol.*, 73, 555, 1983.
199. **El-Sherkawy, M.A., R.S. Loomis and W.A. Williams,** Photosynthetic and respiratory exchanges of carbon dioxide by leaves of the grain amaranth, *J. Appl. Ecol.*, 5, 243, 1968.
200. **Imbaba, S.K. and L.L. Tieszen,** Influence of light and temperature on photosynthesis and transpiration of some C_3 and C_4 vegetable plants from Kenya, *Physiol. Plant.*, 39, 311, 1977.
201. **Sage, R.F. and R.W. Pearcy,** The nitrogen use efficiency of C_3 and C_4 plants. II. Leaf nitrogen effects on the gas exchange characteristics of *Chenopodium album* L. and *Amaranthus retroflexus* L., *Plant Physiol.*, 84, 959, 1987.
201a. **Patterson, D.T.,** C_4 photosynthesis in smooth pigweed, *Weed Sci.*, 24, 127, 1976.
202. **Ku, S.B. and G.E. Edwards,** Oxygen inhibition of photosynthesis in the species *Amaranthus graecizans* L., *Planta*, 147, 277, 1980.
203. **Bjorkman, O., M.R. Badger and P.A. Armond,** Response and adaptation of photosynthesis to high temperature, in *Adaptation of Plants to Water and High Temperature Stress,* Turner, N.C. and P.J. Kramer, Eds., John Wiley & Sons, New York, 1980, 233.
204. **Singh, M., W.C. Ogren and T.M. Widholm,** Photosynthetic characteristics of several C_3 and C_4 plant species grown under different light intensities, *Crop Sci.*, 14, 563, 1974.
205. **Sampath, P. and G. Kulandaivelu,** Photochemical activities and organization of photosynthetic apparatus of C_3 and C_4 plants grown under different light intensities, *Photosynthetic Res.*, 4, 351, 1983.
206. **Bunce, J.A.,** Low humidity effects on photosynthesis in single leaves of C_4 plants, *Oecologia*, 54, 233, 1982.
207. **Bunce, J.A.,** Differential sensitivity to humidity of daily photosynthesis in the field in C_3 and C_4 species, *Oecologia*, 57, 262, 1983.
208. **Patterson, D.T. and E.P. Flint,** Comparative water relations, photosynthesis and growth of soybean (*Glycine max*) and seven associated weeds, *Weed Sci.*, 31, 318, 1983.
209. **Hunt, E.R., J.A. Weber and D.M. Gates,** Effects of nitrate application on *Amaranthus poowellii* Wats. II. Stomatal response to vapor pressure difference is consistent with optimization of stomatal conductance, *Plant Physiol.*, 79, 614, 1985.
210. **Bunce, J.,** Volume and osmotic potential changes in relation to inhibition of photosynthesis by water stress in intact leaves, *Can. J. Bot.*, 64, 557, 1986.
211. **Bunce, J.,** Species-specific responses to water stress of gas exchange parameters mimicked by applied abscisic acid, *Can. J. Bot.*, 65, 103, 1987.
212. **Dubbe, D.R., G.F. Farquhar and K. Raschke,** Effect of abscisic acid on the gain of the feedback loop involving carbon dioxide and stomata, *Plant Physiol.*, 62, 413, 1978.
213. **Hunt, E.R., J.A. Weber and D.M. Gates,** Effects of nitrate application an *Amaranthhus powellii* Wats. III. Optimal allocation of leaf nitrogen for photosynthesis and stomatal conductance, *Plant Physiol.*, 79, 619, 1985.

214. **Sage, R.F., P.W. Pearcy and J.R. Seeman,** The nitrogen use efficiency of C_3 and C_4 plants. III. Leaf nitrogen effects on the activity of carboxylating enzymes in *Chenopodium album* L. and *Amaranthus retroflexus* L., *Plant Physiol.,* 85, 355, 1987.
215. **Carlson, R.W. and F.A. Bazzaz,** Photosynthetic and growth responses to fumigation with SO_2 at elevated CO_2 for C_3 and C_4 plants, *Oecologia,* 54, 50, 1982.
216. **Bazzaz, F.A. and R.W. Carlson,** The response of plants to elevated CO_2. I. Competition among an assemblage of annuals at two levels of soil moisture, *Oecologia,* 62, 196, 1984.
217. **Garbutt, L., W.E. Williams and F.A. Bazzaz,** Analysis of the differential response of five annuals to elevated CO_2 during growth, *Ecology,* 7, 1185, 1990.
218. **Ludlow, M.M.,** Strategies of response to water stress, in *Structural and Functional Responses to Environmental Stresses: Water Shortage,* Krub, K.H., H. Richter, and T.M. Hinckley, Eds., SPB Acad. Publ., The Hague, 1989, 269.
219. **Gamboa, A., E.M. Valenzuela and E. Murillo,** Biochemical changes due to water loss in leaves of *Amaranthus hypochondriacus* L., *J. Plant Physiol.,* 137, 586, 1991.
220. **Bhownik, P.C. and J.D. Doll,** Growth analysis of corn and soybean response to allelopathic effects of weed residues at various temperatures and photosynthetic photon flux densities, *J. Chem. Ecol.,* 9, 1263, 1983.
221. **Menges, R.M.,** Allelopathic effects of palmer amaranth (*Amaranthus palmeri*) and other plant residues in soil, *Weed Sci.,* 35, 339, 1987.
222. **Altieri, M.A. and J.D. Doll,** The potential of allelopathy as a tool for weed management in crops, *Proc. Natl. Acad. Sci., U.S.A.,* 24, 495, 1978.
223. **Munger, P.H., J.R. Abernathy and J.R. Gipson,** The influence of selected plant residues on cotton and sorghum establishment, in *Proc. South Weed Sci. Soc.,* 37, 1984, 320.
224. **Bhowmick, P.C. and J.D. Doll,** Evaluation of allelopathic effects of selected weed species on corn and soybeans, in *Proc. North. Cent. Weed Control Conf.,* 34, 1978, 43.
225. **Hicks, S.K., P.H. Munger, J.R. Abernathy and C.W. Wendt,** Effect of amaranth allelopathic compounds in crops, in *Proc. South. Weed Sci. Soc.,* 39, 1986, 411.
226. **Bradow, J.,** Germination regulation by *Amaranthus palmeri* and *Ambrosia artemisiifolia,* in *The Chemistry of Allelopathy,* Thompson, A.C., Ed., American Chemical Society, Washington, D.C., 1985, 285.
227. **Fischer, N.H. and L. Quijano,** Allelopathic agents from common weeds *Amaranthus palmeri, Ambrosia artemiisifolia* and related weeds, in *The Chemistry of Allelopathy: Biochemical Interactions Among Plants,* Thompson, A.C., Ed., American Chemical Society, Washington, 1985, 133.
228. **Behari, M. and C.K. Andhiwal,** Chemical examination of *Amaranthus spinosus* L., *Curr. Sci.,* 45, 481, 1976.
229. **Roy, S., A.K. Dutta and D.P. Chakraborty,** Amasterol, an ecdysone precursor and a growth inhibitor from *Amaranthus viridis, Phytochemistry,* 21, 2417, 1982.
230. **Bradow, J.M. and W.J. Conmick,** Allelochemicals from Palmer amaranth *Amaranthus palmeri,* S. Wats., *J. Chem. Ecol.,* 13, 185, 1987.
231. **Bradow, J.M. and W.J. Conmick,** Volatile methyl ketone seed-germination inhibitors, *J. Chem. Ecol.,* 14, 1617, 1988.
232. **Bradow, J.M. and W.J. Conmick,** Seed-germination inhibition by volatile alcohols and other compounds associated with *Amaranthus palmeri* residues, *J. Chem. Ecol.,* 14, 1633, 1988.
233. **Conmick, W.J., J.M. Bradow and M.G. Legendre,** Identification and bioactivity of volatile allelochemicals from amaranth residues, *J. Agric. Food Chem.,* 37, 792, 1989.
234. **Flath, R.A., M.A. Altieri and T.R. Mon,** Volatile constituents of *Amaranthus retroflexus* L., *J. Agric. Food Chem.,* 32, 92, 1984.
235. **Mitchell, E.R. and R.R. Heath,** Influence of *Amaranthus hybridus* L. allelochemicals on oviposition behaviour of *Spodoptera exigua* and *S. eridonia* (Lepidoptera, Noctuidae), *J. Chem. Ecol.,* 11, 5, 1985.
236. **Shilling, D.G., R.A. Liebi and A.D. Worsham,** Rye (*Secale cereale* L.) and wheat (*Triticum aestivum* L.) mulch: the suppression of certain broad leaved weeds and the isolation and identification of *phytotoxins,* in *The Chemistry of Allelopathy,* Thompson, A.C., Ed., American Chemical Society, Washington, D.C., 1985, 233.
237. **Neill, R.L. and E.L. Rice,** Possible role of *Ambrosia psilostachya* on patterning and succession in old field, *Am. Midl. Nat.,* 86, 344, 1971.

238. **Pearcy, R.W., N. Tumosa and K. Williams,** Relationships between growth, photosynthesis and competitive interactions for a C_3 and C_4 plant, *Oecologia,* 48, 371, 1981.
239. **Wulff, R.,** Growth responses of soybean (*Glycine max*) and sorghum (*Sorghum bicolor*) to an increase in density of *Amaranthus dubius* plants at two temperatures, *Weed Res.,* 27, 29, 1987.
240. **Weaver, S.E.,** Differential growth and competitive ability of *Amaranthus retroflexus, A. powellii* and *A. hybridus, Can. J. Plant Sci.,* 64, 715, 1984.
241. **Patterson, D.T., E.P. Flint and J.L. Beyers,** Effect of CO_2 enrichment on competition between a C_4 and a C_3 crop, *Weed Sci.,* 32, 101, 1984.

Chapter 5

Biochemistry of Amaranth Proteins

M. Segura-Nieto, A. P. Barba de la Rosa, and O. Paredes-López

CONTENTS

0-8493-5374-2/94/$0.00+$.50

I. INTRODUCTION

Amaranth is one of the few double-duty plants which can supply grains and tasty leafy vegetables of high nutritional quality. It is one of the few nongrass species with potential to become a cereal-like grain crop or source of dietary proteins. Now it is well-known that amaranth grain has proteins with a better balance content of the essential amino acids than that of cereals and legumes.[1-3] Over the past two decades amaranth has become a focus of intense nutritional and agronomical research. This interest began in 1973, when Downton[4] reported that the proteins from amaranth grain contain two to three times the lysine concentration of proteins from wheat, rice, and corn.[4-6] Since then, research has been performed on the biological value of amaranth proteins, on the antinutritional factors in the grain and leaves, on the isolation of protein fractions from grains, and on some biochemical and functional properties of seed proteins.[7-16]

For centuries leaves of amaranth have been used for human consumption in three continents (America, Asia, and Africa); their appearance, texture, and overall nutritional quality from various *Amaranthus* species are comparable to those from spinach. Green parts are rich in protein, minerals, and vitamins, becoming a food or source of protein extracts of high nutritional quality. Because they may contain appreciable amounts of nitrate and oxalate that can cause problems in their consumption, protein concentrates of leaves have been prepared.[2,17]

This chapter will describe in some detail the biochemical and functional properties thus far known about proteins from amaranth species producing edible seeds and leaves. It is pertinent to mention that the terms "seeds" and "grains" are indistinctly used in this work.

II. SEED PROTEINS

A. CHEMICAL COMPOSITION OF AMARANTH SEEDS

There are mainly three species of amaranth that produce nutritive seeds, with potential to become a cereal-like grain crop: *Amaranthus caudatus, Amaranthus cruentus,* and *Amaranthus hypochondriacus.*[1,17] The proximate composition of seed flours are inconsistent among and within these three main species. The range of the reported data is shown in Table 1. The moisture content is lower in *A. cruentus* (6.2 to 8.8%) and highest in *A. caudatus* (9.5 to 11.6%). *A. cruentus* has the lowest crude protein content (13.2 to 18.2%) as compared to *A. caudatus* (17.6 to 18.4%) and to *A. hypochondriacus* (17.9%). Here, it is important to mention that all calculations related to proteins were done with the value 5.85, described as the best conversion factor for amaranth proteins.[18] *A. caudatus* has the highest fiber value (3.2 to 5.8%), and the lowest is found in *A. hypochondriacus* (2.2%). The highest fat content is in *A. caudatus* within a range of 6.9 to 8.1% and the ash, which represents the mineral content in the grain, is higher in *A. caudatus* in the range of 3.1 to 4.4% (Table 1).[17,19-21]

A comparison between the average chemical composition of *A. hypochondriacus* with corn, rice, and wheat is shown in Table 2. Amaranth flour is characterized by a relatively higher concentration of proteins (17.9%) than in cereals (8.5 to 14.0%).[2,19-23] The lipid content is also relatively high (7.7%),[17,21,23-27] as compared to corn (4.5%), rice, or wheat (2.1%), and the starch content (57.0%) is close to the most important cereals (66.9 to 75.5%).[23,25,26]

The essential amino acid composition of different amaranth grains is shown in Table 3. This composition is unusual because its balance is closer to the optimum required in the human diet (FAO/WHO[28]). As noted before, the lysine content is relatively high (3.2 to 6.4%)[3,31-32] as compared with that found among the most common cereals (2.2 to 4.5%), and the sulfur amino acid concentration (2.6 to 5.5%) is higher than that of the most important legumes like pea, beans, soybeans, etc. (1.4%), making this a promising crop as a food or source of dietary proteins.[3,16-17,23,27,29-33] Leucine, isoelucine, and valine, the limiting amino acids, are not considered a serious problem since they are found in excess in most common grains.[23,32,34]

Table 1 **Proximate composition of amaranth seeds species**[a]

Analysis	*A. caudatus*[21] (*A. edulis*)	*A. cruentus*[21]	*A. hypochondriacus*[21]	*A. cruentus* × *A. hypochondriacus*[17]
Moisture	9.5–11.6	6.2–8.8	11.1	—
Crude protein[b]	17.6–18.4	13.2–18.2	17.9	17.4
Fat	6.9–8.1	6.3–8.1	7.7	8.0
Fiber	3.2–5.8	3.6–4.4	2.2	4.3
Ash	3.1–4.4	2.8–3.9	4.1	3.0

[a] Percentage on dry weight basis; [b] N × 5.85.

Table 2 **Comparison of proximate composition between amaranth seeds and some cereals**[a]

Analysis	Amaranth[b]	Corn[17]	Rice[26]	Wheat[17]
Moisture	11.1	13.8	11.7	12.5
Crude protein	17.9[c]	10.3[d]	8.5[d]	14.0[e]
Fat	7.7	4.5	2.1	2.1
Fiber	2.2	2.3	0.9	2.6
Ash	4.1	1.4	1.4	1.9
Carbohydrates	57.0	67.7	75.4	66.9

[a] Percentage on dry weight basis; [b] Values of *A. hypochondriacus* from Table 1; [c] N × 5.85; [d] N × 6.25; [e] N × 5.7.

Table 3 **Essential amino acid composition of seeds from three *Amaranthus* species**[a]

Amino Acid	*A. caudatus*[17]	*A. cruentus*[20,33]	*A. hypochondriacus*[17,20]	FAO/WHO/UNU[28] Adults	Children
Cys	2.3	2.0–3.8	2.0–3.9		
Ile	3.6–4.1	3.4–3.7	2.8–3.8	1.3	4.6
Leu	5.9–6.3	4.8–5.9	5.0–5.8	1.9	9.3
Lys	5.7–6.4	4.8–5.8	3.2–6.0	1.6	6.6
Met	2.4–3.3	1.8–1.6	0.6–1.6		
Phe	3.4–4.0	3.2–4.5	3.8–4.5		
Thr	3.8	3.2–4.2	2.6–4.3	0.9	4.3
Trp	1.1	nd	1.1–4.0	0.5	1.7
Tyr	2.8	2.4–4.0	3.1–4.0		
Val	4.1–4.7	3.9–4.3	3.2–4.2	1.3	5.5
Met + Cys	4.7	3.8–5.4	2.6–5.5	1.7	4.2
Phe + Tyr	6.2	5.6–8.5	6.9–8.5	1.9	7.2

[a] Grams per 100 g protein.

Note: nd = not determined.

Seeds of the three species of *Amaranthus* display some heterogeneity in their color varying from pale yellow to black. In *A. hypochondriacus*, over 95% of the seeds are pale yellow. Differences in color are also reflected in the electrophoretic patterns of total proteins from *A. hypochondriacus* seeds, as shown in Figure 1. In light color seeds there are more polypeptides of low molecular weight (Figure 1, lanes 2 to 4) than those of dark color (lanes 5 to 7).[16] Also, Olivares[35] has observed some differences in the protein bands of the high and low molecular weight polypeptides among varieties. Deeper studies are needed to try to correlate these differences.

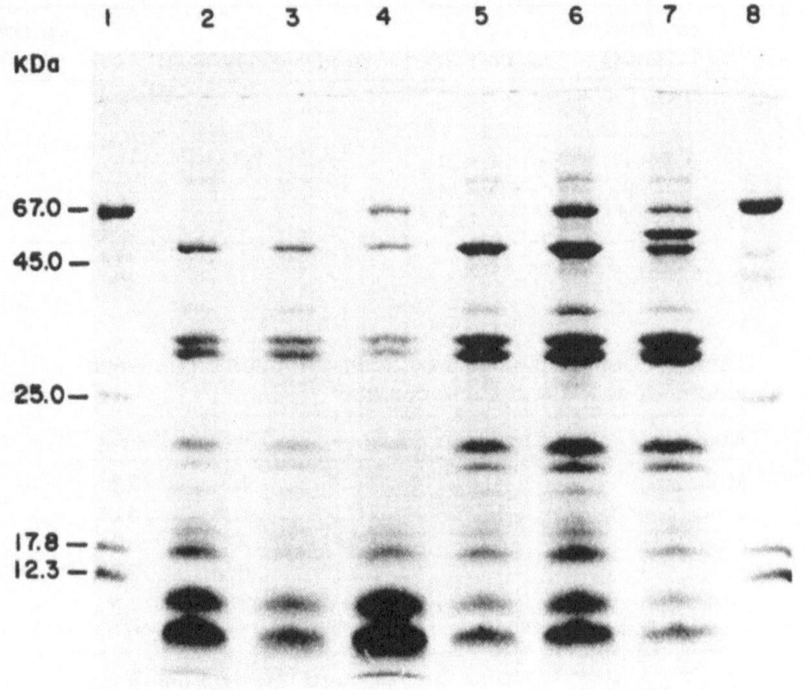

Figure 1 Electrophoretic patterns of *A. hypochondriacus* seeds of different colors. Lanes 1 and 8 molecular weight markers; lanes 2 to 4, light seeds (pale yellow, yellow, and strong yellow, respectively); lanes 5 to 7, dark seeds (light brown, dark brown, and black, respectively).

B. HISTOLOGICAL STUDIES

Histological, chemical, and nutritional studies on *A. cruentus* seeds were used to characterize the effects of successive milling of the seed. In general, nutrients (nitrogen, crude fat, crude fiber, and ash) are more highly concentrated in the seed coat-embryo fraction, 2.3–2.6 times as the original intact seed and it accounted for 26.3% of the total seed weight.[36]

The storage proteins in amaranth seeds, like in other seed storage proteins, are defined by their accumulation in time and space during the maturation of the seeds.[37] There is no available information related to the deposition of storage proteins in protein bodies. Morphological studies on seeds of *A. cruentus* and *A. hypochondriacus* using the scanning electron and light microscopy have shown that the cells of the embryo contain material which has typical properties of lipid and lipid complexes as demonstrated by hexane extraction and by staining with osmium tetroxide.[38,39] Spherical cell bodies embedded in the lipid-containing matrix were partially digested by trypsin, suggesting their proteinaceous nature.[38] In most dicotyledonous plants, the developing embryo utilizes vast portions of the energy reserves of the endosperm, leaving little or no endosperm in the mature seed.[40] In amaranth grain, the endosperm remains mostly as a single cell layer in the region of the embryo attached to the seed coat. At the tips of the cotyledons and root, several cell layers of endosperm remain (Figure 2A). Protein bodies are ubiquitous in embryonic and endosperm tissues; however, cell size and protein body size are variable.[17,36] Thin-walled parenchyma cells of the embryo contain reserves in the form of protein bodies (Figure 2) measuring approximately 3–5 μm in diameter, while, in endosperm cells, they are about 1.5–2.0 um. Protein in the perisperm occurs as very small deposits between starch granules[17,36,39] as shown in Figure 2B.

C. CLASSIFICATION OF AMARANTH SEED PROTEINS

There are several possibilities for the classification of amaranth proteins:

1. On the basis of morphology, they can be grouped into proteins from endosperm, embryo, and perisperm.[38,39]

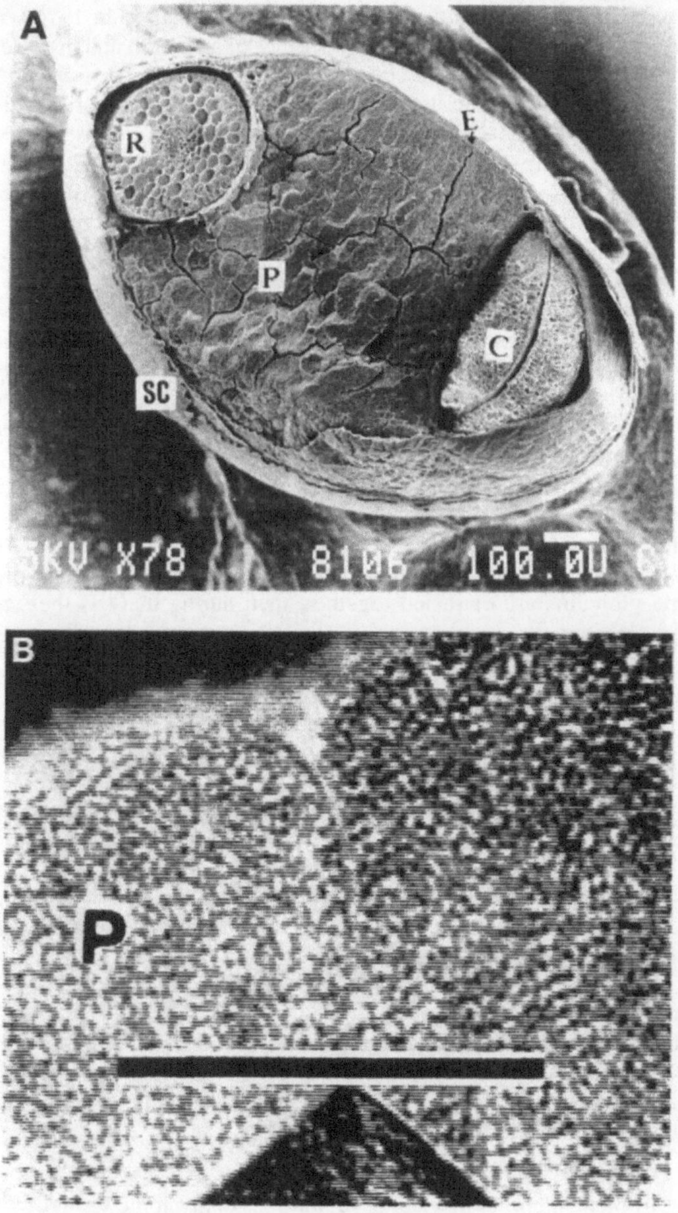

Figure 2 (A) Scanning electron micrograph of *A. hypochondriaucus* seed. R, root, E, endosperm, P, perisperm, Sc, seed coat, C, cotyledons. Bar 100 μm. (B) A. *Cruentus* section of perisperm stained with acid fuchsin and viewed at 500 to 550 nm excitation. Bar 50 μm.

2. On the basis of biological functions, proteins can be classified in two main groups: "housekeeping" proteins (metabolically active enzymes; regulatory, structural, ribosomal proteins, etc.), composed of relatively small amounts of numerous protein species, and "storage proteins", accumulated in significant quantities in the developing seeds and compartmentalized in protein bodies which on germination are rapidly hydrolyzed to provide a source of reduced nitrogen for the early stages of seedling growth.[23,37]
3. Considering the chemical composition, they are defined as: simple or complex proteins (lipoproteins, glycoproteins, nucleoproteins, metalloproteins, etc.).[23,41]
4. According to solubility, Osborne[42] divides them into four classes: albumins soluble in water, globulins soluble in high salt concentration, prolamins soluble in aqueous alcohol, and glutelins soluble in acid or alkaline solutions.

In the course of time, proteins of amaranth seeds have been isolated by many versions of this scheme.[43-45]

D. PROTEIN DISTRIBUTION

Interest about the isolation and characterization of amaranth seed protein fractions started in 1976 with the studies of Abdi and Sahib[11] on the distribution of lysine in different species of amaranth and in protein fractions of *A. hypochondriacus*. This species has high lysine content and the albumin fraction represented 50% of the protein with the highest lysine concentration. In 1985, Konishi et al.[12] described also the protein fractions from the seeds and the physical and chemical properties of globulins from *A. hypochondriacus L.* waxy and nonwaxy types, using the procedure of Landry and Moureaux.[43] Since then, several reports have appeared, describing the isolation and the chemical characterization of protein fractions in amaranth grain.[15,16,30,31,46–48]

Most of the results support that albumins are the main fraction (48.9 to 65%); only Bressani and García-Vela[31] reported values between 8.5 and 19.2% (Table 4). Glutelin fraction is the second in abundance (22.4 to 42.3%), except for Abdi and Sahib[11] and Duarte-Correa[30] who found 8.5 and 7.0%, respectively. Globulins represent the 13.7 to 18.1% of the seed protein, excepting Bressani and García-Vela[31] who found 38.5% (procedure A, Table 4). Most data point out prolamins as the minor fraction with values between 1.0 to 3.2%. Several reports (3 of 11) found higher concentration in the range of 7.2 to 11.0%, as shown in Table 4. The amount of the residue left after protein extraction is probably dependent on the buffers and the extraction time.

The results of these studies varies; some of the differences can be related to the procedure for protein extraction and the type of seeds used (Table 4).

One of the main differences is in the extraction method for albumins and globulins, as pointed out in Table 5. Some procedures are based on a sequential extraction (first albumins, then globulins, and so on), while in other methods, albumins and globulins are extracted together, then during dialysis they are separated as discussed below.[11,12,16,30,31,46,49.]

III. CHARACTERIZATION OF PROTEIN FRACTIONS

A. ALBUMINS

1. Localization

For many years seed albumins were not considered as part of the storage proteins. They were considered as "housekeeping proteins".[50] Youle and Huang[51] demonstrated the presence of albumins in protein bodies. Konishi *et al.*[52] isolated two types of albumin fractions (alb-1 and alb-2) from *A. caudatus* and *A. hypochondriacus* which were biochemically different. Alb-1 disappeared when the flours were predigested with pronase. They suggested that alb-2 were storage proteins that might be associated with protein bodies, inaccessible to the enzymatic digestion.

2. Biochemical and Physicochemical Properties

There are few studies on the biochemical and physicochemical characterization of amaranth albumins.[16,52] Among the criteria used to identify them were solubility, electrophoretic pattern and molecular weight of denatured polypeptides, amino acid composition, and sedimentation coefficient, which will be described in the next sections.

a. Isolation and Purification

The albumins are seed proteins defined by their solubility in water or in solutions of low ionic strength.[16,42] In amaranth grain, they represent one of the major protein groups, as already mentioned. The percentage concentration changes with the extraction procedure used.

The main differences in some of the methods used for the extraction of albumins are pointed out in Table 5 and can be reduced to three main procedures:

1. Albumins and globulins are extracted together, first with high ionic strength (0.4 to 1.0 M) NaCl with or without a second extraction with water; in both cases extracts are dialysed exhaustively against water.[12,46,47,49] After dialysis, the two groups of proteins are separated. Albumins are recuperated in the supernatant and globulins in the pellet. When amaranth proteins are extracted from whole meal, more albumins are obtained than when defatted flour is used.[15,46,49] However, most of the experiments are conducted with defatted meal.
2. Albumins are extracted from defatted flour with water (procedure B, Table 4)[31] or with low ionic strength buffer with proteases inhibitors, without dialysis.[16]

Table 4 **Protein fractionation of *Amaranthus hypochondriacus* seeds extracted with different procedures[a]**

Procedure		Fraction						
		Albumins	Globulins	Alb + Glo	Prolamins	Glutelins	Residue	Ref.
Mitchell (1984)		48.9	13.7		7.2	8.5	21.7	11
Padhye & Salunkhe	w			56.9	1.6	22.4	8.5	12
(1977)	nw			59.4	1.0	29.4	12.6	
Padhye & Salunkhe (1977)		65.0	17.0		11.0	7.0	nd	30
Osborne (1924)	W			39.5	3.2	39.2	18.1	46
	D			29.2	1.3	40.3	29.2	
Landry & Moreaux	A	8.5	38.5		9.1	29.1	14.8	31
(1990)	B	19.2	18.1		2.0	42.3	18.4	
Landry & Moreaux (1990)				57.3	1.4	25.8	12.1	47
Osborne (1924)				51.6	1.2	42.2	2.8	15
		50.0	15.6		2.0	30.4	2.0	16

[a] Percent of extracted protein.

Note: Alb + Glo = albumins + globulins; w = waxy; nw = non-waxy; W = whole flour; D = defatted flour; A = procedure A; B = procedure B.

Table 5 **Distribution of albumins and globulins from *Amaranthus hypochondriacus* seeds extracted under different conditions[a]**

Albumins	Globulins	Alb + Glo	Extraction Conditions	Ref.
48.9	13.7		Information not available	11
		56.9 (w)	0.4 M NaCl + 35 mM K_2HPO_4, o.n.	12
		59.4 (nw)	10 vol (w/v), dialyzed *vs.* water	
65.0	17.0		Information not available	30
		46.8 (W-nd)	0.5 M NaCl twice (1 h each)	46
		39.5 (W-d)		
		47.5 (D-nd)		
		29.2 (D-d)		
		29.2	0.5 M NaCl (10 vol w/v), washing with water, 1 h each step, dialyzed *vs. water*	49
8.5	38.5		0.5 M NaCl, 60 min; water, 30 min	31
19.2	18.1		Water, 60 min; 0.5 M NaCl 60 min	
		57.3	0.5 M NaCl, water twice, dialyzed vs, water 72 h	47
		51.6	Na_2HPO_4 pH 7 1 h, dialyzed *vs.* water	15
50.0	15.6		0.1 M NaCl twice (8 h and 6 h)	16
			1.0 M NaCl twice (8 h and 6 h)	

[a] Percent of extracted protein.

Note: Alb + Glo = albumins + globulins; w = waxy; nw = nonwaxy; o.n. = overnight; W = whole flour; D = defatted flour; nd = nondialyzed; d = dialyzed.

3. Two albumin fractions are extracted. The first albumin fraction (Alb-1) is water-extracted, as in procedure 2 above. Another fraction (Alb-2) is extracted with water after removing Alb-1 and globulins (with high ionic strength). The two albumin fractions are biochemically different.[52]

One observation related to the first albumin extraction method regards the need to critically examine the pellet considered as globulins, and distinguish the amount of the denatured and degraded material (that could include albumins and globulins) from the real globulins.

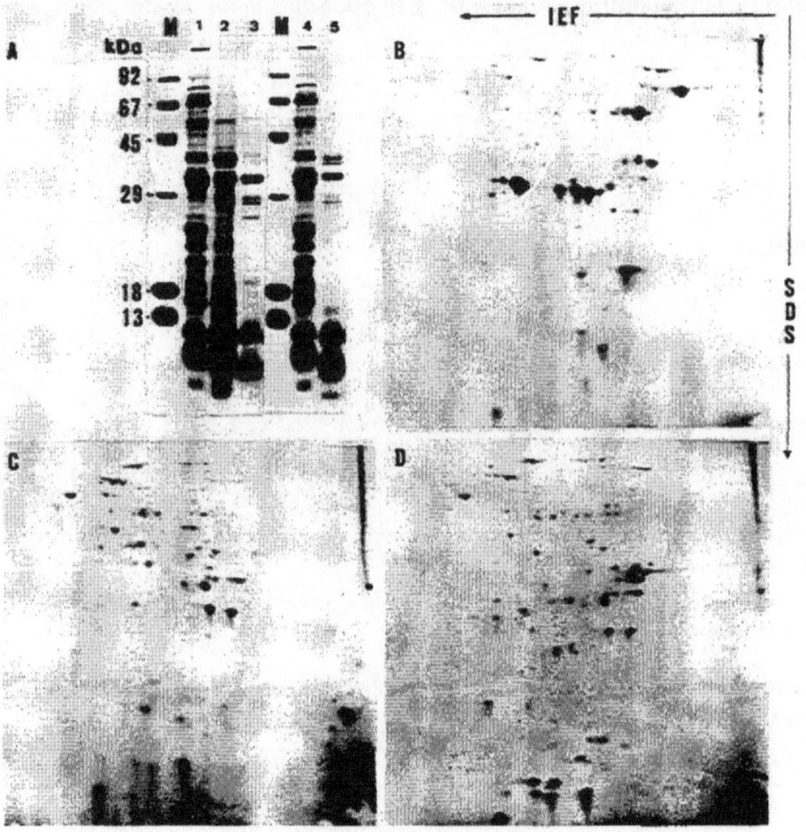

Figure 3 (A) Electrophoretic patterns in one- and two-dimensions of amaranth seed albumins. Lanes 1 and 4, total proteins from defatted flour; lane 2, albumins + globulins; lane 3, albumins extracted by Osborne's method; lane 5, albumins extracted with 0.1 M NaCl. (B) Two-dimensional analysis of total proteins. (C) Albumins extracted by Osborne's method. (D) Albumins extracted with 0.1 M NaCl.

In the second case of low salt concentration for albumin extraction, it will be necessary to determine if there is any contamination with some types of globulins.

Observations for the third method are the same as for method 2; however, it is difficult to determine the real ionic strength used for the Alb-2 extraction.

b. Electrophoretic Pattern in One- and Two-Dimensions

The electrophoretic patterns of the albumin fractions isolated by the Osborne protocol (supernantant of dialysis) and with 0.1 M NaCl shown in Figure 3A (lanes 3 and 5, respectively), reveal subtle differences.[16,42] In lane 2 we see the electrophoretic pattern of the mixture of albumins and globulins before dialysis. In both albumin preparations, there are two groups of polypeptides. The main group has a molecular weight below 18 kDa and the second between 42 and 25 kDa, as compared to molecular weight markers in lane M. Polypeptides from albumins fraction (lanes 2 and 5, Figure 3A) can be detected among total proteins from crude extracts of defatted meal in lanes 1 and 4. More studies are needed on the albumins fraction in order to identify the storage from the housekeeping proteins.

Two-dimensional analysis of total proteins and from albumins extracted by the two methods above are shown in Figure 3 (B, C, and D, respectively). These studies show that albumins have a complex polypeptide pattern. There are several polypeptides with the same molecular weight but different isoelectric points, and some common polypeptides can be detected among the total proteins extracted from defatted meal in panel 3B. It is known that during dialysis of albumins and globulins, some degradation could take place if there is no protection against proteolytic digestion.[16,53]

Table 6 **Amino acid composition of albumins from different types of *Amaranthus hypochondriacus* seeds extracted under various conditions**[a]

Amino Acid	(30)	(31) v-4	(31) v-5	(49)	(47)	(15)	(16)
Cys	nd	2.1	1.9	2.4	1.8	5.7	1.4
Ile	3.6	4.4	4.0	3.2	3.0	3.6	3.3
Leu	4.8	7.1	6.1	4.4	5.5	5.5	5.1
Lys	7.7	6.7	6.2	9.4	7.6	7.4	6.8
Met	1.8	2.3	2.3	2.7	2.6	4.0	2.7
Phe	3.4	4.7	4.5	3.2	3.2	5.0	3.6
Thr	3.3	5.0	4.0	2.9	4.8	3.8	2.9
Trp	nd	1.5	1.2	nd	0.8	nd	nd
Tyr	3.6	2.4	3.7	3.2	3.8	3.2	3.7
Val	3.9	5.1	4.6	4.0	4.0	4.4	3.4
Ala	3.3	4.5	4.1	3.7	5.6	4.9	3.9
Arg	10.6	7.1	6.8	8.3	0.9	7.8	11.0
Asx[b]	7.8	10.7	9.6	7.0	8.3	6.0	8.9
Glx[c]	22.8	19.4	19.6	22.6	18.8	17.0	20.9
Gly	7.3	4.9	4.3	8.3	8.3	6.0	5.6
His	2.1	2.1	2.1	1.8	2.5	2.4	2.6
Pro	3.5	nd	nd	3.4	2.6	3.6	4.0
Ser	4.4	5.1	4.5	4.6	10.4	4.6	4.8
Met + Cys	nd	4.4	4.2	5.1	4.4	9.7	4.1
Phe + Tyr	7.0	7.1	8.2	6.4	7.0	8.2	7.3

[a] Grams per 100 g protein; [b] Mixture of aspartic acid and asparagine; [c] Mixture of glutamic acid and glutamine.

Note: ( ) = reference; v-4 = variety 4; v-5 = variety 5; nd = not determined.

c. Amino Acid Composition

The amino acid composition of albumins extracted under different conditions are shown in Table 6. The most striking differences are in their lysine content (in the range of 6.2 to 9.4% grams per 100 g) and sulfur amino acids (4.1 to 9.7%).[15,16,30,31,47,49]

d. Sedimentation Coefficient

Albumins from many dicot seeds are also identified by their low sedimentation coefficient around 1.7S to 2.0S.[54] In amaranth albumins, the sedimentation pattern reveals two sets of components. A major one between 1.4S to 2.0S (X = 1.7S) and a minor one with a sedimentation coefficient of 4.6S are shown in Figure 4A.[16] The electrophoretic pattern of albumin fractions isolated by sucrose gradient is shown in Figure 4B. The main set of albumins sedimenting between 1.4S and 2.0S consists of polypeptides, with molecular weights ranging from 4 to 20 kDa (see fractions 4 to 7 in Figure 4B). These values are the same as those reported for many other 2S albumins.[54-57] These 1.7S albumins are one of the major group of polypeptides in the amaranth grain. The same holds for some oil seed albumins.[54] In panel 4B, each lane in the SDS-PAGE contains about the same amount of protein, so it does not reflect the real protein concentration in the gradient (Figure 4A).[16]

3. Methionine Rich Proteins

Methionine rich proteins (MRPs) are a group of important proteins found in different types of seeds. In dicots, it has been found that in some oil seeds (like sunflower), one protein of 10 kDa contains 15.5% (in mols percent) methionine; in Brazil nut, there is a 12 kDa polypeptide, with 18.8% methionine.[55,58,59] In monocots like corn and rice, the proteins rich in methionine are a group of prolamins. In corn, there are two polypeptides with 10 and 15 kDa that have 22.5 and 11.3% methionine, respectively.[60,61] In rice, the 10 kDa polypeptide exceptionally rich in sulfur amino acids has 20% methionine and 10% cysteine.[55] In amaranth, there is a group of 2S polypeptides with 16 and 18% methionine content.[62]

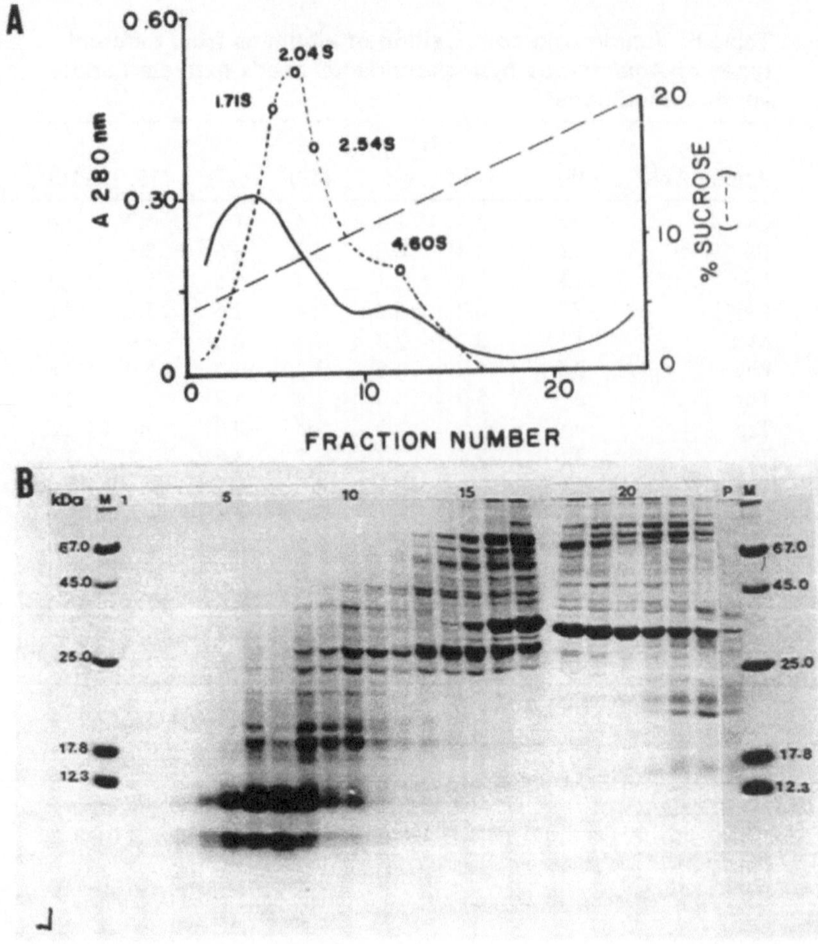

Figure 4 (A) Sedimentation coefficients of albumins. The absorbance of each fraction from the sucrose gradient was measured at 280 nm; (——), sedimentation pattern of protein markers of known sedimentation coefficient; (______) sedimentation pattern of albumins. (B) Electrophoretic pattern of every fraction from the sucrose gradient in (A).

a. Isolation and Purification

In amaranth albumin fractions, there is a small group of 2S MRPs. These polypeptides were isolated from total albumins by reverse phase chromatography on a C_{18} column, as shown in Figure 5A. The amino acid composition of each HPLC-fraction (data not shown) revealed that fractions 7 and 8 (Figure 5A), had 10 and 13% methionine, respectively, and a molecular weight in a range of 18 to 12 kDa, shown in lanes 7 and 8 in Figure 5B. The purification of two MRPs from these two fractions was performed with the same chromatographic system as before.[62]

b. Electrophoretic Pattern and Molecular Weight

In Figure 5C, the electrophoretic pattern of the mixture of MRPs in lane 1 is shown. Both polypeptides have a molecular weight of 18 kDa. These polypeptides are susceptible to degradation under acidic solutions.[53,62]

c. Amino Acid Composition

The amino acid composition of two MRPs is shown in Table 7. The methionine content of MRP-1 is 18.6% (in % moles) and 16.6% MRP-2. These values are similar to that detected in the Brazil nut albumin.[62,63] The MRPs also contain a relatively high lysine concentration (6.1 to 6.4%). Other essential amino acids are in lower concentration as compared to casein.[64]

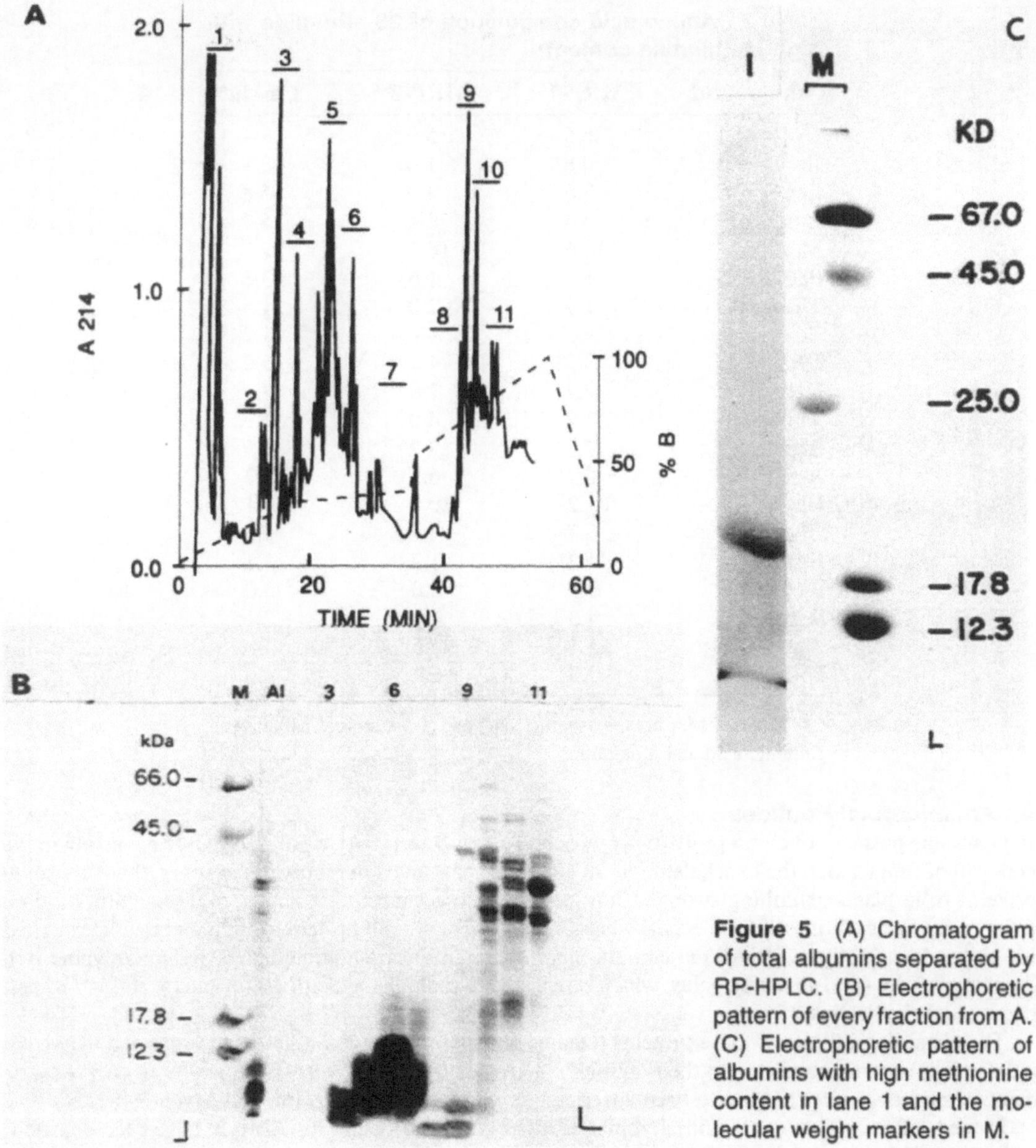

Figure 5 (A) Chromatogram of total albumins separated by RP-HPLC. (B) Electrophoretic pattern of every fraction from A. (C) Electrophoretic pattern of albumins with high methionine content in lane 1 and the molecular weight markers in M.

d. N-terminal Sequence

The N-terminal amino acid sequence of two MRPs was determined on performic acid-oxidized proteins as shown below. The N-terminal of these polypeptides did not show homology with other methionine rich polypeptides.[62]

	1	2	3	4	5	6	7	8	9	10
MRP-1	P	L	F	E	M	G	L	Q	A	L
MRP-2	P	I	F	M	E	G	K	M	E	R

	11	12	13	14	15	16	17	18	19	20
MRP-1	K	L	Q	M	K	C	P	N	M	T
MRP-2	Q	L	M	N	K	E	P	N	Y	M

	21	22	23	24	25
MRP-1	E	E	C	A	M
MRP-2	E	E	Y	M	M

More studies are in progress trying to complete these sequences.[53,62]

Table 7 **Amino acid composition of 2S albumins with high Methionine content**[a]

Amino Acid	MRP-1[62]	MRP-2[62]	Casein[64]
Cys	6.5	5.9	—
Ile	0.8	1.6	3.8
Leu	3.6	4.0	5.8
Lys	6.4	6.1	5.4
Met	18.6	16.6	2.0
Phe	1.3	1.6	3.6
Thr	1.9	2.0	1.5
Trp	—	—	1.2
Tyr	3.9	4.5	5.0
Val	2.2	2.8	3.3
Ala	3.2	4.0	1.9
Arg	7.3	7.2	2.9
Asx[b]	6.7	6.3	5.2
Glx[c]	21.2	20.5	15.1
Gly	6.7	7.1	1.5
His	1.2	1.2	2.1
Pro	3.9	4.0	5.0
Ser	4.3	4.2	2.1
Met + Cys	25.1	22.5	—
Phe + Tyr	5.2	6.1	8.6

[a] % Moles; [b] Mixture of aspartic acid and asparagine; [c] Mixture of glutamic acid and glutamine.

4. Antimicrobial Peptides

In plants, the presence of chitin [poly (b-1,4-*N*-acetyl-*D*-glucosamine)], a polysaccharide occurring in the cell wall of fungi and in the exoskeleton of invertebrates has not been reported. However, there are some proteins from plants exhibiting strong affinity to this polysaccharide. Examples of these chitin-binding proteins are basic chitinases from beans, wheat, and tobacco; a small protein from rubber tree latex called hevein, and some others.[65-68] All these chitin-binding proteins share a homologous cysteine/glycine-rich domain of about 40 to 43 amino acids, which is repeated twofold in the nettle lectin and fourfold in wheat, barley, and rice lectins.[69]

The physiological role of these proteins remains uncertain; however, it has been suggested they have a defense-related function due to their antibiotic activities that all have been shown by *in vitro* experiments. Antifungal properties have been adscribed to chitinases, nettle lectin, and hevein.[66-68]

In seeds of *A. caudatus*, two antimicrobial peptides (Ac-AMP-1 and Ac-AMP-2) have been detected, isolated, and characterized, which are homologous to the cysteine/glycine-rich domain of chitin binding proteins.[70]

a. Isolation and Purification

Antimicrobial peptides (Ac-AMPs) are isolated from a crude soluble extract of amaranth meal with 100 mM KCl in the presence of protease inhibitors with 10/15 mM Na/Na_2PO_4 buffer by a 4 step procedure. In the first, ammonium sulfate fractionation of proteins precipitates in an interval of 30 to 75% relative saturation. In the second step, the basic protein fractions (pI > 9) are isolated by ion exchange chromatography at pH 9. The protein not retained by the column has the antifungal activity and is further separated in a third step by cation exchange chromatography at pH 7.0. The active peptides are eluted in two separated fractions. Finally, these two components are purified by reverse phase chromatography and named as Ac-AMP1 and Ac-AMP2, respectively.[70]

b. Characterization and Amino Acid Sequence

The amaranth antimicrobial pure peptides were analyzed by SDS-PAGE before and after reduction with dithiothreitol. The two nonreduced peptides migrated with an apparent molecular weight of 4 and 5.5 kDa, respectively. Under reduced conditions, Ac-AMP1 and Ac-AMP2 both migrated as single bands with an apparent molecular weight of about 3 kDa. These results suggest that Ac-AMPs are homodimers. On the basis of fast atom bombardment mass spectroscopy, Ac-AMP1 and Ac-AMP2 have monoisotopic molecular weights of 3025 and 3181, respectively. The pI values of Ac-AMP1 is 10.3 and Ac-AMP2 is 10.6.[70]

The unreduced Ac-AMPs did not react with thiol reagent 5,5'-dithiobis (nitrobenzoic acid), whereas the reduced peptides did, indicating that all cysteine residues share disulfide bonds.[70]

The amino acid sequence of both polypeptides were identical, as shown below, except that Ac-AMP2 has an additional arginine residue at the C-terminal. This is the reason why it is more basic.[70]

	1	2	3	4	5	6	7	8	9	10
AC-AMP1/2[a]	V	G	E	C	V	R	G	R	C	P
	11	12	13	14	15	16	17	18	19	20
	S	G	M	C	C	S	Q	F	G	Y
	21	22	23	24	25	26	27	28	29	30
	C	G	K	G	P	K	Y	C	G	R[b]

[a] Includes both polypeptides; [b] R just in Ac-AMP2.

The amino acid sequence of Ac-AMP2 shows striking similarity to the cysteine/glycine-rich domain of chitin-binding chitinases, chitin-binding lectins, and hevein.[65-68]

c. *Activity*

These amaranth peptides inhibit the growth of fungi to a much higher extent than other previously characterized antifungal, chitin-binding proteins.[69,70] The antifungal activity was not abolished by treatment with proteinase K, pronase E, chymotrypsin, or trypsin. This unique property of Ac-AMPs as potent inhibitors of fungal growth *in vitro* suggests that they may play a role in the defense of seeds against invasion of fungal organisms. However, it remains to be proven if Ac-AMPs can exert antifungal effects in its natural microenvironment.[69,70]

5. Proteinaceous Inhibitors

Proteins that form complexes with proteases and inhibit their proteolytic activity are widespread in nature. In addition to their role regulating proteolytic activities, they are important for protecting fluids or tissues for degradation by unwanted or foreign proteolytic activities.[71] The term protease includes both endopeptidases and exopeptidases, whereas the term proteinase is used to describe only endopeptidases.[72,73]

Several nonhomologous families of protease inhibitors are recognized among the animal, microorganism, and plant kingdoms.[73] In plants, ten possible protease inhibitors families have been recognized.[72-75] The inhibitor families have been found that are specific for each of the four mechanistic classes of proteolytic enzymes: serine, cysteine, aspartic, and metalloproteases. By far, the majority of known protease inhibitors are specific for serine proteinases. Seven of the ten families inhibit enzymes of the serine class of endopeptidases and a single family that is specific for either cysteine, aspartyl, or metallo-classes of proteases.[72]

Protein inhibitors have been particularly well-studied in important food crops such as cereals, legumes, and potatoes, since most of the initial work in this field came from those interested in animal nutrition concerned about the possible deleterious effects exerted by the inhibitors.[74]

In the case of amaranth, there are very few studies regarding protein inhibitors of proteinases, and these are members of the serine class.[76] Saunders and Becker[17] were one of the first reporting the presence of small amounts of heat-labile trypsin inhibitors detected in *A. cruentus*. They reported a small amount of trypsin inhibitors that was not significant from the nutritional point of view.[17]

a. *Isolation and Characterization*

So far, there are two reports on the isolation and characterization of *A. hypochondriacus* trypsin inhibitors.[10,76] One of them reported that the inhibitor was isolated by affinity chromatography in a trypsin-Sepharose 4B column.[76] The molecular weight was not determined, but the electrophoretic analysis in nondissociating conditions revealed 13 trypsin inhibitors. One notable observation is that the inhibitor was very thermostable, retaining 20% of its original activity after 7 h at 100°C.[76] In the other report, the authors also mention the existence of several trypsin inhibitors, but they selected the most abundant one named amaranth trypsin inhibitor (ATI). Its purification was done through ion exchange chromatography, gel filtration, and RP-HPLC. In SDS-PAGE, ATI appears as a single band with a molecular weight around 7 kDa. This is a basic inhibitor with a pI around 7.5, and its amino acid composition shows two cysteins and high valine content.[10]

b. N-Terminal Sequence

The N-terminal sequence of amaranth trypsin inhibitor (ATI), is highly acidic, with 5 glutamic acids in the first 26 amino acid residues. This partial sequence shows some homology with potato I inhibitor family.[10]

	1	2	3	4	5	6	7	8	9	10	11	12	13
ATI	A	R	E	C	P	G	K	Q	E	W	P	E	L
	14	15	16	17	18	19	20	21	22	23	24	25	26
	V	G	E	Y	G	Y	K	A	A	A	I	I	E

c. Specificity

In *A. cruentus,* the whole grain flour contains 0.094 trypsin inhibitor activity (TIU) per milligram (3.18 TIU per milligram of N), the bran-germ fractions 0.14 TIU/per milligram (2.08 23 TIU/per milligram of N), and the perisperm fraction 0.076 TIU per milligram (6.18 TIU per milligram of N) as compared to whole wheat flour, which contains 0.54 TIU per milligram (27.9 TIU per milligram of N). These amounts in amaranth flour probably are not nutritionally significant.

In *A. hypochondriacus,* there are several trypsin inhibitors.[10,76-77] Koeppe et al.[76] reported that 3 of 13 inhibitors also have antichymotrypsin activity. In another study with the purified inhibitor, a typical trypsin-like inhibitory activity against proteases from fungi, stored insects, and amaranth germinated grains was found.[10] More studies are needed on the identification and characterization of proteinase inhibitors in different species of amaranth seeds.

6. Lectins

Lectins are carbohydrate-binding proteins (glycoproteins) with no enzymatic activity.[78] Due to their specific binding properties, they have been used in studies of cell surface architecture, blood group typing, and isolation and characterization of oligosaccharide structures.[79-81] Lectins are widely used in immunology, cell biology, and cancer research due to their biological effects such as immunosuppression, mitogenicity, and cytotoxicity.[81-83] Lectins are common to all kingdoms, and in the case of plants, they have been isolated from legumes, solanaceae, and cereal families for commercial purposes.[79,80] From the Amaranthaceae, lectins of three *Amaranthus* species have been isolated: *caudatus, hypochondriacus,* and *cruentus.* Some of their properties are described in Table 8.[77,83-86]

a. Isolation and Purification

Lectins can precipitate polysaccharides and glycoproteins in a specific manner. This property is being used increasingly for their purification. Affinity chromatography is a common technique used for amaranth lectins purification from the three species. Different adsorbents used for this purpose are shown in Table 8. In the case of *A. caudatus,* lectin was purified using Synsorb-T beads, while *A. cruentus* lectin was purified in a fetuin-Sepharose column. In the case of *A. hypochondriacus,* a column packed with stroma from blood group "A" red cell was utilized.

A lectin of 45 kDa per subunit has been isolated from *A. hypochondriacus* seeds. This glycoprotein contains 10% carbohydrate (w/w) containing 6 *N*-acetyl D-glucosamines, 4-D-galactoses, 1 D-glucose, and traces of xylose residues for each 3 D-mannose residues per molecule. A lectin isolated from *A. cruentus* (ACL) is a dimer with an estimated molecular weight of 66 kDa (gel filtration) and each subunit of 35 kDa (electrophoresis). ACL contains 2.2% of neutral carbohydrates, while in amaranthin, a lectin from *A. caudatus* is a homodimer of 33 to 36 kDa (electrophoresis) per subunit; the native molecule of 61.4 kDa by low-angle laser light scattering and 62.9 kDa by equilibrium sedimentation and does not contain covalently linked carbohydrate.[83,86]

These three lectins are similar in several aspects: the three agglutinate A, B, and O red cells equally well, interact with fetuin, are inhibited by D-GalNAC, and do not seem to require divalent cations for activity (Table 8).

Table 8 **Characteristics of lectins from amaranth seeds**

	A. caudatus[86] (Amaranthin)	*A. cruentus*[84] (ACL)	*A. hypochondriacus*[83]
Adsorbent for chromatography	Synsorb-T beads (Gal1,3GalNAca-O-(CH2)8CO	Fetuin-agarose	Blood group A Stroma-column
Native Mr.	61.4–62.9 kDa	66.0 kDa	n.r.
(subunits)	(33.0–36.0 kDa)	(35 kDa)	(45 kDa)
Hemagglutinin	ABO human	ABO human	ABO human
Blood type		Rabbit, sheep steer	
Cytotoxic	(+)	n.r.	(-)
Mitogenic activity	(-) Hog lymphocytes	n.r.	(+) Mouse spleen lymphocytes
Carbohydrates	n.d.	neutrals: 6*	d
Gal	n.d.		4
Man	n.d.		3
Glc	n.d.		1
Xyl	n.d.		traces
GlcNac	n.d.		6
Thomsen-Friedenreich-Antigen	(+)	n.r.	n.r.
Inhibition:			
agglutination	Galb1, 3GalNAca-O-R" R = -OH, -CH3; D-GalNAc	D-GalNAC	D-GalNAc
Hemagglutination	Fetuin	Fetuin	Fetuin
Reaction Ab *vs.* *A. cruentus* lectin	Identity with seed extract	Identity with seed extract	Identity with seed extract

Note: One hemagglutination unit (HU) = minimal amount of lectin to completely agglutinate cells and is equivalent to lectin concentration; d = per three mannose residues; d-GalNAc = *N*-acetyl-D-galactosamine; n.d. = no detected; n.r. = not reported, *,6 approximately corresponds to 2.2% of neutral hexoses originally reported by Koeppe and Rupnow in Reference 84.

b. Amino Acid Composition

The amino acid composition of amaranthin, ACL, and *A. hypochondriacus* lectins is shown in Table 9 and reveals an unusual amount of essential amino acids such as methionine, lysine, and tryptophan, as well as acidic residues (aspartic and glutamic acid). Amaranthin contains nine tryptophan residues per subunit.[86]

Polyclonal antibodies against purified ACL gave a reaction of identity to amaranth seed extracts from various *Amaranthus* species (*caudatus*, *cruentus*, *hypochondriacus*, *hybridus*, and *retroflexus*).[86] Thus, all of the amaranth species tested contain a presumable lectin with the same antigenic determinants (or very similar to ACL) which could not be distinguished by the ACL antibody.[86]

In the case of *A. hypochondriacus* lectin, there are notable differences in molecular weight, mitogenic activity, and amino acid composition from ACL and amaranthin as shown in Table 8. An important property of amaranthin is its high specificity for the Thomsen-Friedenreich antigen (T- and cryptic T-antigen) that has been proposed as a specific carcinoma marker.[85,86]

Further studies are needed on carbohydrate binding sites of *A. hypochondriacus* and *A. cruentus* lectins before additional comparisons with amaranthin can be made.

B. GLOBULINS

The globulins represent the main storage protein group in legumes. There are two major classes of these proteins, named 7S and 11S according to their sedimentation coefficients. Each class of proteins is preserved across species, and even certain non-legumes species possess such proteins, e.g., oats.[87-89] It is not known why there are two main classes of storage globulins in seeds. There is some evidence linking environmental factors with the variation in the 7S:11S ratio.[89]

Table 9 **Amino acid composition of amaranth lectins from different species**[a]

Amino Acid	*A. caudatus*[86] (Amaranthin) Res/molecule (36 kDa)	*A. cruentus*[84] (ACL) Res/molecule (33 kDa)	*A. hypochondriacus*[83] Res/molecule (45 kDa)
Cys	5.4	4.8	3.0
Ile	16.9	14.5	16.0
Leu	27.6	15.1	24.0
Lys	21.9	10.4	8.0
Met	5.7	6.0	7.0
Phe	17.0	9.8	13.0
Thr	16.1	14.5	23.0
Trp	9.0	nd	2.0
Tyr	17.0	6.2	12.0
Val	20.1	12.1	20.0
Ala	13.3	20.9	38.0
Arg	10.6	7.6	11.0
Asx[b]	44.5	29.0	37.0
Glx[c]	30.6	41.6	37.0
Gly	30.2	44.9	40.0
His	7.0	3.8	7.0
Pro	11.5	10.5	2.0
Ser	32.8	47.5	24.0
Total	337.2	298.7	324.0
Met + Cys	11.1	10.8	10.0
Phe + Tyr	34.0	16.0	25.0

[a] Residues per subunit; [b] Mixture of aspartic acid and asparagine; [c] Mixture of glutamic acid and glutamine.

Various trivial names exist to call the 7S and 11S globulins contained in legume seeds: e.g., phaseolin (kidney bean 7S), vicilin (pea 7S), B-conglycinin (soybean 7S), legumin (pea 11S), glycinin (soybean 11S), arachin (groundnut 11S), and excelsin (Brazil nut 11S).[88-93]

The 7S globulins are composed of three main subunits called α (57 to 68 kDa), α′ (57 to 72 kDa), and β (42 to 52 kDa). They are bound by noncovalent linkages to form a trimer, giving a molecular wight from 170 to 200 kDa. Each subunit has one or two N-linked glycosyl groups. The 11S globulin molecules of legumes are hexamers with molecular weight from 300 to 400 kDa. The hexamers are composed of acidic polypeptides (acidic pI) with molecular weight of 30 to 40 kDa and basic polypeptides (basic pI) of 18 to 22 kDa. The acidic and basic polypeptides are linked together by a disulfide bond.[91,93,94]

The widespread occurrence of 11S and 7S type storage globulins in angiosperm seeds has been recognized. Recent studies are conducted on less conventional systems, such as rapeseed, sunflower, oats, lupin, cucurbits, and amaranth.[95-98]

There are several studies describing the isolation and purification of amaranth globulins (Tables 4 and 5).[12,15-16,47,99] This special attention is perhaps due to the fact, that in legumes, globulins play an important role as functional agents in food products.[99,100-102] Now it is known that amaranth globulins, in addition to their adequate nutritional value, are composed by both 11S- and 7S-like proteins and present some good functional properties such as emulsifying capacity.[13,15,16,103]

1. Biochemical Characterization

a. *Isolation and Purification*

As shown in Table 5, different extracting agents have been used for the amaranth globulins extraction. In some cases, globulins are extracted together with albumins, and after dialysis the precipitate is recovered as globulins fraction or recuperated after extraction of albumins.[16,30,47,49] Saline solutions of 0.5 to 1 M have been used, and different buffer containing salts.[15,16,49,47] Barba de la Rosa et al.[15] have tested different extracting agents and suggested 0.1 M phosphate buffer (pH 7) as the best agent for globulin extraction.

With the use of different extraction procedures, some differences exist among globulin yields. Most of the percentages are in a range of 13.7 to 18.1% except for Bressani and García-Vela[31] which obtain a yield of 38.5% (Table 5).[11,12,15,16,30,47-49] The ratio between albumins and globulins varies as follows: 3.6;[11] 0.48;[12] 2.5;[15] 3.2;[16] 1.1;[31] 1.7 to 2.0.[49]

b. *Electrophoretic Pattern in One- and Two-Dimensions*

The amaranth globulin fractions analyzed by SDS-PAGE can be classified as hetero-oligomers composed of nonequimolar ratio of subunits; these subunits can be divided into four categories according to their size, as shown in Figure 6A (lane 2) and in Table 10: group I corresponds to the high molecular weight (73 to 50 kDa) that might be related to 7S type; group II with polypeptides around 31.5 to 38.0 kDa and group III with lower molecular weight (20 to 27 kDa). Both groups II and III can be related to the 11S type. The low molecular weight globulins or group IV is constituted of polypeptides smaller than 18 kDa.[12,15,16,49,53]

The electrophoretic profile of globulins does not change when they are extracted from whole or defatted flours.[15] Gorinstein et al.[47] have reported that amaranth globulins are constituted only by major bands of 18 to 14 kDa. The abundance of low molecular weight polypeptides and the absence of the high molecular weight ones, can be partly due to the extraction procedure used.

Charge heterogeneity of amaranth globulin was investigated by two-dimensional analysis.[16] There are various globulin components (Figure 6B) and the four groups can be identified according to their charge and molecular weights.[16] The group of high molecular weight components can be related to the α, α', and β subunits of 7S-like proteins, as previously reported (Figure 6A, lane 2 and 6B).[104,105] The second group (around 31 to 38 kDa), and the third group (of molecular weight around 20 to 27 kDa) can correspond to 13S globulins made up of acidic and basic subunits, like soybean and oat globulins and rice glutelins.[106-109]

Krochko and Bewley[110] reported that in alfalfa the true globulins proteins of the endosperm have near neutral isoelectric points so they separate almost entirely on both the IEF and NEPHG (nonequilibrium pH gradient). Segura-Nieto et al.[16] have also described in amaranth globulins a group of low molecular weight polypeptides (25 to 12 kDa) with a sedimentation coefficient of 1.9S (Figure 7A). Their electrophoretic pattern is different from the other globulins as shown in the gel of Figure 7B, lanes 1 to 5. More studies are needed to characterize this polypeptide group.

c. *Amino Acid Composition*

In terms of essential amino acids, globulins are relatively rich in lysine (3.7 to 7.6%) compared to cereals and legumes (Table 11).[23,87] All reports showed good levels of sulfur amino acids (3.1 to 7.1.0%)[12,16,49,31] except for one that reported values of 2.6.[14]

In terms of nonessential amino acids, amaranth globulins are higher than other species on arginine, glutamic acid + glutamine, and aspartic acid + asparagine. Such high contents of amino acids appear to be a dominant trait in common seed storage proteins.[14-16,31,49,88]

2. Physicochemical Properties

Konishi et al.[13] and Mora-Escobedo et al.[49] have reported that one amaranth globulin fraction is eluted (on Sephacryl S-300 column) at the void volumen with a strong UV-absorbance, but it contains a small amount of protein when assayed by the dye-binding Bradford method. This suggests that this fraction contains mostly nucleic acid and some amount of high molecular weight aggregates of protein, as reported by Madhusudhan and Singh.[98] Many discrepancies have been noted on the molecular weight of amaranth globulins, reporting values of 440, 398, and 260 kDa.[12,15,49,111] Kitamura et al.[112] noted that gel filtration was not useful for the purification of 7S and 11S globulins; they found that, depending on the eluting buffer, 7S globulin was capable of 7S—9S aggregation or interaction with 11S like-globulins and eluted in an earlier position than that of 11S. In some cases, 11S-like globulins can dissociate, giving molecular weights lower than 7S-like globulins. These phenomena can be responsible for the differences found by some authors.[12,15,111]

a. *Sedimentation Coefficients*

The studies of seed globulins have been centered on the ultracentrifugal investigations. Sedimentation coefficient values of 1.9S, 8S, and 13S of amaranth globulins have been reported.[12,16,103,111,113] In Figure 7A, the sedimentation pattern of amaranth globulins (continuous line) are shown with their corresponding sedimentation coefficient values already mentioned. The sedimentation pattern of protein markers of

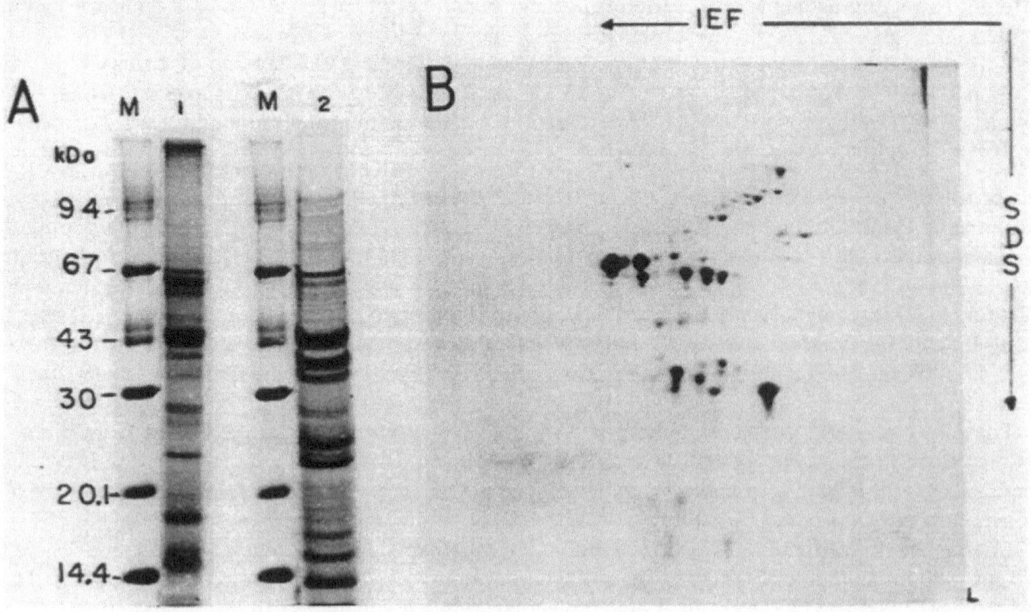

Figure 6 (A) Electrophoretic patterns of globulin fractions; lane M, molecular weight markers; lane 1, globulins without reduction; lane 2, with reduction conditions. (B) Two-dimensional analysis of globulins under reduced conditions.

Table 10 **Molecular weight of globulin subunits**[a]

Group				
I	II	III	IV	Ref.
—	36.0–32.0	20.0	18.0	12
—	36.1–32.9	21.2–14.4	12.5	49
67.0	37.5–31.5	26.5–20.5	14.5	15
73.0–50.0	38.0–34.0	27.0–21.0	14.0	53

[a] kDa.

known sedimentation coefficient are labeled with dashed lines (Figure 7A). After electrophoresis of the main globulin fraction, it can be observed that the peak with 1.9S shows only bands of low molecular weight, suggesting that this peak can be an albumin contamination, although the globulin preparation was previously dialysed, and the pellet solubilized in high ionic strength before loaded on the sucrose gradient (Figure 7A).[16] The peak with 8S has similar electrophoretic pattern (lanes 8 to 12 in Figure 7B) than those of 7S soybean fraction, and the 13S shared the same acidic and basic subunits characteristic of 11S soybean globulin fraction (Figure 6B and lanes 17 to 20 from Figure 7B).[16,103,111,113]

b. Molecular Association/Dissociation

One characteristic common to most, if not all globulins, is that they present the phenomena of association/dissociation under nondenaturing conditions.[88,89] Amaranth globulins, like other species, show these phenomena. Freeze-dried globulins cause aggregation of 8S to 10S.[103] Konishi et al.[12] have suggested that amaranth globulins can dissociate the oligomeric protein into monomers at alkaline pH solutions. Amaranth globulins also present the phenomena of cryoprecipitation.[12,103] On the other hand, Barba de la Rosa et al.[103] have shown that protein concentration changes the sedimentation coefficient dramatically. Finally, Marcone and Yada[99] have reported that disulfide groups are the responsible of sulfydryl-disulfide interchange reaction causing the polymerization of amaranth globulins.

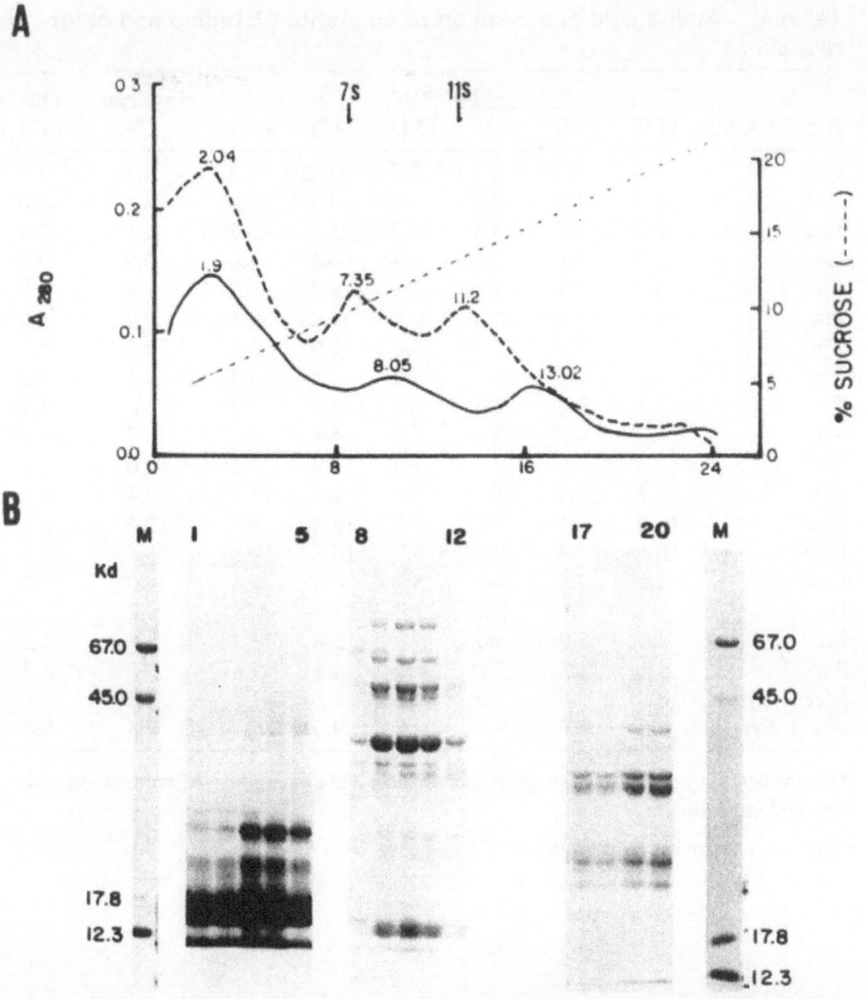

Figure 7 (A) Globulins sedimentation coefficients. The absorbance in both sucrose gradients with standards (——) and globulins (____) was measured at 280 nm. (B) Electrophoretic pattern of some fractions from the globulin peaks in A.

c. Hydrophobicity

Due to the importance of hydrophobic interactions on the stability and protein function, a great effort has been made to calculate the protein hydrophobicity. This value has been estimated according to the polarity of its amino acid composition.[114] However, this calculation does not seem to represent the "effective hydrophobicity" which depends on the flexibility and conformation of the protein molecule. The fluorescence techniques using *cis*-parinaric acid (CPA) and anilino-naphtalenesulfonate (ANS), which determine aliphatic and aromatic hydrophobycity, respectively, have been proposed.[115-117] Marcone and Yada[99] have reported that hydrophobicity using CPA probe showed two maxima, one at pH 5.5 and the second above pH 9.5 (Figure 8). It is known that in acidic and basic pHs, the hydrophobic areas of a polypeptide are exposed outward, leading to uncovering all aromatic residues. Then, the maximum hydrophobicity observed in amaranth globulins may have been caused by a combination of changes in tertiary structure exposing the more hydrophobic core.[118-119] It is known that in the a-subunits of legumins, the most hydrophobic group is probably buried in the inner part of the protein, when they are under its native conformation.[89,120] Konishi and Yoshimoto,[13] using ANS fluorescent probe, reported a value of 15,800 for amaranth globulins being higher than those of soy protein isolate (SPI), wheat gluten, and casein (9600, 10,100, 5100, respectively). Bovine serum albumin (BSA) had the highest value (63,400). Although only aromatic hydrophobicity has been correlated with protein

Table 11 **Amino acid composition of amaranth globulins and other globulins**[a]

Amino Acid	Amaranth (12)	(49)	(31)	(14)	(15)	(16)	Soybean (29)	Oat (87)
Cys	5.5	2.4	4.1	0.7	3.8	1.3	1.4	1.0
Ile	4.6	4.0	2.6	4.6	4.1	3.8	4.5	4.6
Leu	6.3	6.8	3.8	6.3	5.5	5.7	6.8	7.1
Lys	3.7	4.6	5.7	4.4	6.5	7.6	5.9	2.8
Met	1.2	3.1	2.9	1.9	3.3	1.8	1.4	0.9
Phe	6.1	4.3	4.5	8.1	4.8	2.4	4.4	5.0
Thr	3.2	3.6	2.2	3.6	4.0	3.4	3.8	4.7
Trp	nd	nd	0.6	nd	nd	nd	1.1	nd
Tyr	3.4	3.6	3.4	2.8	4.2	3.9	3.4	3.4
Val	4.7	4.3	2.8	5.1	4.5	4.1	4.7	6.1
Ala	3.4	3.7	2.7	3.3	3.9	3.2	4.0	5.8
Arg	9.7	10.4	11.7	13.8	9.2	12.3	7.4	6.3
Asx[b]	11.6	8.1	8.4	8.2	8.4	9.0	10.6	8.8
Glx[c]	16.5	17.2	29.6	15.7	16.8	18.8	18.5	18.3
Gly	4.1	5.6	4.5	3.7	6.4	6.6	4.0	7.2
His	3.0	3.1	2.0	2.9	1.1	2.6	2.2	2.1
Pro	4.0	5.2	nd	5.0	3.8	3.3	5.5	4.7
Ser	5.8	5.3	3.2	4.9	4.8	5.6	5.0	6.7
Met + Cys	6.7	5.5	7.0	2.6	7.1	3.1	2.8	1.9
Phe + Tyr	9.5	7.9	7.9	10.9	9.0	6.3	7.8	8.4

[a] Grams per 100 g protein; [b] Mixture of aspartic acid and asparagine; [c] Mixture of glutamic acid and glutamine.

Note: ( ) = References; nd = not determined.

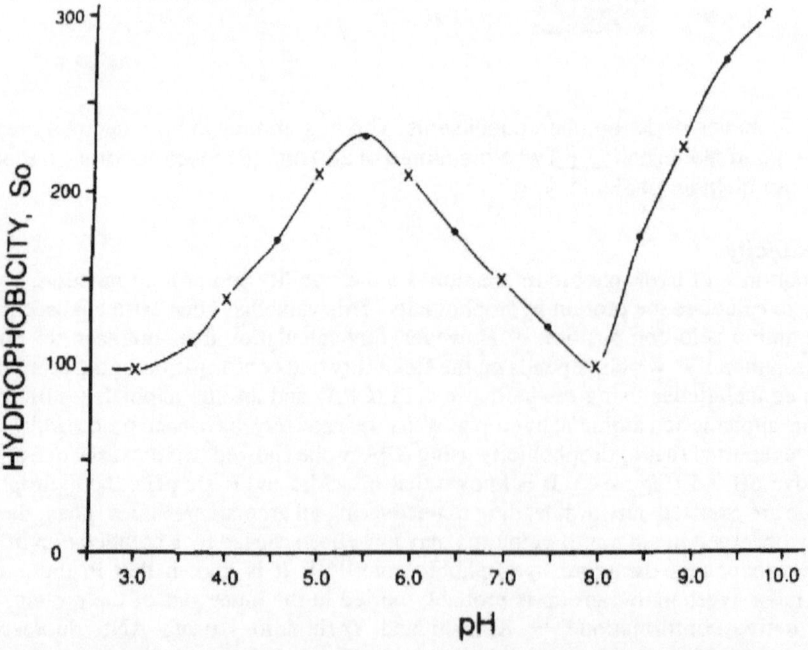

Figure 8 *Cis*-parinaric acid (CPA) surface hydrophobicity of globulins at differing pH values.

insolubility, it is interesting to note that CPA measurements correlate well with protein functionality, i.e., solubility and emulsification activity.[13,99,117,121]

More studies are needed on the correlation between globulin functionality and aliphatic and aromatic hydrophobicity.

3. Functional Properties

The functional and physical properties, rather than the nutritional value of proteins, will largely determine their acceptability as ingredients in prepared foods.[121,122] The ability of proteins to bind and adsorb other food components (such as water, lipids, and flavors) is an important attribute exploited in formulations of many foods such as sausages, cheeses, and custards.[121] Binding ability of proteins is related to adhesion, film formation, fiber formation, and viscosity.[121,122] Moreover, the ability of proteins to bind fats is a very important functional property, principally because it enhances flavor retention and improves mouthfeel.[121]

a. *Solubility*

Globulins show a characteristic precipitation in the pH range 4.5 to 6.8.[12,99] Further examination revealed that the maximum insolubility occurs in the pH range 4.5 to 6.0 and that partial precipitation begins between pH 6 and 6.8. This partial precipitation can be attributed to protein existing in different ionizable states, as noted previously in the ionic-exchange behavior.[111]

b. *Emulsifying Activity*

Emulsion stability and fat-binding capacity of the proteins can be predicted using the hydrophobicity and solubility of protein together.[123,124] The hydrophobic proteins easily interact with oil, and the effective hydrophobicity gives an idea of the capacity of the proteins in the stabilization of the protein-lipid interaction.[116,117] Konishi and Yoshimoto[13] have found that emulsion stability (ES) and emulsion activity (EA) of amaranth globulin are about twice those of SPI and lower than casein. The maximum ES is obtained at pH 6.5 in the acidic side, and pH 11 in the alkaline region (Figure 9). Please note that Figure 9 only shows ES and not EA.

Amaranth globulin appears relatively heat-stable and is expected to be useful as a good emulsifier with a high nutritional value.

C. PROLAMINS AND GLUTELINS

1. Biochemical Properties

In cereals there are two major protein fractions; one of them is found in wheat, barley, and maize as the largest fraction. This fraction is soluble in aqueous ethanol and is known as prolamins.[23,125] The other fraction, called glutelins, requires extreme pH solutions for its extraction and is found as the main protein fraction in rice.[23] Further analysis suggested that prolamins were not particularly soluble in aqueous ethanol at room temperature and adding reducing agents improved their extraction. For this reason, there is a problem in the evaluation of techniques for prolamin extraction methods. The glutelins represent the more insoluble protein fraction from the seed meal and are extracted with acidic or basic solutions, generally using borate buffer at pH 10 and, in some cases, with detergents and/or reducing agents.[16,126]

In amaranth, the amount of alcohol-soluble proteins is low, as in oat and rice where their main fractions are globulins and glutelins, respectively.[23] The glutelins are the second most important fraction in amaranth seeds as shown in Table 4. Some works have focused on the extraction of amaranth prolamins and glutelins under different conditions.[15,16,30,31,126,127]

a. *Isolation*

Prolamins have been extracted using different solvents as well as different rates of solvent/amaranth meal (Table 12). Gorinstein et al.[126] reported that the highest prolamin extraction of 2.1% was obtained using 55% isopropanol + 5% (v/v) of 2-mercaptoethanol; Bressani and García-Vela,[31] with the same concentration of alcohol without reducing agent, reported a 7.3% of prolamins. In general, 70% isopropanol is the most common solvent used for prolamin extraction. Using 0.01 M K_2HPO_4 in 60% propanol and extraction for 14 h did not improve the prolamins yield (2%).[16] The highest value of prolamins, 11%, was reported by Duarte-Correa *et al.*[30] (Table 12).

There are great discrepancies on glutelins yield extraction; it ranges from 7 to 42.3% (Table 4). It appears that when more albumins + globulins are extracted, glutelins fraction is decreased and *vice versa*. Many buffers have been used as extracting agents, as shown in Table 13, but, in general, borate buffer

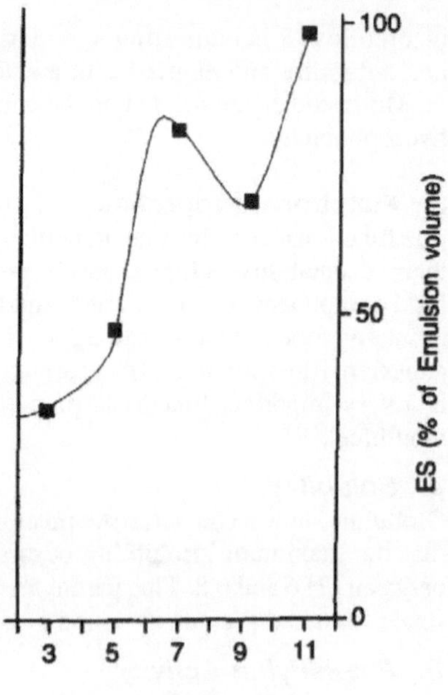

Figure 9 Emulsion stability (%) of globulins at different pH values.

Table 12 **Systems used to extract prolamins from *Amaranthus hypochondriacus***

Yield (%)	Extracting Agent	Extraction Conditions	Ref.
11	Not available		30
7.3	55% 2-PrOH (A)	Differences in alb + glo extraction.	31
0.8	55% 2-PrOH (B)	Extraction time 60 + 30 min	
1.8	55% 2-PrOH + 5% 2-ME (A)	Extraction 30 + 30 min	
1.5	55% 2-PrOH + 5% 2-ME (B)		
1.2	55% 2-PrOH		126
2.1	55% 2-PrOH + 5% 2-ME	After extraction, prolamins were precipated with acetone	
0.9	70% 2-PrOH		
1.9	70% 2-PrOH + 5% 2-ME		
0.7	45% EtOH		
0.7	70% 2-PrOH (W)	After extraction, prolamins were dialyzed vs. acetic acid and freeze-dried	15
1.2	70% 2-PrOH (D)		
2.0	60% 1-PrOH + 0.01M K_2HPO_4 pH 7.5	Extraction time 6 + 8 h After extraction, prolamins were precipitated with acetone and redissolved with 0.1 N NaOH	16

is most commonly used in combination with SDS and/or reducing agents (2-mercaptoethanol). The highest yield value reported was 42.2% from defatted flour[15] and the lowest was of 7%.[30]

b. Electrophoretic Patterns

Gorinstein et al.[126] reported that alcohol soluble proteins showed the main subunit in the region of 10 to 14 kDa and a very faint line of about 20 kDa. Segura-Nieto et al.[16] reported three main bands at 16, 18, and 22 kDa and two lower bands below 12 kDa. Some minor polypeptides were at 30 to 40 kDa, as shown in Figure 10. Barba de la Rosa et al.[15] found that when the electrophoresis is performed without reducing agents, prolamins are composed of bands of high molecular weight which, in reduction conditions, become 67, 34, and 22 kDa. Other studies have shown that the electrophoretic pattern of prolamins changes depending on the solvent used to defat the amaranth flour. When hexane is used, the protein pattern behavior shows bands of high molecular weight, but with acetone the prolamins show mainly bands of low molecular weight.[127]

Table 13 **Systems used to extract glutelins from *Amaranthus hypochondriacus***

Yield	Extracting agents (%)	Extraction Conditions	Ref.
7.0	Not available		30
7.1	Sodium carbonate pH 10 + 0.6% (v/v) 2-mercaptoethanol[a]	Defatted flour with acetone	47
18.7	Sodium carbonate pH 10 + 0.6% 1-MetOH + 0.5% SDS		
7.3	Borate buffer pH 10 + 0.6% 2-ME – 0.5 M NaCl	Defatted flour with hexane	31
8.3			
21.8	Borate buffer pH 10 + 0.6% 2-ME + 0.5% SDS	Defatted flour with hexane	15
34.0		Defatted flour with hexane	
23.8	0.1 M NaOH	Whole flour	
40.6		Defatted flour with hexane	
22.5	0.1 M borate buffer pH 10	Whole flour	
37.5		Defatted flour	
26.2	0.1 M borate buffer pH 10 + 1% SDS	Whole flour	
41.6		Defatted flour	
23.3	0.1 M borate buffer pH 10 + 0.6% 2-ME	Whole flour	
38.8		Defatted flour	
28.5	0.1 M borate buffer pH 10 + 1% SDS + 0.6% 2-ME	Whole flour	
42.2		Defatted flour	
31.1	0.01 M boric acid pH 9 + 1% 2-ME + SDS	Defatted flour with hexane	16

[a] Abbreviated 2-ME.

There are no differences in glutelin electrophoretic patterns when they are extracted with borate with or without NaOH.[15] But there are some changes when they are obtained in the presence of SDS with or without 2-mercaptoethanol.[31,15]

The electrophoretic pattern of glutelins in one-dimension is shown in Figure 11A, lane 1. Glutelins are made up of 6 main sets of polypeptides, 2 with a molecular weight around 55 to 67 kDa, 2 between 35 to 38 kDa and the smaller ones around 22 to 23 kDa.[15,16] Two-dimension electrophoretic pattern (Figure 11B) shows four defined groups of polypeptides. The manipulation of glutelins, like freeze and thawing, tends to degrade the 67 kDa and the 23 kDa glutelins.[53] In Figure 11B, the 67 kDa is missing. Glutelins are the most abundant polypeptides in amaranth grain, as can be seen in the two-dimensional gel electrophoretic pattern of defatted meal from amaranth grain in Figure 3, panel B, where the most abundant spots of polypeptides correspond to glutelins (compare Figure 3B with 11B).

Further studies are needed on the characterization of this abundant group of amaranth seed proteins.

c. Amino Acid Composition

The amino acid composition of prolamins shows relatively high values in the essential amino acids leucine (5.5 to 9.8%) and threonine (3.1 to 6.4%).[16,31] Like other amaranth fractions, prolamins have high aspartic acid + asparagine (6.0 to 9.6%), glutamic acid + glutamine (11.0 to 16.6%), arginine (5.3 to 9.1%), alanine (2.6 to 7.2%) and glycine (4.3 to 8.5%) as compared to wheat protein shown in Table 14.[23]

Glutelins have high leucine content (5.9 to 10.2%), as well as tyrosine + phenylalanine (6.7 to 10.3%) as presented in Table 15. Among the nonessential amino acids, arginine and aspartic acid + asparagine are in higher concentration than in wheat.[23]

IV. VEGETABLE AMARANTHS

The leaves of all 50 to 60 species of amaranths are edible. For centuries the most widely grown species have been: *A. tricolor* and *A. dubius* grown in tropical regions of India, China, and Southeast Asia; and *A. cruentus* extensively grown in humid tropical regions as well as other species like *A. retroflexus* and *A. graecizans*.[2,5,33]

The leaves and the softest portions of the shoots are usually boiled in several changes of water and then separated from the cooking liquid.[128] Their appearance, texture, flavor, and overall eating quality of

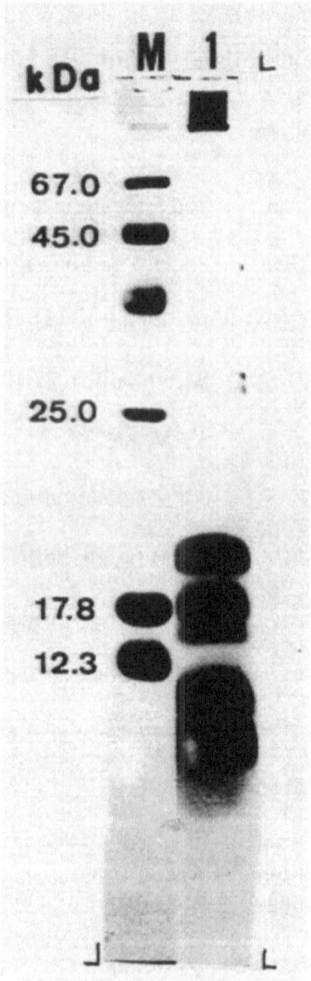

Figure 10 Electrophoretic pattern of prolamins in reduced conditions. Lane M, molecular weight markers; lane 1, prolamins fraction.

amaranth species are comparable with spinach. Their high yield and nutritional quality make vegetable amaranth a promising food source for human nutrition.[1,2,5]

A. CHEMICAL COMPOSITION

Amaranth leaves contain 27.8 to 48.6% dry matter as crude protein as shown in Table 16. The fat content ranges from 2.9 to 7.1% and contains linoleic acid as the major concentration of unsaturated fatty acid (46%) and palmitic acid as saturated fatty acid (18 to 25%).[129] Amaranth leaves and stems are a very good source of fiber, 11.1 to 23.2%. Minerals such as potassium, iron, magnesium, and calcium also exist in significant concentrations, giving a total of 33.1 to 40.7% of ash content (Table 16). Vitamins C and A (B-carotene) are in concentrations of nutritionally significant levels, averaging 420 and 250 ppm, respectively.[2,17,25,130]

Amaranth leaves contain somewhat more dry matter than spinach, vinespinach, and chard, mainly because of their high levels of fiber and ash (Table 17). The presence of large amounts of oxalates, ranging from 0.2 to 11.4% (dry weight), may limit availability of the nutrients.[2,17,128-130] For this reason, leaf protein isolates have been prepared and are the protein source successfully utilized.

B. LEAF-PROTEIN ISOLATES

Amaranth leaf protein is highly extractable, and approximately one third of the protein in the plant is obtained in the protein isolate. Compared with other species, amaranth leaf protein has the highest protein extraction yields. In the preparation of protein isolates, most of the harmful compounds are eliminated in the press cake (e.g., oxalates, nitrites). The green cheeselike coagulum is washed with water slightly acidified with dilute acetic acid (vinegar) to further reduce the amounts of possible antinutritional

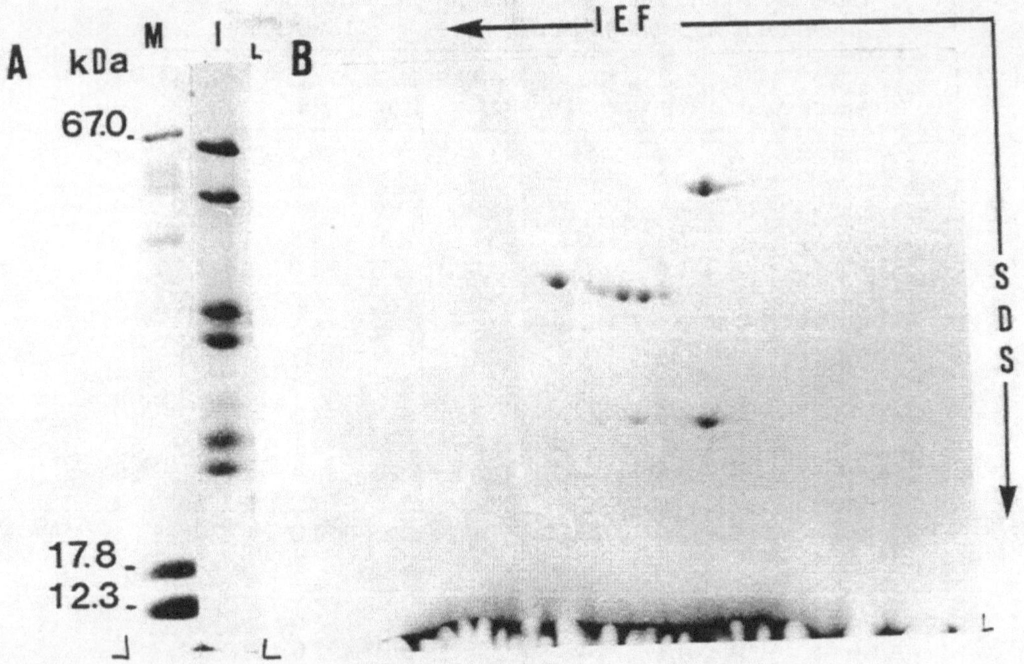

Figure 11 (A) Electrophoretic pattern of glutelins fraction. Lane M, molecular weight markers; lane 1, glutelins. (B) Two-dimensional analysis of glutelins.

Table 14 **Amino acid composition of prolamins from *Amaranthus hypochondriacus***[a]

Amino Acid	Amaranth (30)	(47)		(31)	(15)	(16)	Wheat (23)
Cys	nd	1.5	0.9	nd	6.3	0.8	2.3
Ile	4.5	2.6	3.3	5.8	5.9	4.4	3.9
Leu	6.3	6.9	6.1	8.8	5.5	9.8	6.2
Lys	5.3	2.8	4.2	6.5	4.1	7.3	0.5
Met	2.0	5.2	1.0	1.4	7.2	1.9	1.2
Phe	3.6	3.2	4.5	5.8	8.7	4.8	4.8
Thr	5.0	6.1	4.1	5.4	3.1	6.4	1.9
Trp	nd	1.3	1.1	0.8	nd	nd	5.5
Tyr	4.5	2.3	3.3	3.2	3.9	4.0	2.3
Val	7.4	6.0	3.8	5.8	2.6	4.0	3.8
Ala	7.2	6.5	4.7	2.6	4.5	5.7	1.9
Arg	5.4	8.8	8.8	7.1	9.1	8.9	2.3
Asx[b]	7.6	7.6	9.6	8.5	6.0	7.9	2.9
Glx[c]	15.1	13.0	16.6	17.9	13.0	11.0	34.5
Gly	7.2	8.5	7.5	6.2	4.3	6.0	1.6
His	1.8	2.0	2.8	2.4	1.1	2.1	1.9
Pro	5.7	5.2	5.1	nd	4.6	3.9	12.7
Ser	4.2	5.5	7.8	6.6	5.0	6.3	4.4
Met + Cys	nd	6.7	1.9	nd	13.5	2.7	3.5
Phe + Tyr	8.1	5.5	7.8	6.7	12.6	8.8	7.1

[a] Grams per 100 g protein; [b] Mixture of aspartic acid and asparagine; [c] Mixture of glutamic acid and glutamine.

Note: ( ) = References.

Table 15 **Amino acid composition of glutelins from *Amaranthus hypochondriacus*[a]**

Amino Acid	Amaranth (30)	(31)	(47)	(15)	(16)	Wheat (23)
Cys	nd	nd	1.2	6.0	0.4	0.8
Ile	3.7	4.0	3.9	5.6	4.8	3.1
Leu	6.1	7.7	5.9	10.2	7.8	6.4
Lys	5.8	6.1	3.1	4.5	4.5	1.6
Met	0.6	nd	1.3	3.0	2.9	1.5
Phe	4.1	4.8	5.0	6.6	5.2	3.8
Thr	4.2	3.4	3.2	8.3	6.5	2.6
Trp	nd	1.9	0.1	nd	nd	1.3
Tyr	3.0	1.9	4.1	3.7	4.0	3.8
Val	5.5	4.7	3.5	3.7	4.3	4.1
Ala	5.7	4.4	3.5	3.5	4.1	2.5
Arg	6.8	9.4	12.8	2.6	10.9	3.0
Asx[b]	10.6	10.3	8.8	5.9	7.1	2.6
Glx[c]	18.4	20.7	17.8	12.8	12.0	34.8
Gly	7.9	5.5	5.6	4.7	5.7	5.0
His	1.6	4.9	3.9	4.6	2.7	1.7
Pro	4.3	nd	5.7	4.6	5.0	11.1
Ser	4.7	4.8	5.4	5.1	7.0	4.4
Met + Cys	nd	nd	2.5	9.0	3.3	2.3
Phe + Tyr	7.1	6.7	9.1	10.3	9.2	7.6

[a] Grams per g protein; [b] Mixture of aspartic acid and asparagine; [c] Mixture of glutamic acid and glutamine.

Note: ( ) = References.

Table 16 **Proximate composition of leaves from some *Amaranthus* species[a]**

Analysis	*cruentus*[2]	*edulis*[2]	*graecizans*[130]	*tricolor*[2]	*hybridus*[2]	*spinosus*[2]	*caudatus*[2]
Protein[b]	46.5	33.3	37.1	32.7	27.8	44.5	48.6
Fat	6.8	2.9	3.8	7.1	3.9	4.6	3.3
Fiber	11.1	22.9	23.2	—	17.5	14.1	12.8
Ash	35.4	40.7	34.7	—	—	33.1	35.2

[a] Percentage on dry weight; [b] Crude protein (%N × 6.25)

Table 17 **Comparison of proximate composition between amaranth and selected vegetable leaves[a]**

Component	Amaranth[b]	Spinach[17]	Chard[17]	Collard[17]
Protein[c]	46.5	34.4	26.9	32.6
Fat	6.8	3.2	3.4	5.4
Fiber	11.1	46.2	51.7	51.0
Ash	35.4	16.1	18.0	10.9

[a] Percentage on dry weight; [b] *A. cruentus* from Table 16; [c] Crude protein (%N × 6.25).

factors.[17,25,128,130] The resulting leaf nutrient concentrate is specifically useful for food for young children. The leaf concentrate is especially high in protein, vitamin A, and iron. The fibrous pulp remaining after extracting the amaranth greens is a suitable feedstuff.[130]

Table 18 **Essential amino acids content of protein isolates from some *Amaranthus* and other species**[a]

Amino Acid	*retroflexus*[130]	*hybridus*[130]	*graecizans*[130]	Alfalfa[25]	Spinach[25]	Cabbage[25]	Corn[17]
Cys	—	1.0	—	—	—	—	—
Ile	4.8	5.4	5.3	5.6	4.8	3.7	3.7
Leu	7.5	8.6	7.3	9.3	8.5	5.4	12.5
Lys	6.2	6.8	6.6	5.9	6.4	3.8	2.7
Met	—	1.5	—	2.3	1.8	1.0	—
Phe	3.8	4.4	3.7	5.9	5.3	3.2	4.9
Thr	3.0	3.5	3.0	5.1	4.6	3.5	3.6
Val	4.8	4.6	3.9	6.3	5.6	4.6	7.5

[a] Grams per 100 g protein.

The protein in the isolate may well be of high nutritional value due to a high lysine content (6.2 to 6.8%), higher than alfalfa, cabbage, and corn (Table 18) and to relatively good amounts of other essential amino acids, comparing well with spinach.

Currently, because many people in developing countries utilize vegetative amaranth as a major part of their diet, they could also utilize such protein isolates for increasing their protein intake.

V. CONCLUDING REMARKS

The above mentioned composition, properties, and future applications of amaranth proteins demonstrate the food potential of this under-utilized crop.

There is a lack of information on the molecular biology of synthesis, processing, and deposition of storage proteins in seeds and leaves. Earlier attempts to improve cereal and legume seed proteins by breeding have been only partially successful, but the basic studies generated around these topics have furthered the understanding of the regulation of the expression and the structure of seed proteins that determine quality and their genes. This information provides the basis for present and future attempts to manipulate protein quality using molecular biology techniques. Thus, both approaches, traditional and modern breeding techniques, complement each other

Many more studies are needed on biochemical, physicochemical, and functional properties of proteins from grains and leaves of different amaranth species.

Amaranth is an important, cheap source of proteins of high nutritional quality, either from seeds or leafy vegetables. The leaves are also an inexpensive source of several nutrients, in particular, the carotenoids; however, the main constraint to their nutritional exploitation is the presence of some antinutritional and toxic elements, such as nitrate, oxalate, and saponin in some amaranth types. Selection of lines and development of varieties with low antinutritional factors and higher nutritional qualities will increase the consumption of amaranth. Further studies are needed to correlate nutritional and antinutritional factors.

In the food processing area, research and development work is needed on shelf life, on the functionality of proteins from grain and from protein concentrates of grain and leaves, as well as on the effects of processing on functional and nutritional qualities.

The main challenge is to incorporate amaranth proteins into existing food formulations, to modify their functional and nutritional quality, and to create new products from grain and vegetable amaranth.

REFERENCES

1. National Academy of Science, *Amaranth: Modern Prospects for an Ancient Crop,* National Academy Press, Washington, D.C., 1984, 1.
2. **Teutonico, R. A. and Knorr, D.,** Amaranth: composition,properties, and applications of a rediscovered food crop, *Food Technol.,* 39, 49, 1985.
3. **Sánchez-Marroquín, A.,** Dos cultivos olvidados, de importancia agroindustrial: el amaranto y la quinua, *Arch. Latinoam. Nutr.,* 33, 11, 1983.
4. **Downton, W. J. S.,** *Amaranthus edulis:* A high lysine grain amaranth, *World Crops,* 25, 20, 1973.

5. **Grubben, G. J. H. and Van Sloten, D. H.,** Genetic resources of amaranthus, International Board for Plant Genetic Resources, *Food and Agricultural Organization,* Rome. 1981
6. **Tucker, J. B.,** Amaranth: the once and the future crop, *Bioscience,* 36, 9, 1986.
7. **Cheeke, P. R. and Bronson, J.,** Trials with *Amaranthus* grain, forage and leaf protein concentrates, in *Proc. 2nd Amaranth Conf.,* Rodale Press, Emmaus, PA, 1980, 5.
8. **Pedersen, B., Hallgren, I., and Eggum, B. O.,** The nutrititve value of amaranth grain (*Amaranthus caudatus*). I. Protein and minerals of raw and processed grain, *Plant Foods Human Nutr.,* 36, 309, 1987.
9. **Afalobi, A. O., Oker, O. L., and Umoh, I. B.,** Preliminary studies on the nutritive value of some cereal-like grains, *Nutr. Rep. Int.,* 24, 389, 1981.
10. **Valdés-Rodríguez, S., Segura-Nieto, M., Chagolla-López, A., Ververy Vargas-Cortinas, A., Martínez-Gallardo, N., and Blanco-Labra, A.,** Purification, characterization, and complete amino acid sequence of a trypsin inhibitor from amaranth (*Amarathus hypocondriacus*) seeds, *Plant Physiol.,* 103, 1407, 1993.
11. **Abdi, H. and Sahib, M. K.,** Distribution of lysine in different legumes and some species of Amaranthus seeds, *J. Food Sci. Technol.,* 13, 237, 1976.
12. **Konishi, Y., Fumita, Y., Ikeda, K., Okuno, K., and Fuwa, H.,** Isolation and characterization of globulin from seeds of *Amaranthus hypochondriacus* L., *Agric. Biol. Chem.,* 49, 1453, 1985.
13. **Konishi, Y. and Yoshimoto, N.,** Amaranth globulin as a heat- stable emulsifying agent, *Agric. Biol. Chem.,* 53, 3327, 1989.
14. **Marcone, M. F. and Yada, R. Y.,** Isolation, purification, and characterization of the oligomeric seed globulin from *Amaranthus hypochondriacus, Agric. Biol. Chem.,* 55, 2281, 1991.
15. **Barba de la Rosa, A. P., Gueguen, J., Paredes-López, O., and Viroben, G.,** Fractionation procedures, electrophoretic characterization, and amino acid composition of amaranth seed protein, *J. Agric. Food Chem.,* 40, 931, 1992.
16. **Segura-Nieto, M., Vázquez-Sánchez, N., Rubio-Velázquez, H., Olguín-Martínez, L R., Rodíguez-Nester, C., and Herrera-Estrella, L.,** Characterization of amaranth (*Amaranthus hypochondriacus* L.) seed proteins, *J. Agric. Food Chem.,* 40, 1553, 1992.
17. **Saunders, R. M. and Becker, R.,** *Amaranthus:* a potential food and feed resource, in *Advances in Cereal Science and Technology,* Pomeranz, Y., Ed., American Association of Cereal Chemist, St. Paul, MN, 1984, chap. 6.
18. **Carlsson, R.,** Amaranth species and related species for leaf protein concentrate production, in *Proc. 2nd Amaranth Conf.,* Rodale Press, Emmaus, PA., 1980, 83.
19. **Bressani, R., González, J. M., Zúñiga, J., Breuner, M., and Elías L. G.,** Yield, selected chemical composition and nutritive value of 14 selections of amaranth grain representing four species, *J. Sci. Food Agric.,* 38, 347, 1987.
20. National Academy of Science, Underexploited tropical plants with promising economic value, Report of the Adhoc Panel of the Advisory Committee on Technology Innovation (BOSTID), Washington, D.C., 1975, 1.
21. **Singhal, R. S. and Kulkarni, P. R.,** Composition of the seeds of some Amaranthus species, *J. Sci. Food Agric.,* 42, 325, 1988.
22. **Bressani, R.,** The proteins of grain amaranth, *Foods Rev. Int.,* 5, 13, 1989.
23. **Lásztity, R.,** *The Chemistry of Cereal Proteins,* CRC Press, Boca Raton, FL, 1984.
24. **Pedersen, B., Kalinowski, L. S., and Eggum, B. O.,** The nutritive value of amaranth grain (*Amaranthus caudatus*). Protein and minerals of raw and processed grain, *Plant Foods Human Nutr.,* 36, 309, 1987.
25. **Singhal, R. S., and Kulkarni, P. R.,** Review: Amaranths — an underutilized resource, *Int. J. Food Sci. Technol.,* 23, 125, 1988.
26. **Paredes-López, O., Barba de la Rosa, A. P., Hernández-López, D., and Cárabez-Trejo, A.,** Amaranto — Características Alimentarias y Aprovechamiento Agroindustrial. Secretaría de la Organización de los Estados Americanos, Washington, D.C., 1990, 1.
27. **Prakash, D. and Pal, M.,** Seed protein, fatty acid profile of *Amaranthus* species, *J. Sci. Food Agric.,* 58, 145, 1992.
28. FAO, Energy and protein requirements, in Nutrition Meeting Report Series, 52, Food and Agriculture Organization, Rome, 1973.
29. **Smith, A. K. and Circle, S. S.,** Chemical composition of the seed, in *Soybeans: Chemistry & Technology,* Vol. 1, AVI Publishing, Westport, CT, 1981, chap. 3.

30. **Duarte-Correa, A., Jokl, L., and Carlsson, R.,** Amino acid composition of some Amaranthus sp. grain proteins and of its fractions, *Arch. Latinam. Nutr.,* 26, 466, 1986.
31. **Bressani, R. and García-Vela, L. A.,** Protein fractions in amaranth grain and their chemical characterization, *J. Agric. Food Chem.,* 38, 1205, 1990.
32. **Becker, R., Wheeler, E. L., Lorenz, K., Stafford, A. E., Grosjean, O. K., Betschart, A. A., and Saunders, R. M. A.,** Compositional study of Amaranth grain, *J. Food Sci.,* 46, 1175, 1981.
33. **Schmidt, D.,** Grain amaranth: A look at some potentials, in *Proc. 1st Amaranth Sem.,* Rodale Press, Emmaus, PA, 1977, 121.
34. **Bressani, R., Elías, L. G., and García-Soto, A.,** Limiting amino acids in raw and processed amaranth grain protein from biological test, *Plant Foods Hum. Nutr.,* 39, 223, 1989.
35. **Olivares, R.,** personal communication, 1993.
36. **Betschart, A. A., Irving, D. W., Shepherd, A. D., and Saunders, R. M.,** *Amaranthus cruentus:* milling characteristics, distribution of nutrients within seed components, and the effect of temperature on nutritional quality, *J. Food Sci.,* 46, 1181, 1981.
37. **Higgins, T. J. V.,** Synthesis and regulation of major proteins in seeds, *Annu. Rev. Plant Physiol.,* 35, 191, 1984.
38. **Irving, D. W., Betschart, A., and Saunders, R. M.,** Morphological studies on *Amaranthus cruentus, J. Food Sci.,* 46, 1170, 1981.
39. **Irving, D. W. and Becker, R.,** Seed structure and composition of potential new crops, *Food Microstruc.,* 4, 43, 1985.
40. **Esau, K.,** *Plant Anatomy,* 2nd ed., John Wiley & Sons, New York, 1965, 767.
41. **Goodwin, T. W. and Mercer, E. I.,** Nitrogen fixation, amino acid biosynthesis and proteins, in *Introduction to Plant Biochemistry,* Pergamon Press, Oxford, 1983, chap. 9.
42. **Osborne, T. B.,** The vegetable proteins, in *Monographs in Biochemistry,* 2nd ed., Longmans, Green and Co., New York, 1924, 25.
43. **Landry, J. and Moreaux, T.,** Hétérogénéite des glutélins du grain de mais: extraction sélective et composition en acides aminés des trois fractions isolée, *Bull. Soc. Chim. Biol.,* 52, 1021, 1970.
44. **Padye, V. W. and Salunkhe, D. K.,** Biochemical studies on black gram (*Phaseolus mungo*): solubilization and electrophoretic characterization of the protein, *J. Food Biochem.,* 1, 111, 1977.
45. **Byers, M., Miflin, B. J., and Smith, S. S.,** A quantitative comparison of the extraction of protein fractions from wheat grain by different solvents, and of the polypeptide and amino acid composition of the alcohol-soluble proteins, *J. Sci. Food Agric.,* 34, 447, 1983.
46. **Paredes-López, O., Mora-Escobedo, R., and Ordorica-Falomir, C.,** Isolation of amaranth proteins, *Lebensm. Wiss. Technol.,* 27, 59, 1988.
47. **Gorinstein, S., Moshe, R., Greene, L. J., and Arruda, P.,** Evaluation of four amaranthus species through protein electrophoretic patterns and their amino acid composition, *J. Agric. Food Chem.,* 39, 851, 1991.
48. **Soriano-Santos, J., Iwabuchi, S., and Fujimoto, K.,** Solubility of amaranth seed proteins in sodium sulphate and sodium chloride: the main factor in quantitative extraction for analysis, *Int. J. Food Sci. Technol.,* 27, 337, 1992.
49. **Mora-Escobedo, R., Paredes-López, O., and Ordorica-Falomir, C.,** Characterization of albumins and globulins from amaranth, *Lebensm. Wiss. Technol.,* 23, 484, 1990.
50. **Derbyshire, E. and Boulter, D.,** Isolation of a legumin-like protein from *Phaseolus aureus* and *Phaseolus vulgaris, Phytochemistry,* 15, 411, 1976.
51. **Youle, R. J. and Huang, A. H. C.,** Albumin storage proteins in the protein bodies of castor bean, *Plant Physiol.,* 61, 13, 1978.
52. **Konishi, Y., Horikawa, O. Y., Azumaya, J., and Nakatani, N.,** Extraction of two albumin fractions from amaranth grains: comparison of some physicochemical properties and the putative localization in the grain, *Agric. Biol. Chem.,* 55, 2745, 1991.
53. **Segura-Nieto, M.,** unpublished data, 1993.
54. **Youle, R. J. and Huang, A. H. C.,** Ocurrence of low molecular weight and high cysteine containing albumin storage proteins in oilseeds of diverse species, *Am. J. Bot.,* 68, 44, 1981.
55. **Altenbach, S. B. and Simpson, R.,** Manipulation of methionine rich protein genes in plant seeds, *Trends Biotechnol.,* 8, 156, 1990.
56. **Crouch, M. L. and Sussex, I. M.,** Development and storage- protein synthesis in *Brassica napus* L. embryos *in vivo* and *in vitro, Planta,* 153, 74, 1981.

57. **Krebbers, E., Herdies, L., De Clercq, A., Seurinck, J., Leemans, J., Van Damme, J., Segura, M., Gheysen, G., Van Montagu, M., and Vandekerckhove, J.**, Determination of the processing sites of an Arabidopsis 2S albumin and characterization of the complete gene family, *Plant Physiol.*, 87, 859, 1988.
58. **Lilley, G. G., Caldwell, J. B., Kortt, A. A., Higgins, T. J., and Spencer, D.**, Isolation and primary structure for a novel methionine-rich protein from sunflower seeds (*Helianthus annus,* L.), in *Proc. World Congr. Vegetable Protein Utilization in Human Foods and Animal Feedstuffs,* Applewhite, T. H., Ed., American Oil Chemists Society, 1989, 497.
59. **De Castro, L. A. B., Lacerda, Z., Aramayo, R. A., Samparo, J. A. M., and Gander, E. S.**, Evidence for a precursor molecule of Brazil nut 2S seed proteins from biosynthesis and cDNA analysis, *Mol. Gen. Genet.*, 206, 338, 1987.
60. **Pedersen, K., Argos, P., Naravana, S. V. L., and Larkins, B.**, Sequence analysis and characterization of a maize gene encoding a high sulphur zein protein of Mr. 15,000, *J. Biol. Chem.*, 261, 6279, 1986.
61. **Kirihara, J. A., Hunsperberg, J. P., Mahoney, W. C., and Messing, J. W.**, Differential expression of a gene for a methionine rich storage protein in maize, *Mol. Gen. Genet.*, 211, 477, 1988.
62. **Segura-Nieto, M., Rodríguez-Néster, C., and Olguín-Martínez, L.**, Identification of methionine rich proteins from *Amaranthus hypochondriacus* seed, presented at *Int. Symp. Biotechnology for Crop Improvement in Latin America,* November 1 to 7, 1992, CQI 10.
63. **Ampe, C., Van Damme, J., De Castro, L. A. B., Sampaio, M. J. M., Van Montagu, M., and Vandekerkhove, J.**, The amino acid sequence of the 2S sulfur rich proteins from seeds of Brasil nut (*Bertholletia excelsa,* H. B. K.), *Eur. J. Biochem.*, 159, 597, 1986.
64. **Kinsella, J. E., Whitehead, D. M., Brady, J., and Bringe, N. H.**, Milk proteins: possible relationships of structure and function, in Developments in Dairy Chemistry-4, Fox, P. F., Ed., Elsevier Applied Science, 1989, chap. 2.
65. **Boller, T., Gehri, A., Mauch, F., and Vogeli, U.**, Chitinase in bean leaves: induction by ethylene, purification, properties, and possible function, *Planta,* 157, 22, 1983.
66. **Molano, J., Polacheck, I., Duran, A., and Cabib, E.**, An endochitinase from wheat germ, *J. Biol. Chem.*, 254, 4901, 1979.
67. **Shinshi, H., Mohnen, D., and Meins, F.**, Regulation of a plant pathogenesis-related enzyme: inhibition of chitinase and chitinase mRNA accumulation in cultured tobacco tissues by auxin and cytokinin, *Proc. Natl. Acad. Sci. U.S.A.*, 84, 89, 1987.
68. **Van Parijs, J., Broekaert, W. F., Goldstein, I. J. and Peumans, W. J.**, Hevein: an antifungal protein from rubber-tree (*Hevea Brasiliensis*) latex, *Planta,* 183, 258, 1991.
69. **Schlumbaum, A., Mauch, F., Vögeli, U., and Boller, T.**, Plant chitinases are potent inhibitors of fungal growth, *Nature,* 324, 365, 1986.
70. **Broekaert, W. F., Mariën, W., Terras, F. R. G., De Bolle, M. F. C., Proost, P., Van Damme, J., Dillen, L., Claeys, M., Rees, S., Vanderleyden, J., and Cammue, B. P. A.**, Antimicrobial peptides from *Amaranthus caudatus* seeds with sequence homology to the cysteine/ glycine-rich domain of chitin-binding proteins, *Biochemistry,* 31, 4308, 1992.
71. **Neurath, H.**, Evolution of proteolytic enzymes, *Science,* 224, 350, 1984.
72. **Richardson, M.**, Protein inhibitors of enzymes, *Food Chem.*, 6, 235, 1980.
73. **Travis, J. and Salvesen, G. S.**, Human plasma proteinase inhibitors, *Annu. Rev. Biochem.*, 52, 655, 1983.
74. **Ryan, C. A.**, Protease inhibitors in plants: genes for improving defenses against insects and pathogens, *Annu. Rev. Phytopathol.*, 28, 425, 1990
75. **Laskowski, Jr. and Kato, I.**, Protein inhibitors of proteinases, *Annu. Rev. Biochem.*, 49, 593, 1980.
76. **Koeppe, S. J., Rupnow, J. H., Walker, C. E., and Davis, A.**, Isolation and heat stability of trypsin inhibitors in amaranth (*Amaranthus hypochondriacus*), *J. Food Sci.*, 50, 1519, 1985.
77. **Duarte-Correa, A., Jokl, L., and Carlsson, R.**, Chemical constituents in vitro protein digestibility, and presence of antinutritional substances in amaranth grains, *Arch. Latinoam. Nutr.*, 36, 319, 1986.
78. **Goldstein, I. J., Hughes, R. C., Monsigni, M., Osawa, T., and Sharon, N.**, What should we call a lectin?, *Nature,* 285, 66, 1980.
79. **Sharon, N. and Lis, H.**, Glycoproteins: research booming on long ignored ubiquitous compounds, *Mol. Cell Biochem.*, 42, 167, 1982.
80. **Goldstein, I. J. and Hayes, C. E.**, The lectins: carbohydrate binding proteins of plants and animals, *Adv. Carbohydr. Chem. Biochem.*, 35, 127, 1978.

81. **Lis, H. and Sharon, N.,** Lectins: their chemistry and application to immunology, in *The Antigens,* Vol. 4, Sela, M., Ed., Academic Press, New York, 1977.
82. **Pardoc, G. I., Bird, G. W. C., Uhlenbruck, G., Sprenger, I., and Heggen, M.,** Immunotaestforsh, *Allerg. Klin. Imunol.,* 140, 374, 1970.
83. **Zenteno, E. and Ochoa, J. L.,** Purification of a lectin from *Amaranthus leucocarpus* by affinity chromatography, *Phytochemistry,* 27, 313, 1988.
84. **Koeppe, S. J. and Rupnow, J. H.,** Purification and characterization of a lectin from the seeds of amaranth (*Amaranthus cruentus*), *J. Food Sci.,* 53, 1412, 1988.
85. **Rinderle, S. J., Goldstein, I. J., and Remsen, E. E.,** Physicochemical properties of amaranthin, the lectin from Amaranthus caudatus seeds, *Biochemistry,* 29, 10555, 1990.
86. **Rinderle, S. J., Goldstein, I. J., Matta, K. L., and Ratclliffe, R. M.,** Isolation and characterization of amaranthin, a lectin present in the seeds of *Amaranthus caudatus,* that recognizes the T- (or cryptic T) - antigen, *J. Biol. Chem.,* 264, 16123, 1989.
87. **Peterson, D. M.,** Subunit structure and composition of oat seed globulin, *Plant Physiol.,* 62, 506, 1978.
88. **Fukushima, D.,** Structures of plant storage proteins and their functions, *Food Rev. Int.,* 7, 353, 1991.
89. **Wright, D. J.,** The seed globulins, in *Developments in Food Proteins - 5,* Hudson, B. J. F., Ed., Elsevier Applied Science, 1987, chap. 3.
90. **Bhattacharyya, S. P. and Biswas, B. B.,** Purification and characterization of high-salt-soluble vicilin from mung bean (*Vigna radiata*), *Biochem. Int.,* 21, 667, 1990.
91. **Weller, D. L.,** The 7S protein from brazil nut, *J. Food Biochem.,* 13, 353, 1989.
92. **Mimouri, B., Robin, J. M., and Azanza, J. L.,** Comparative studies of 11S globulin constituents of *Brassica napus* L. and of its related species *Brassica campestris* L. and *Brassica oleraceae* L., *Plant Sci.,* 67, 183, 1990.
93. **Brooks, J. R. and Morr, C. V.,** Current aspects of soy protein fractionation and nomenclature, *J. Agric. Oil Chem. Soc.,* 62, 1347, 1985.
94. **Peng, I. C., Quass, D. W., Dayton, W. R., and Allen, C. E.,** The physicochemical and functional properties of soybean 11S globulin — a review, *Cereal Chem.,* 61, 480, 1981.
95. **Krishnan, H. B., White, J. A., and Pueppke, S. G.,** Characterization and localization of rice (*Oryza sativa* L.) seed globulins, *Plant Sci.,* 81, 1, 1992.
96. **Robert, L. S., Nozzolillo, C., and Altosaar, I.,** Characterization of oat (*Avena sativa* L.) residual proteins, *Cereal Chem.,* 62, 276, 1985.
97. **Esnault, M.-A., Merceur, A., and Citharel, J.,** Characterization of globulins of yellow lupin seeds, *Plant Physiol.,* 29, 573, 1991.
98. **Madhusudhan, K. T. and Singh, N.,** Studies on linseed proteins, *J. Agric. Food Chem.,* 31, 959, 1983.
99. **Marcone M. F. and Yada, Y.,** Study of the charge profile and covalent subunit association of the oligomeric seed globulin from *Amaranthus hypochondriacus, J. Agric. Food Chem.,* 40, 365, 1992.
100. **Saio, K. and Watanabe, T.,** Food use of soybean 7S and 11S proteins, extraction and functional properties of their fractions, *J. Food Sci.,* 38, 1139, 1973.
101. **Swanson, B. G.,** Current status of soy protein functionality in food systems, *J. Agric. Oil Chem. Soc.,* 67, 265, 1990.
102. **Morr, C. V.,** Pea and lentil protein extraction and functionality, *J. Agric. Oil Chem. Soc.,* 67, 276, 1990.
103. **Barba de la Rosa, A. P., Paredes-López, O., and Gueguen, J.,** Characterization of amaranth globulins by ultracentrifugation and chromatographic techniques, *J. Agric. Food Chem.,* 40, 937, 1992.
104. **Lei, M.-G., Tyrell, D., Bassette, R., and Reeck, G. R.,** Two- dimensional electrophoretic analysis of soybean proteins, *J. Agric. Food Chem.,* 31, 963, 1983.
105. **Komatsu, S., Kajiwara, H., and Hirano, H.,** Soybean seed 34- kDa oil body-associated protein separated by two-dimensional gel electrophoresis, *Plant Sci.,* 81, 21, 1992.
106. **Robert, L. S., Matlashewski, G. J., Adeli, K., Nozzolillo, C., and Altosaar, I.,** Electrophoretic and developmental characterization of oat (*Avena sativa* L.) globulins in cultivars of different protein content, Cereal Chem., 60, 231, 1983.
107. **Snow, S. D. and Brooks, J. R.,** Fractionation of rice glutelin polypeptides using gel filtration chromatography, *J. Food Sci.,* 54, 730, 1989.
108. **Guo, Y.-J., Bishop, R., Ferhnstrom, H., Yu, G.-Z., Lian, Y.- N., and Huang, S.-D.,** Classification of Chinese rice variety by electrofocusing, *Cereal Chem.,* 63, 1, 1986.
109. **Wen, T.-N. and Luthe, D. S.,** Biochemical characterization of rice glutelin, *Plant Physiol.,* 78, 172, 1985

110. **Krochko, J. E. and Bewley, J. D.,** Use of electrophoretic techniques in determining the composition of seed storage proteins in alfalfa, *Electrophoresis,* 9, 751, 1988.
111. **Barba de la Rosa, A. P. and Paredes-López, O.,** Biochemical characterization of globulins from amaranth seed, presented at VII PAABS Cong. XIX Congr. Mexican Soc., 1992, 091T.
112. **Kitamura, K., Okubo, K., Shibasaki, K.,** The purification of soybean 11S globulin with ConA-Sepharose 4B and Sepharose 6B, *Agric. Biol. Chem.,* 38, 34, 1974.
113. **Olguín-Martínez, L. and Segura-Nieto, M.,** Globulins from amaranth seeds (*Amaranthus hypochondriacus* L.), presented at 1st Amaranth Int. Symp. Mexico, September 1 to 5, 1991.
114. **Bigelow, C. C.,** On the average hydrophobicity of proteins and the relation between it and protein structure, *J. Theor. Biol.,* 16, 211, 1967.
115. **Kato, A., Matsuda, T., Matsudomi, N., and Kobayashi, K.,** Determination of protein hydrophobicity using a sodium dodecyl sulfate binding method, *J. Agric. Food Chem.,* 32, 284, 1984.
116. **Kato, A. and Nakai, S.,** Hydrophobicity determined by a fluorescence probe method and its correlation with surface properties of proteins, *Biochim. Biophys. Acta,* 624, 13, 1980.
117. **Hayakawa, S. and Nakai, S.,** Relationship of hydrophobicity and net charge to the solubility of milk and soy proteins, *J. Food Sci.,* 50, 486, 1985.
118. **Tsutsui, T., Li-Chang, E., and Nakai, S.,** A simple fluorometric method for fat-binding capacity as an index of hydrophobicity of proteins, *J. Food Sci.,* 51, 1268, 1986.
119. **Matsudomi, N., Mori, H., Kato, A., and Kobayashi, K.,** Emulsifying and foaming properties of heat-denatured soybean 11S globulins in relation to their surface hydrophobicity, *Agric. Biol. Chem.,* 49, 915, 1985.
120. **Gueguen, J.,** Relation between conformation and surface hydrophobicity of pea (*Pisum sativum* L.) globulins, *J. Agric. Food Chem.,* 37, 1236, 1989.
121. **Kinsella, J. E. and Phillips, L. G.,** Structure: function relationship in food proteins, film and foaming behavior, in *Food Proteins,* Kinsella, J. E., and Soucie, W. G., Eds., American Oil Chemists Society, 1989, chap. 4.
122. **Cherry, J. P.,** Peanut protein and product functionality, *J. Agric. Oil Chem. Soc.,* 67, 293, 1990.
123. **Voutsina, L. P., Cheung, E., and Nakai, S.,** Relationships of hydrophobicity to emulsifying properties of heat denatured proteins, *J. Food Sci.,* 48, 26, 1983.
124. **Keshavarz, E. and Nakai, S.,** The relationship between hydrophobicity and interfacial tension of proteins, *Biochim. Biophys. Acta,* 576, 269, 1979.
125. **Shewry, P. R., Hill, J. M., Pratt, H. M., Leggatt, M. M., and Miflin, B. J.,** An evaluation of techniques for the extraction of hordein and glutelin from barley seed and a comparison of the protein composition of Bomi and Riso 1508, *J. Exp. Bot.,* 29, 677, 1978.
126. **Gorinstein, S., Nue, I. A., and Arruda, P.,** Alcohol-soluble and total proteins from amaranth seeds and their comparison with other cereals, *J. Agric. Food Chem.,* 39, 848, 1991.
127. **Leyva-López, N.,** personal communication, 1993.
128. **Carlsson, R.,** Amaranth species and related species for leaf protein concentrate production, in *Proc. 2nd Amaranth Conf.,* Rodale Press, Emmaus, PA, 1977, 83.
129. **Fernando, T. and Bean, G.,** A comparison of the fatty acids and sterols of seeds of weedy and vegetable species of Amaranthus spp., *J. Am. Oil Chem. Soc.,* 62, 89, 1985.
130. **Hill, R. M. and Rawate, P. D.,** Evaluation of food potential, some toxicological aspects, and preparation of protein isolate from the aerial part of amaranth (pigweed), *J. Agric. Food Chem.,* 30, 465, 1982.

Chapter 6

Amaranth Carbohydrates

Mercedes G. López, Luis Arturo Bello-Pérez, and Octavio Paredes-López

CONTENTS

0-8493-5374-2/94/$0.00+$.50

I. AN OVERVIEW

Not unexpectedly, the last few years have witnessed an increasing interest in the chemical structure of carbohydrates. They possess an enormous potential to fill diverse chemical and structural roles: this potential stems from the various ways each sugar can be linked together. A fundamental step in understanding the relationship between structure and function of carbohydrates is to determine which type(s) of linkages are present.

Among amaranth carbohydrates, the main interest is unquestionably starch, since very little is known about the molecular organization of the principal constituents of starch: amylose and amylopectin. Lower molecular weight carbohydrates are also present in numerous amaranth varieties, but have been found only in small or trace amounts.

II. INTRODUCTION

The study of carbohydrates is of considerable theoretical and practical interest; their biological roles are determined by their specific chemical, functional, and biochemical behavior. These observations may be considered in terms of biomolecular evolution: during the sequential development of aerobic and anaerobic glycolysis as well as photosynthesis, glucose rather than any other sugar has emerged as the universal energy source.

The availability of many polysaccharides in large quantities makes them the polymer of choice for renewable resources; their properties have been exploited mainly by food, pharmaceutical, textile, paper, and fuel industries.

It is generally recognized that the useful and unique properties of carbohydrates stem, essentially, from the molecular conformation or configuration of the polymer. Therefore, in order to understand the fundamental basis of such properties, it is imperative that we improve our knowledge of the molecular character of these polymers.

III. MONOSACCHARIDES AND RELATED SUBSTANCES

A. GENERAL CONSIDERATIONS AND QUANTITIES

Sugar aldehydes or ketones are simple carbohydrates which possess reducing power due to their carbonyl group; this group is also responsible for their reactivity. Many of these reducing sugars are the principal constituents of foodstuffs, hence a great number of reports dealing with this class of molecules have been published.

In the case of amaranth seeds, Becker et al.[1] reported only trace amounts of monosaccharides (glucose, fructose, and perhaps galactose) in seven out of ten different amaranth samples tested. The amounts found in the species studied are listed in Table 1. Among the monosaccharides reported by Lorenz and Gross,[2] glucose was the principal free sugar (ranging from 0.12 to 0.67% on dry basis) obtained from an ethanol extract, followed by fructose (0.05 to 0.13% dry weight).

Singhal and Kulkarni[3] reported that the concentration of reducing sugars in different amaranth species was greater than the levels found in other cereals and legumes; the levels reported ranged from 2.47 to 9.03 g kg^{-1}. However, they did not differentiate between individual sugars, or between mono- or oligosaccharides present in the reducing mixture. In general, the levels of simple carbohydrates present in amaranth varieties range from 0.05 to 0.67%. These levels are quite high compared with the amounts found in wheat (0.01 to 0.04%)[4] and maize (0.02 to 0.09%).[5]

Some of the most common related compounds of monosaccharides are alditols and sugar acids. Among the former, inositol generally occurs in the size member ring form. Becker et al.[1] studied the inositol levels (0.06 to 0.10%) in different amaranth samples; the concentrations found by these workers are similar to the amounts found in other sources.[6,7] Thus, except for galactose, all the amaranth varieties studied showed higher concentrations of sugars and comparable values of inositol. As far as we were able to determine, no attempts have been made to examine the presence of sugar alditols other than the cyclitol (inositol) under similar conditions. It seems possible that the presence of higher monosaccharide concentrations in amaranth seeds, compared with wheat and maize, could be related to the biosynthesis of starch in the amaranth granules.

Table 1 **Saccharides in amaranth seeds**

Species	S	F	G	R	ST	M	I	Ref.
A. hypochondriacus	1.95	—	—	0.76	0.10	0.18	0.10	1
A. cruentus	1.83	—	—	0.82	0.13	0.02	0.10	1
A. edulis	1.35	—	—	0.45	0.02	0.22	0.10	1
A. cruentus	0.74	0.05	0.57	0.28	—	—	—	2
A. cruentus	0.59	0.05	0.12	0.42	—	—	—	2
A. hypochondriacus	0.87	0.13	0.67	0.27	—	—	—	2
A. hybridus	0.41	0.00	0.00	0.13	—	—	—	2

Note: S: sucrose; F: fructose; G: glucose; R: raffinose; ST: stachyose; M: maltose; I: inositol.

B. IDENTIFICATION AND ANALYSIS

1. Derivatization

The use of simple carbohydrate analysis are as limitless as the information obtained from them. There is much literature of interest regarding analysis of monosaccharides.

The presence of very polar hydroxyl groups makes monosaccharides and related compounds generally nonvolatile. They must be derivatized to volatile species prior to analysis — usually via gas chromatography. Monosaccharides can be converted to acetates,[8] trifluoroacetates,[9,10] trimethylsilylates (TMS),[11] methylates,[12] etc. The reducing group of sugars can be selectively modified to oximes[13] or to per-O-acetylaldononitriles.[14]

Ahmed et al.[15] proposed that TMS derivatization presents the disadvantage of forming several derivatives instead of only one. This raises potential uncertanties regarding their separation, identification, and especially their quantitation.

Since alditols are quantitatively converted to their acetate form, this procedure has been used most extensively among the derivatization procedures listed above. This excellent alternative for the quantitation of simple sugars requires a chemical modification step (reduction with $NaBH_4$) prior to derivatization.

2. Quantitation

The introduction of capillary columns to carbohydrate analysis allows separation of complex mixtures of carbohydrates in an effective fashion and in a reproducible manner. Quantitative analysis of simple sugars can be very precise; however, quantitation of complex mixtures is often more difficult due to overlapping chromatographic peaks.

With respect to the monosaccharide concentrations in amaranth seeds reported by Beckerr et al.,[1] it is possible that the observed trace amount levels could have been distorted by the derivatization procedure, since these workers used TMS derivaties with the concomitant disadvantages mentioned in Section III.B.1. Even so, the values obtained were still significantly higher than those reported by Lorenz and Gross[2] using high performance liquid chromatography (HPLC). Indeed, reported values in the literature vary widely even among the same species of amaranth examined.

IV. OLIGOSACCHARIDES

A. TYPES OF OLIGOSACCHARIDES AND QUANTITIES

Generally, oligosaccharides are not found in plants; however, oligosaccharides with a low degree of polymerization have been found in wheat[4] and maize.[5] Singhal and Kulkarni[3] reported significant quantities of reducing sugars. These authors claimed that their values were higher than those found in cereals and legumes and are in excellent agreement with the values previously reported by Lorenz and Gross.[2] However, the former investigators did not report sucrose levels, which appears to be the predominant oligosaccharide in amaranth species. Becker et al.[1] reported that sucrose was the predominant free disaccharide present in nine amaranth samples. They pointed out that the levels found were much higher than those reported for grains such as wheat, rye, triticale, and millets. Besides sucrose, the amaranth samples tested contained raffinose, stachyose, and maltose; the amounts found are listed in Table 1.[1,2] After sucrose, raffinose was the principal α-galactoside in most amaranth species, followed by stachyose.[16] In many cases, oligosaccharides are accompanied by monosaccharides and the two must be analyzed together; this may decrease the sensitivity of the method employed for oligosaccharide determination.

Lorenz and Gross[2] reported raffinose as the most abundant oligosaccharide in eight different amaranthus species studied by them. While maize[5] contains aproximately 4 to 8% by weight oligosaccharides, amaranth contains 1.65% and wheat as low as 0.6%.

B. IDENTIFICATION

The identification and quantitation can be achieved in the same manners described for monosaccharides (Sections III.B.1 and III.B.2).

V. STARCH

A. GENERAL ASPECTS

Starch is the major dietary component in all human populations. The consensus of recent opinion on healthy eating habits favors an increase in the proportion of polymeric plant carbohydrates (including starch) in the daily diet, with a proportionately reduced intake of "refined" sugars and fats. However; in our culture, the main purpose of starch utilization in foods remain aesthetic rather than nutritional. This biopolymer constitutes an excellent raw material to modify food texture and consistency. Not only is the amount of starch important for the texture of a given food product, but starch type is equally critical.[17] The more important potential sources of starches are: cereals (40 to 90% starch, dry weight), tubers (65 to 85%), and legumes (30 to 70%). World production of starch has been estimated at 18 million metric tons, which are extracted principally from maize (10 million tons) and potato (2.5 to 3.0 million tons); of these amounts, 50% are used in foods products.[18]

Starch, a homopolysaccharide of α-D-glucose, is the major storage carbohydrate of all higher plants. In its native state, starch exists as water insoluble, partially crystalline granules whose morphology, chemical composition, and supermolecular structure are characteristic of each particular plant species.[19-21] The variation in size of these granules is especially remarkable (between 0.5 and 100 μm); the largest granules are present in the potato (15 to 100 μm) and the smallest in amaranth species (0.8 to 2.5 μm)[22-26] (Table 2).[3,26-30] The particle size, including size distribution, is one of the characteristics most influencing the functional properties of starch granules. It has been reported that smaller granules have a higher swelling power, a lower solubility, a greater water-binding capacity, and a lower sustceptibility to α-amylase.[31]

Granule composition, morphology, and supermolecular organization are, to a certain extent, under genetic control, i.e., they are dictated by the biochemistry of the starch-synthesizing organelle (amyloplast) as well as the physiology of the plant.[17] However, the exact mechanism(s) by which plant genetic information is translated into specific granule composition, morphology, and organization is still unknown. The exact roles and *in vivo* regulation of various enzymes involved in starch biosynthesis (glucosyl transferases, branching enzyme, and phosphorylase) are also not fully understood.[32,33] Once such information is available it may be possible to manipulate the amount, composition, and other properties of starch using genetic engineering techniques.[17]

Starch granules are composed essentially of α-D-glucan polymers and small amounts of noncarbohydrate consituents, particularly lipids, proteins, and minerals (Table 3);[1,28-30,34] the latter, although a minor component, may influence the functional properties of starch.[4,34] In amaranth species, starch levels, ash, protein, and fat are higher than levels found in other materials such as potato, wheat, and waxy maize. Among all the amaranth species, *A. hypochondriacus* possesses the highest fat content. This high fat level could be partially responsible for the special functional properties observed for amaranth starch. Amaranth starch *per se* has been classified as "waxy-type". Waxy starches are characterized by their high amypopectin content. This composition confers very useful and unique applications, such as high viscosity and gelatinization at high temperatures.

In general, starch owes its functionality (e.g., as a thickening agent, flavor carrier, and binder in food related applications) to its two high molecular weight components, amylose and amylopectin, as well as to the physical organization of these macromolecules into the granule structure. Therefore, the starch structure needs to be considered at two distinct levels: (1) at a molecular level which refers to the amount, fine structure, size, and shape of its molecular components and (2) at the supermolecular structure of the granule.[17,21,37,38]

Table 2 **Characteristics of amaranth starch granules**

Species	Size (mm)	Shape	Ref.
A. hypochondriacus	1.0	Polygonal	27
A. hypochondriacus	1.0	Polygonal	29
A. hypochondriacus	1.0	Polygonal	28
A. cruentus	1.0	Lenticular	26
A. paniculatus	1.5–2.5	Circular	3
A. polygamous	1.1–1.9	Circular	3
A. gracilis	1.1–2.1	Circular	3
A. spinosus	1.2–1.5	Circular	3
A. teunifolious	0.8–2.3	Circular	3
A. paniculatus	1.5–2.0	Elliptical	30

Table 3 **Proximate composition of amaranth starch**

Species	Moisture	Ash[a]	Protein[a,b]	Fat[a]	Amylose[a]	Ref.
A. hypochondriacus	6.16	1.39	0.12	0.39	4.90	28
A. hypochondriacus	—	—	0.49	1.10	7.20	1
A. hypochondriacus	—	—	0.97	1.80	2.10	29
A. retoflexus	—	0.18	0.42	0.14	—	34
A. hypochondriacus	—	—	0.10	0.62	2.00	35
A. paniculatus	9.50	0.80	0.92	1.00	11.50	30
A. hypochondriacus	2.70	0.11	0.35	—	1.98	36

[a] Dry basis; [b] Nitrogen × 5.85.

B. AMYLOSE

Amylose is a linear macromolecule (Figure 1), which consists of α-D-glucopyranoside residues bound by α-(1 → 4) linkages with a degree of polymerization (DP) of some hundreds of glucose residues anywhere from 500 to 2000 units.[25] The perception of amylose as a linear molecule is based upon the quantitative conversion to maltose by β-amylase, an exo-enzyme which hydrolyses at the α-(1 → 4) linkages. In the case of amylose, this hydrolysis is incomplete, suggesting the presence of branching points. Branching, however, occurs with sufficient rarity that amylose behavior is essentially that of a linear polymer.[4]

Table 4 comprises some amylose properties from different sources and underlines the incomplete β-amylolysis.[18] Some other properties of amylose are noted in Table 5.[17] Many of these properties can be explained in the context of amylose's ability to adopt different molecular conformations. In neutral aqueous solutions the normal conformation is a random coil (Figure 2). Amylose has the capacity to interact with iodine, which produces a helicoidal inclusion complex, having about six glucose molecules per turn, in which the iodine molecule is located in the central cavity of the helicoidal polysaccharide. This complex gives a blue color with maximum absorption at wavelengths between 620 to 680 nm. Amylose has a strong effect on the gelatinization and retrogradation of starch,[17] and since amaranth starch does not contain a large amylose amount these phenomena are rarely observed in amaranth starches.

Amylose can form complexes with lipids in the superficial regions of the granule. It is known that this type of complex inhibits starch degradation by enzymes such as phosphorylase, α-amylase and β-amylase. Other researchers have postulated that amylose and lipids coexist independently inside the granule and only form complexes once gelatinization takes place.[39]

Based on the amylose content of starches, these can be classified into different groups: the waxy starches containing very small amounts of amylose (e.g., amaranth starch), around 1%; normal starches with amylose contents between 17 and 24%; and high amylose starches with 70% or more.[40] Starches with above 1% amylose are also known as nonwaxy type.

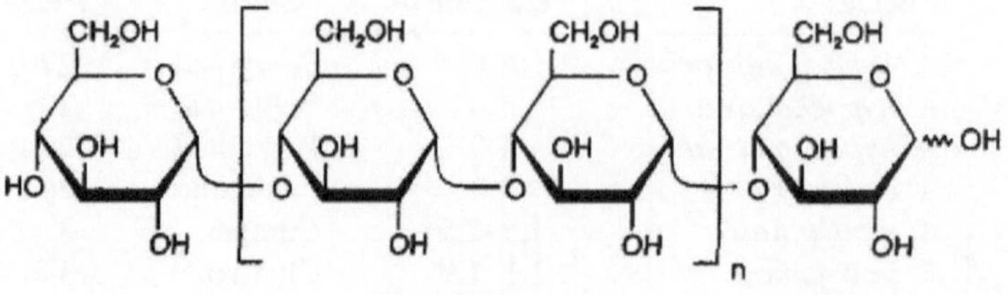

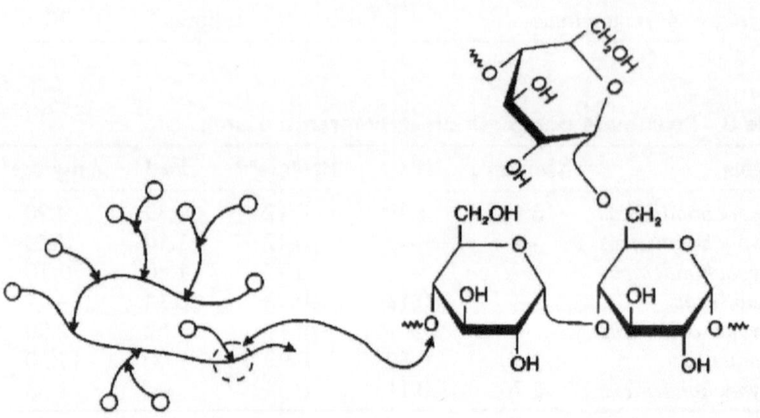

Figure 1 Structure of amylose and amylopectin.

Table 4 **Amylose properties**

Starch	IBC[a]	β-amylolysis limit[b]	η-(ml/g)[c]	DP[d]
Cereals				
Wheat	19.1–19.7	68–72	250–330	2100
Maize	19.0	—	127	940
Amylomaize	192	77	180	1300
Barley	19.0–19.6	73–74	250–355	1850
Oat	19.2–19.5	77–78	180–435	1300
Legumes				
Pea	19.1	78	264	1300
Tubers				
Potato	19.5–20.5	74–76	410–430	3200
Fruits				
Banana	19.9	82	—	—
Mango	19.3	77	400	—

[a] Iodine binding capacity (mg of iodine per 100 mg of starch); [b] Maltose base; [c] Intrinsic viscosity in 1 M KOH at 25°C; [d] From intrinsic viscosity; DP = polymerization degree.

Table 5 **Structure and physicochemical properties of starch polymers**

Property	Amylose	Amylopectin
Molecular structure	Essentially linear α-(1→4)	Branched α-(1→4) and α-(1→6)
Degree of polymerization	$\sim 10^3$	$\sim 10^4$–10^5
Molecular weight	1.5×10^5–10^6	(50–500) × 10^6
Average chain length	$\sim 10^3$	~20–25
Iodine complex, λ_{max} (color)	640–600 nm (blue)	530–550 nm (purple)
Digestibility (%)		
β-amylase	~70	~55
β-amylase and debranching enzymes	~100	~100
Stability of dilute aqueous solutions	Unstable (retrogrades)	Stable
Gel texture	Stiff, irreversible (Tm > 100°C)	Soft (thermally reversible at <100°C)
Film properties	Strong, coherent	Brittle

[a] Tm is melting temperature.

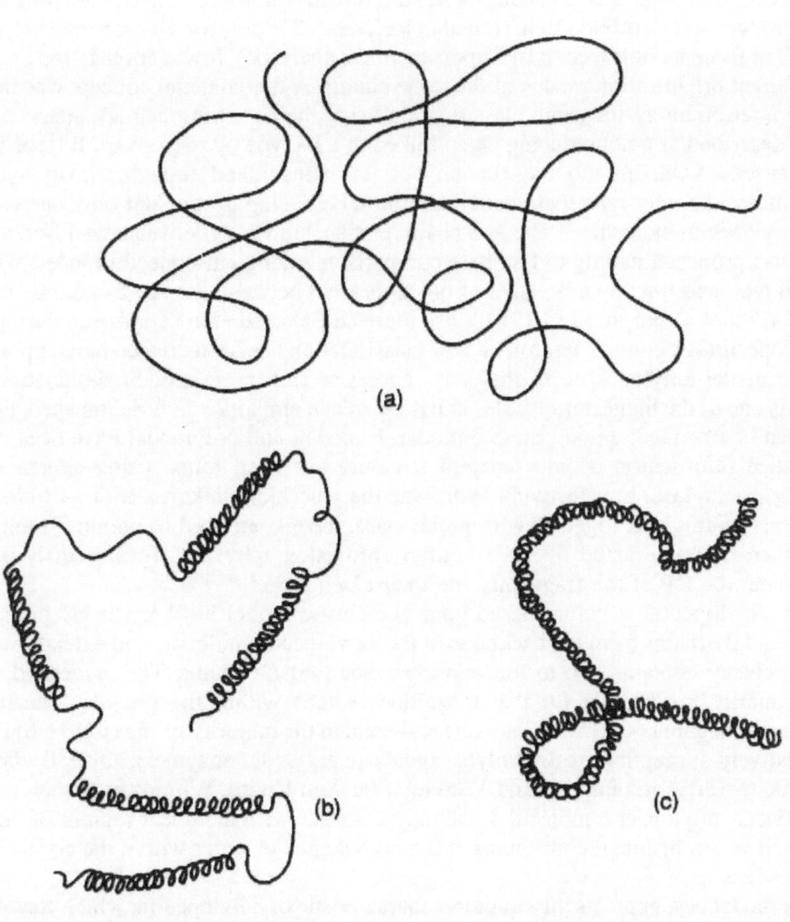

Figure 2 Models proposed for the amylose molecule in aqueous solution. (a) Random coil; (b) interrupted helix; (c) deformed helix.

C. AMYLOPECTIN

1. General Aspects

Amylopectin is the branched component of starch (Figure 1). It is formed by chains of α-D-glucopyranoside residues (between 17 and 25 units, Table 5) principally bound through α-(1 → 6) linkages at the branch points.[31] The amylopectin can be degraded by β-amylase cleavage at α-(1 → 4) linkages to produce β-limit dextrins (that are the chain residues which contain the branch points) and, thereafter, by pullulanase or isoamylase which act on the α-(1 → 6) bounds yielding maltose. The molecular weight of amylopectin varies between 50 and 500×10^6. These variations are influenced by the botanical origin of the starch, the method employed for fractionation of starch to amylose and amylopectin, and the method used to determine the molecular weight. The large size of this molecule creates some problems in molecular weight assessments[41] which have been overcome with improvements in the more newly developed techniques (e.g., light-scattering techniques, ultracentrifugation).[42]

2. Structural Characterization

Because the structure and properties of amylopectin contribute notably to granule composition, therefore, this review pays attention to various aspects including molecular size, branching, and lengths of internal and external chains of these macromolecules.[43]

Our knowledge about the amylopectin structure progresses from the report of Haworth et al. (cited in Reference 37) who, in 1937, proposed a laminated structure. In the same year, Staudinger and Husemann (cited in Reference 37), suggested a herring-bone structure in which a principal chain possessed all branch points. Later, Meyer and Bernfeld (1940) (cited in Reference 37) proposed a randomly branched structure (Figure 3). All of these models were based upon chemical analysis.[37] It was emphasized that these models contained different organization modes of the same chains. A fundamental concept was introduced later on, involving a terminology for chain classification using the following capital letters: A-, B-, and C-. Type A- was described as a nonreducing chain linked to a B- type by α-(1 → 6). B- type chains contain either one or several A- chains and may contain B- type chains linked through primary hydroxyl groups. Type C- chains are the only type that contain a terminal reducing group, and only one C- chain can be found in an amylopectin molecule.[37] The A:B chain relationship varies between the different models. This terminology was proposed mainly to facilitate comparison among amylopectin models where the chain length of each type was unknown. In spite of the agreement between the A-, B-, and C- models and the cluster model (Figure 4) proposed in 1974,[37,38,44] there still existed some confusion during the interpretation of amylopectin structure. Therefore, a new classification has been created based upon the polymerization degree of the amylopectin. In this way, an easier and more explicit designation is available. Amylopectin is one of the biggest molecules in nature. When amylopectin is debranched, different modal distribution can be obtained, among these bimodal, trimodal, and polymodal have been published.[45,46]

More detailed information on amylopectin structure has been found using debranching enzymes (pullulanase or isoamylase) to selectively hydrolyze the glucosidic linkages α-(1 → 6) localized on the branching points. β-amylase, together with pullulanase, have been used to obtain β-limit dextrin. This product can then be fractionated by gel filtration chromatography, and further analysis will provide information about the DP of the fragments and their chain ratio.[45,47,48]

A different amylopectin structure based upon the cluster model has been lately proposed.[46] In this model, the A- and B- chains form the backbone of the amylopectin molecule and extend over two or more clusters. Each cluster contains two to four closely associated A- chains. The associated clusters of A-chains are primarily responsible for the crystalline regions within the granule. The intercrystalline (amorphous) areas occur at 60 to 70 Å intervals and contain the majority of the α-(1 → 6) linkages; these bonds are relatively susceptible to hydrolytic agents (e.g., acids, enzymes). Overall, the amylopectin molecule is 100 to 150 Å in diameter and 1200 to 4000 Å in length. Within the granules, amylose may be located between amylopectin molecules and may associate with the linear regions of the amylopectin molecules. Such an arrangement contributes to the high degree of order within the crystalline regions of amylopectin molecules.[37]

The cluster model best explains the structural characteristic of amylopectin, which contains crystalline regions that are very organized and amorphous regions with a high degree of disorganization (Figure 4).[44] This description agrees very well with the high relative viscosity of amylopectin.[49] This model has been supported by Hizukuri,[46] who found that chain length distrubutions were inconsistent with Haworth, Staudinger, or Meyer models, but consistent with the cluster model. The A:B chain relationship (between 2.6:1.0 and 1.0:1.0) has been used as a pattern to verify or refute different models of the amylopectin molecule.[47] X-ray diffraction studies have also been used to confirm the cluster model and the role of

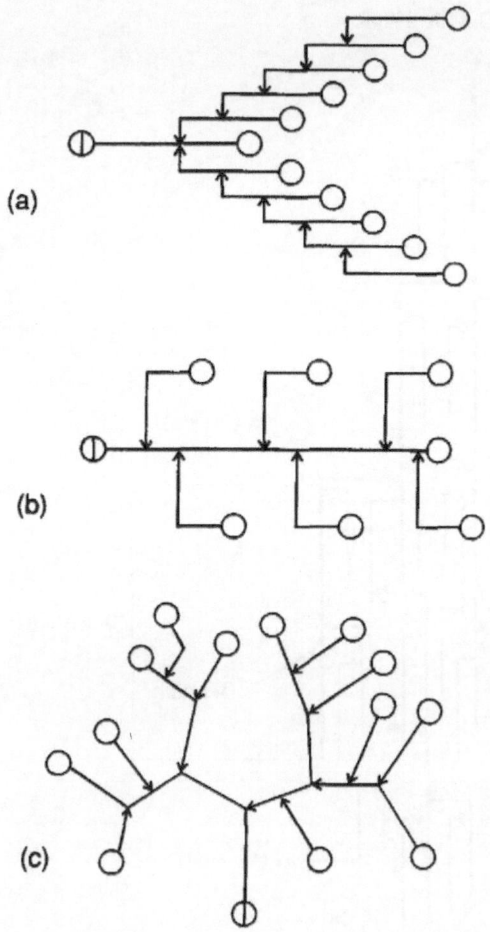

Figure 3 Proposed models for amylopectin. (a) Laminated structure; (b) herring-bone structure; (c) randomly branched structure.

amylopectin in starch crystallinity. The degree of amylopectin crystallinity depends on the average molecular weight and chain length. For example, A-, B-, and C- types present a polymerization degree of 26 (short); 36 (long), and 28 (medium) chain length, respectively.[43] These chain lengths seem to be specific for different botanical sources. Recently, it has been suggested that the production of crystalline varieties is determined exclusively by the chain lengths of amylopectin.[17] Based on this, it is now known that the chain length of amylopectin starches type A- are shorter than the chain length of type B-. In addition, amylodextrins with short and long chains have a relatively high trend to crystallize into the A- and B- types, respectively.[17] It has been concluded that the average length of amylopectin chains depends principally upon type B-. This suggests a fine control mechanism, which differs slightly among different plant species. At the same time, this control mechanism can be implicated in amylopectin biosynthesis. It can be stated that the chain length of amylopectin is an intrinsic factor which plays a determinant role in the crystalline structure of starch granules.[17] The regular variation of the amylopectin molecular structure is of interest to understand the functional properties of amaranth starches.

There are several reports on the structural characteristics of amylopectin. Kobayashi et al.[48] worked with ten varieties of wheat amylopectin. They debranched the amylopectin with pullulanase and found that the fraction profile obtained by high pressure size exclusion chromatography did not show significant differences. This result suggests that the structure of isolated amylopectins did not greatly affect the functional properties of wheat starches. MacGregor and Morgan[45] worked with isolated amylopectin from small and large granules of normal and waxy barley. They found that all the amylopectins presented identical structures, and the results agree extremely well with the cluster model. However, Taki et al.[50] who worked with maize starches (normal, waxy, and high amylose) and potato starch reported that waxy maize amylopectin structure presented a different pattern from all the other amylopectins studied. Later on, Hizukuri et al.[51] reported large differences among amylopectins extracted from different rice starches which contain different amylose and amylopectin levels in the granule.

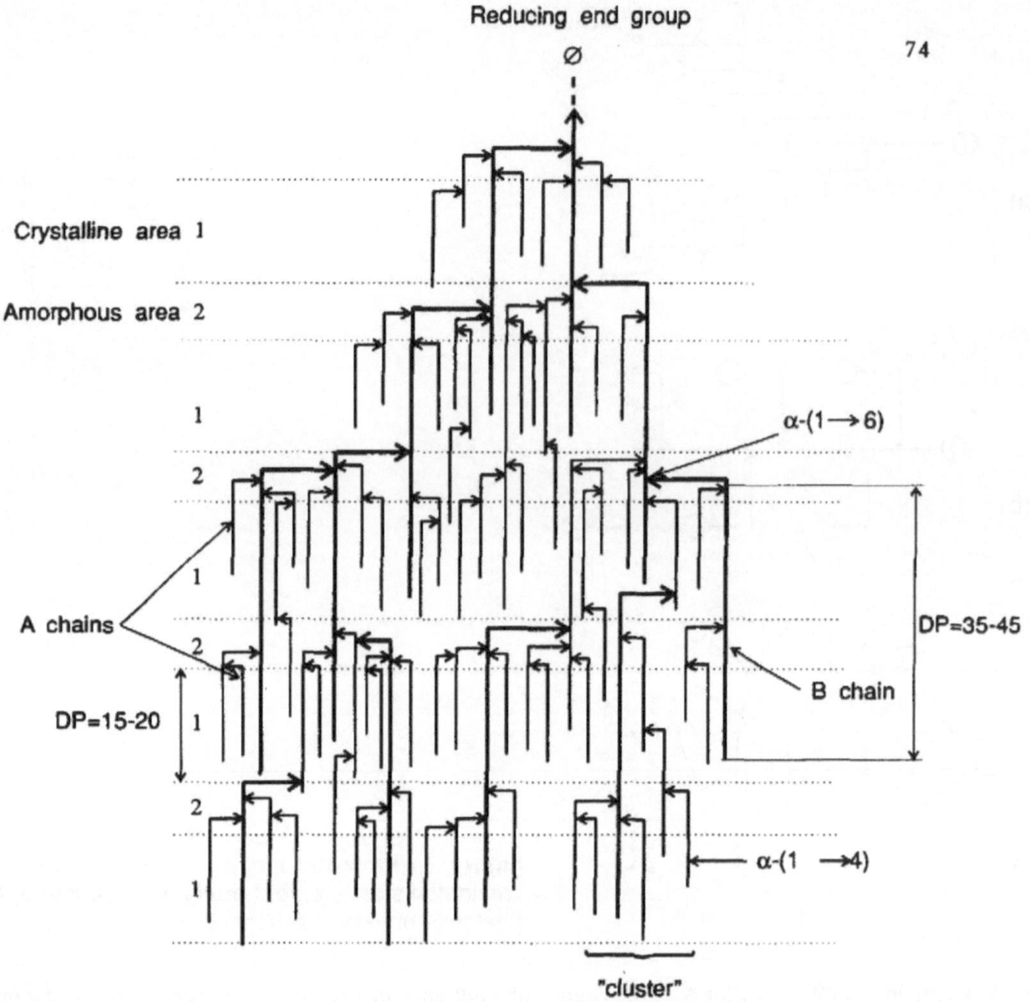

Figure 4 Cluster model for amylopectin. (From Robin, J.P., Mercier, C., Chabonniere, R., and Guilbot, A., *Cereal Chem.*, 51, 389, 1974. With permission.)

VI. STARCH ISOLATION, PURIFICATION, AND CHARACTERIZATION

A. STARCH ISOLATION

Amaranth starch isolation can be performed by different methods, which basically involve the use of sodium hydroxide, mercuric chloride, and sulfur dioxide. Among these methods, the most extensively used on amaranth starch isolation have been the alkaline and the mercuric chloride methods.

1. Alkaline Method

The alkaline procedure for the isolation of amaranch starch was first introduced by MacMasters et al.[52] Overall, this method is a series of washing, centrifuging, and rinsing; therefore, the method turns very tedious and time consuming. In spite of this, the alkaline method has been widely used by Sugimoto et al.,[27] Tomita et al.,[28] and Konishi et al.[53] for different amaranth species such as *A. hypochondriacus* and *A. caudatus.* Yañez and Walker[29] modified the above method. This modification involves the elimination of two steps: soaking in sodium metabisulfite and extracting with sodium chloride. These authors have substituted these two steps by a singular step using 0.25% sodium hydroxide. The advantage conferred by the modified method is mainly time-saving. However, the starch obtained by this procedure presents higher impurity levels than those reported in the original MacMasters method.

Paredes-López et al.[54] have also used an alkaline method for starch isolation from amaranth samples; this method is basically the same as the MacMasters method. However, it presents the advantage of obtaining a lyophilized starch; therefore, starch is more easily dispersed in polar solvents such as water and dimethylsulfoxide. However, solubility still varies considerably among isolated starches.

2. Mercuric Chloride Method

This method is usually performed by dipping a sample in a $HgCl_2$ solution (0.01 M) for 1 h to inhibit the α-amylase activity. The sample is then milled in a commercial blender. The starch obtained at this point must be filtered and, if necessary, more $HgCl_2$ may be added during filtration. This last step can be repeated until only a small amount of starch passes through the filter. The starch is allowed to settle, then separated from the supernatant, resuspended in distilled water, and washed a few times (until the supernatant is clear). At this point, the starch is highly pure and can be dried in a forced air oven.

Wankhede et al.[30] have also used $HgCl_2$ as an enzyme deactivator; however, they first immersed the grains in distilled water (about 100 g per 0.5 l) containing a few milliliters of toluene for 12 h at 5 or 6°C. After the recommended time the grains are homogenized in a Waring blender with $HgCl_2$ (0.01 M) for few hours. The starch obtained at this stage can be further purified by repeatedly shaking the suspension in sodium chloride:toluene (1:0.2 v/v). This method gives very good yields, approximately 60 to 62% of starch from the whole grain. Other authors, Adkins and Greenwood,[55] Lorenz and Gross,[2] and Hernández-López,[56] have used the same method with slight variations. The sample is first submerged in $HgCl_2$:acetate buffer (0.01 M:0.02 M), close to neutral pH (6.5) at 5°C for 24 h to inhibit the enzymatic activity. Then, the sample is washed with distilled water, wet-milled in a Waring blender, and, finally, centrifuged for 15 min at 5000 × g and dried. The only drawback that this method presents vs. the previous procedures described above is the relatively high protein content in the isolated starch.

B. STARCH FRACTIONATION

1. Amylose and Amylopectin Separation

Separation of starch into amylose and amylopectin can be achieved by the addition of butanol or thymol; these chemicals have the property of forming insoluble amylose complexes. Sugimoto et al.[27] and Kintner[57] have been able to isolate amaranth starch inducing the generation of these types of complexes. Tomita et al.[28] reported a genetic strain of *A. hypochondriacus* containing 22% amylose. On the other hand, the amylopectin fraction can be found in the supernatant liquor remaining after removal of the amylose complex. This fraction can be concentrated and then treated with n-butyl alcohol to remove amylose residues.

2. Amylose Separation by Gel Filtration

Amylose content can be determined by gel-permeation column chromatography, as described by Banks and Greenwood,[58] using columns packed with Sepharose CL-2B gel. The column can be run in the ascending or descending mode. Sargent[59] also fractionated amylose samples using a Sepharose CL-6B column.

3. Amylopectin Debranching and Separation

Debranching of amylopectin is usually performed by a specific debranching enzyme, pullulanase, which cleaves the α-(1 → 6) (branching points). The enzyme is prepared in an aqueous 20% dimethylsulfoxide solution. The molecular size and the average chain length (CL) of debranched amylopectin is determined by dividing the total carbohydrate by its reducing value (total carbohydrate analyzed by the procedure published by Dubois et al.[60]), and such value is measured by a modified Park-Johnson procedure (Hizukuri and Maehara[61]). Distributions of the amylopectin branch chain lengths are usually analyzed by gel-permeation column chromatography.

$$CL = \frac{\text{Total carbohydrate (as glucose)}}{\text{Reducing capacity after debranching (as glucose)}}$$

4. High Performance Liquid Chromatography

Hizukuri[46,62] was the first investigator that developed an HPLC technique for the quantitation of amylopectin starches from different sources. The use of this technique presents some advantages over the previous technique, such as better separation, higher sensitivity, smaller amount of sample, etc. Furthermore, the results drawn from this type of analysis confirm that all the amylopectin fractions fit extremely well with the cluster model described in Section V.C.2. On top of that, HPLC is also a tool that can be utilized to withdraw information on the relationship between the molecular structure and the crystalline form of the branched component.

C. PHYSICOCHEMICAL CHARACTERIZATION

1. Viscosimetry

a. Viscoamylography

The viscosity created after stirring and heating an aqueous starch suspension, followed by holding the temperature to any desired time and then cooling to a defined rate, can be measured by different viscoamylographs such as the Brabender type and reporting the readings in Brabender Units (B.U.). It should be kept in mind that sample manipulation involves the disruption of some intra- and intermolecular hydrogen bonds, as well as the generation of new ones; as a result of this, the structure of the colloid collapses as the integrity of the starch granule is lost, which generates changes in its functional properties too. Amaranth viscosities have also been measured by other techniques, such as those reported by Barba de la Rosa et al.[63] and Mora-Escobedo et al.[64]

b. Photopastegraphy

The degree of gelatinization temperature of different amaranth species can be assessed by various methods. Using the Kainuma et al.[65] method, which involves photopastegraphy, Sugimoto et al.[27] recorded gelatinization temperature of two different types of *A. hypochondriacus*. The gelatinization behavior of the amaranth samples was similar to that previously reported by Kainuma et al.[65] Okuno and Sakaguchi[66] also studied the gelatinization temperatures by photopastegrams; however, they used the whole amaranth grain during their determinations.

2. Differential Scanning Calorimetry

Zobel et al.[67] have widely used calorimetric techniques to study the physical transitions of starches in aqueous solutions when these are heated. Donovan,[68] Paredes-López et al.,[54] and Hernández-López[56] are some of the several investigators who have used differential scanning calorimetry (DSC) techniques to measure the gelatinization endothermic transitions and enthalpy of amaranthus species. DSC is based on the measurement of the degradation of a very small sample (~1 mg) which decreases the thermal lag within the system. The instrument should be calibrated with a metal, which is usually indium (ΔH fusion 6.8 cal/g, melting point 156.4°C). The sample is placed in an aluminum pan. After sealing, the pan is left to equilibrate for about an hour and scanned at a rate of 10°C/min. Usually an empty pan is used as an inert reference. Gelatinization energy (ΔH) in cal/g is determined by measuring the peak area. Degree of gelatinization (expressed as percent) is determined from gelatinization energy.

3. X-Ray Diffractometry

Sugimoto et al.[27] studied the X-ray diffraction pattern of normal and waxy amaranth starch. The diagrams obtained were of an A- type crystalline structure; this type of structure is typical of the diagrams for rice and maize starches. In the same year, Tomita et al.[28] used the same procedure to investigate the structure of amaranth starch and several millets which also presented an A- type pattern; in this study, the diagrams were akin to normal maize starch. The procedure and parameters used by both research groups were basically the following: fully moistured starch granules were placed in a spectrodiffractometer following Hizukuri's method.[69] The parameters used during the performance were CuKalfa (Ni filtered), voltage of 35 kV, electric current 12.5, time constant 4 s, scanning speed of the goniometer, 1°/min, and chart speed of 10 mm/min.

4. Scanning Electron Microscopy

Scanning electron microscopy (SEM) has been used by Lorenz[70] to investigate the surface of amaranth seeds. He reported that *A. hypochondriacus* granules are angular and polygonal compared to the oval to round shapes of granules of wheat starch. Amaranth native starch was examined by Wankhede et al.;[30] the starch granules showed that they are mostly spherical and very small in size (1.5 to 2.0 μm). Barba de la Rosa et al.[63] used SEM to study the microstructure of raw, gelatinized, and hydrolyzed amaranth flour. They observed that hydrolysis temperature changed considerably the structural appearance of flour particles. Using SEM, Paredes-López et al.[54] demostrated that starch from *A. hypochondriacus* Mercado type granules presented a polygonal shape slightly more rounded than other varieties reported previously.[70,71,72] The overall description of the method is as follows: a cross-sectioned cut seed is mounted on circular specimen holders using an adhesive, then coated first with carbon and then with gold. The thickness of the total coat must be around 150 to 180 Å, then the samples are scanned under the electron microscope, examined, and photographed.

VII. FUNCTIONAL PROPERTIES AND STRUCTURE-FUNCTION RELATIONSHIP

A. FUNCTIONAL PROPERTIES

The structure, phase transitions, and interactions of starch with food constituents are topics stimulating considerable research interest, since they are related to the functionality of this plant polymeric carbohydrate. Recent advances in the physical chemistry and structure of starch point to the nonequilibrium (metastable) character of aqueous starch systems, and also to the need to study more extensively the phenomena of annealing, gelatinization, gelation, and retrogradation from the perspective of starch as a partially crystalline polymer plasticized by water.[19,73,74]

A considerable number of reports have also recognized the similarities between synthetic polymers and starch; among these similarities the following can be listed: the supermolecular organization (relative arrangement of macromolecules in the solid state), the effect of plasticizing solvents (i.e., water) on the kinetics of state transformations, and the stability of thermally processed starch materials.[74]

1. Gelatinization

a. Basic Aspects

The starch granule is organized in a ringed structure. Amylopectin molecules are perpendicular to these rings and directed toward the outer surface of the granule. That is, they are aligned along an imaginary axis extending from the hilum of the granule radially towards the edge of the granule. Portions of these long, branched molecules are in crystallites, which are also perpendicular to the rings. It is the regular orientation of the amorphous and crystalline regions that gives the granule its characteristic birefringent pattern, known as the maltese cross. The noncrystalline, or amorphous regions, contain the amylose molecules and sections of amylopectin molecules not involved in crystallites. Absorbed water hydrates the amorphous areas being able to move rather freely within the granule.[75]

When starch is placed in cold water, the granules swell slightly (10 to 20%), owing to diffusion and absorption of water into the amorphous regions, but this swelling is a reversible process upon drying. However, when granules are heated in water to progressively higher temperatures, a point is reached where granule swelling becomes irreversible and structural order disappears as the crystals melt,[76] and optical birefringence is lost. Phenomena are examinable via observation of losses in the X-ray diffraction pattern. As the granules continue to expand, amylose leaches out into the aqueous intergranular phase. These molecular changes lead to a substantial increase in the sample viscosity, which can be measured using the Brabender viscoamylograph (Figure 5).

When comparing maize and amaranth starches in a cycle of heat, followed by holding and cooling steps, the amaranth samples present a higher viscosity peak at higher gelatinization temperatures. This behavior is mostly attributable to the high levels of amylopectin in amaranth starch. Moreover, during the cooling step, amaranth normally presents a low set back — a typical characteristic of high amylopectin starches.[24]

b. Thermal Analysis

Overall, the disruption of granular structure, swelling and hydration, and solubilization of starch molecules are collectively described by the term gelatinization.[77,78] Because gelatinization is an endothermic process, thermal analysis methods have been widely used to assess some of the involved phenomena. In particular, DSC has attracted much interest during the last decade for studies of phase transitions in aqueous starch systems. DSC can provide the characteristic temperatures and transition enthalpies, being accomplished over a wide range of starch concentrations. A typical DSC thermal curve of starch is illustrated in Figure 6. The endothermic transitions E_1 and E_2 are thought to reflect disorganization (heat/moisture-mediated) of starch crystallites, whereas the high-temperature endotherm, E_3, has been attributed to dissociation of amylose-lipid complex. The latter transition is thus absent in defatted or waxy-type starch.[35,68,79] Endothermic transitions beginning at higher temperatures (Table 6)[23,53,63] are due to a high amylopectin content; this high content is also reflected in gelatinization enthalpies, whose values are higher compared with other starches. Starches from different sources have been reported to exhibit variations in the values obtained for the highest transition temperatures. This is most likely due to variations in the rate of heating in the calorimeter, or in the concentrations of water in the sample.

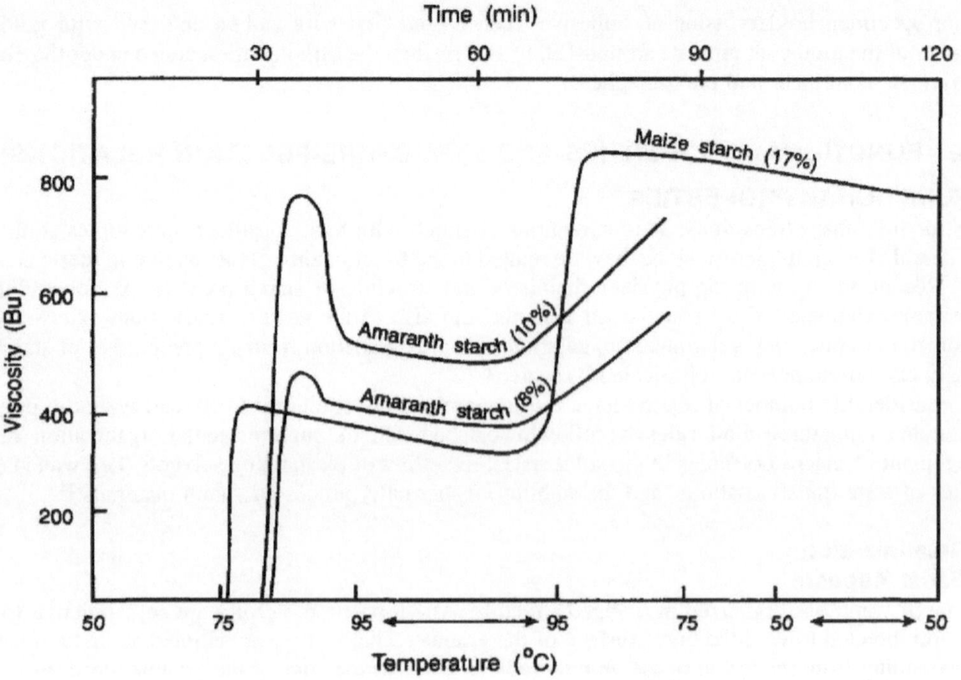

Figure 5 Viscoamylograms of amaranth and maize starches.

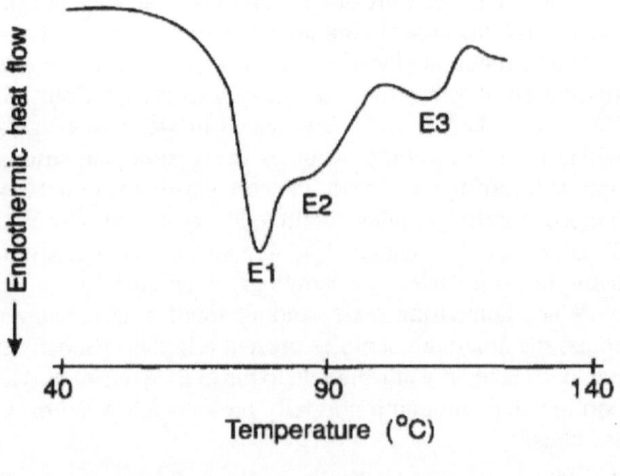

Figure 6 Typical differential scanning calorimetry (DSC) curve of starch.

Table 6 **Differential scanning calorimetry of amaranth starch**

	Endothermic Transitions[a] (°C)			ΔH	
Species	**To**	**Tp1**	**Tc**	**(cal/g)**	**Ref.**
A. hypochondriacus	67.7	72.6	78.1	3.2	63
A. hypochondriacus	66.0	75.0	82.0	2.6	53
A. hypochondriacus	68.0	73.0	81.0	3.4	53
A. caudatus	53.0	57.0	63.0	1.2	53
A. caudatus	51.0	56.0	64.0	1.6	53
A. hypochondriacus	62.0	64.0	68.0	1.7	23

[a] To = onset temperature; Tp1 = peak temperature; Tc = completion temperature.

Table 7 **Influence of water on gelatinization temperature of amaranth starch**

Moisture Content of Starch[a] (g/g total)	Transition Temperature[b] (°C) To	Tpl	Tc
0.85	61.3	64.8	73.7
0.77	63.4	67.0	75.0
0.69	63.0	67.7	77.7
0.61	62.0	66.0	84.0
0.53	64.2	67.8	88.9
0.48	64.0	67.5	89.5
0.06	106.5	111.7	—

[a] 1 mg starch was used in all experiments; [b] To = onset temperature; Tpl = peak temperature; Tc = completion temperature.

Amaranth starches do not show the E_3 transition. This is fundamentally due to the absence of an amylose-lipid complex resulting from the low amylose content of amaranthus species.

By analogy with other synthetic polymers, gelatinization is presently viewed as a melting process of starch crystallites. Therefore, starch-water mixtures containing more than 60% (w/w) water show a single symmetrical endotherm. While the size of this transition decreases progressively with decreasing water content, there is a concomitant development of the second transition, E_2. Moreover, there is a shift of E_2 to higher temperatures as the moisture content decreases. The influence of water on gelatinization temperatures for amaranth starch is shown in Table 7.[35] When the water level is down to 0.48 g water per gram total, a significant increase is observed in onset (T_o) and peak (T_{p1}) temperatures. Completion temperature (T_c) increased significantly as well. The intermediate transition temperature (T_{p2}) usually appears at 0.61 g water per gram total and below.[80] These observations have led to a thermodynamic treatment of melting data for starches using the Flory equation, which describes phase transition behavior of polymer-diluent mixtures; it is based on assumptions of equilibrium thermodynamics.[19]

Donovan[68] suggested that there are two distinct mechanisms by which ordered regions of starch undergo phase transitions. He postulated that the low temperature endotherm (E_1) reflects "stripping" and disorganization of polymer chains from the crystallites, processes that are facilitated by the swelling action of water on the amorphous regions. This mechanism predominates in excess of water. As the water content is reduced and becomes insufficient for the previous process to be completed, the remaining crystallites melt at a higher temperature (E_2) which is dependent on the volume fraction of diluent (water), as predicted by the Flory theory.

After so many years of viewing granular starch in terms of crystals or crystallite assemblies, the focus is now shifting to amorphous regions and their role in determining the phase transition behavior of this biopolymer.

Slade and Levine[73] have suggested that a glass $\rightarrow$ rubber transition precedes the melting endotherms of starch crystallites and that gelatinization is controlled by the mobility of the amorphous material surrounding the crystallites. As with partially crystalline synthetic polymers, crystallite melting cannot commence unless the characteristic glass transition temperature (T_g) of the glassy regions is exceeded. The technological importance of the glass transition in amorphous polymers and the characteristic temperature at which it occurs (T_g) is a keystone of synthetic polymer science. It was reported that the glass transition is perhaps the most important single parameter which one needs to know before deciding on the application of the many noncrystalline (synthetic) polymers that are now available. Especially in the last years, a growing number of food scientists have increasingly recognized the practical significance of the glass transition as a physicochemical event that can govern food processing, product properties, quality, safety, and stability.[73] In this respect, gelatinization has been described as a nonequilibrium process, since melting of crystallites (at their equilibrium T_m) is kinetically constrained by the immobile glass at temperatures below T_g. It has also been pointed out that the T_g of the glassy regions of starch granules, as is known for other hydrophilic or water-compatible glassy polymers, is highly sensitive to the water content.[19] The water acts as a plasticizer. In this field, the classical definition of a plasticizer indicates that it is a material incorporated into a polymer to increase the polymer's workability, flexibility, or extensibility. Plasticization, and its modulating effects upon the temperature location of the glass transition, is another key technological aspect of synthetic polymer science. It has become well-documented, in large part through DSC studies, that plasticization by water results in a depression of the T_g (and of the melt viscosity and elastic modulus) of completely amorphous and partially crystalline food ingredients, and that this T_g depression may be advantageous or disadvantageous to product processing, functional properties, and storage stability.[73]

c. *Influence of Solutes and Annealing*

The gelatinization of starch is also influenced by the presence of small molecular weight solutes (nonionic and electrolytes) as well as by hydrophilic hydrocolloids. Understanding these effects is an important step for a better process control and for improving the texture and other quality attributes of starch-based foods. Increasing concentrations of sugars and other polyhydroxy compounds are known to impede granule swelling, raise the temperature of gelatinization, and delay the increase in viscosity (pasting).[77]

This effect of solutes on gelatinization was studied using amaranth starch. The influence of sucrose on gelatinization temperature is shown in Table 8. It was found that gelatinization temperature increased when sucrose was added to the system; the three transition temperatures were increased. Several theories have been proposed to explain this effect, including competition between starch and sucrose for available water, sucrose inhibition of granular hydration, and sucrose-starch interactions. When solutes such as sucrose are added to the aqueous phase, less water is available to act as plasticizer and, consequently, the temperatures associated with gelatinization increase.[77,80] This effect of sugars on the gelatinization and pasting behavior of native and modified starches is also well known. It is important to the processing and properties of foods produced as baked goods, in that this effect influences the extent of wheat starch gelatinization, its retardation, or even inhibition during baking of, for example, high sugar cookie doughs and cake batters.[73] Another solute, sodium chloride, has a different and more complex effect on amaranth starch gelatinization (Table 9) then sucrose.[80] Changes in salt concentration increased all three transition temperatures T_o, T_{p1}, and T_c from 63.0, 67.7, and 75.9°C to 70.4, 74.6, and 87.2°C, respectively, when the salt concentration increased from 0 to 0.066 g sodium chloride per gram total. With subsequent increases in salt level up to 0.20 g per gram total, the 3 transition temperatures fell to about the initial values. Changes in salt level also affected the degree of gelatinization. This latter parameter fell from 76%, which was the control value, to a minimum of 31% at a salt concentration of 0.20 g per gram total. Sodium chloride appears to exhibit a maximum inhibitory effect on starch gelatinization at 6 to 9%.

It has also been suggested that starch acts as a weak acid ion exchanger and that cations tend to protect and to stabilize the granule structure while anions become the gelatinizing agents by rupturing hydrogen bonds. Paredes-López and Hernández-López[80] hypothesized that when sodium chloride is added to a starch suspension, some hydroxyl groups in the starch granule are converted to sodium hydroxylate groups. These compounds are more easily dissociated, thus causing a rise in the Donnan potential, which more effectively excludes the chloride ions from the granules. However, the absorption of sodium ions is rather limited, hence, the increase in gelatinization temperature. Chloride ions assist gelatinization by breaking hydrogen bonds between starch chains. The results of a study on amaranth starch suggested that the specific effects of electrolytes may be dependent on the particular starch and salt under study.[80] Another important factor affecting gelatinization is annealing. Annealing, defined as the heating of starch in excess water at subgelatinization temperatures, narrows the temperature range and raises both the peak temperature and enthalpy of gelatinization.

The effect of annealing was studied in amaranth starch.[80] It was found that transition temperatures T_o and T_{p1} (Table 10) progressively increased after increasing the incubation temperature of starch from 25 to 60°C . T_c did not change significantly; consequently, the treatment changed and narrowed the temperature range for gelatinization. Interestingly, the values for the three transition temperatures were statistically the same at the incubation temperatures of 60 and 65°C. As the temperature of annealing exceeded a certain limit, there was a reduction in the enthalpy values, presumably due to partial melting of the less thermostable crystallites. The X-ray diffraction patterns showed that the crystalline structure decreased as the holding temperature increased, and only the amorphous portion remained after holding the temperature at 75°C for 22 h.[80] It was found that annealing caused an increase in enthalpy and also narrowed the gelatinization temperature range for all maize starches. The extent of these changes was inversely proportional to the degree of crystallinity of the native starch, this being a function of amylopectin content. Since starch is a semicrystalline material consisting of amorphous and crystalline regions, holding a starch suspension just below the gelatinization temperature would be expected to give rise to more perfectly ordered crystals with higher melting points.[81]

Table 8 **Influence of sucrose on gelatinization temperature of amaranth starch**

Sucrose Content[a,b] (g/g total)	Transition Temperature[c] (°C)		
	To	Tpl	Tc
0 (0)	63.1	67.7	78.6
0.10 (15)	67.1	71.3	84.3
0.20 (30)	75.6	80.1	95.2
0.30 (45)	81.7	86.2	100.2

[a] All experiments were performed with 1 mg starch + 2 mg water; [b] Values in brackets represent sucrose concentration in the aqueous phase; [c] To = onset temperature; Tpl = peak temperature; Tc = completion temperature.

Table 9 **Influence of sodium chloride on gelatinization temparature of amaranth starch**

Sodium Chloride Content (g/g total)	Transition Temperature[b] (C)		
	To	Tpl	Tc
0	63.1	71.2	77.0
0.033	70.8	74.6	88.5
0.066	70.8	76.2	88.5
0.10	70.4	75.4	89.2
0.133	68.8	73.5	85.4
0.166	65.4	72.3	84.3
0.20	61.9	68.8	83.1

[a] All experiments were carried out with 1 mg starch + 2 mg water; [b] To = onset temperature; Tpl = peak temperature; Tc = completion temperature.

Table 10 **Influence of annealing temperature on gelatinization temperature of amaranth starch**

Annealing Temperature[a,b] (°C)	Transition Temperature[c] (°C)		
	To	Tpl	Tc
25	60.8	64.9	78.6
45	62.8	66.1	79.8
50	63.7	67.1	79.5
55	65.5	68.3	79.6
60	71.2	73.8	79.0
65	70.8	73.4	78.8

[a] All experiments were performed with 1 mg starch + 3 mg water; [b] Samples were treated at indicated temperatures for 12 h.c.; c. To = onset temperature; Tpl = peak temperature; Tc = completion temperature.

2. Gelation and Retrogradation

After gelatinization or "pasting" of starch, the amylose and amylopectin can be considered "dissolved". The entangled amylose molecules have a strong tendency to associate through the formation of hydrogen bonds with other adjacent amylose molecules when the solution is cooled, or upon standing for long periods of time.[4]

Swelling degree and granule disintegration, as well as exudation of amylose, depend on the type and concentration of starch, temperature, presence of other solutes, and the shear or agitation applied during heating.[17] Upon cooling, a dispersion of gelatinized starch is converted into a turbid viscoelastic paste or, at sufficiently high starch concentration (>6% w/w), into an opaque elastic gel.[82]

At concentrations above the critical level (6% w/w), a three-dimensional network is formed where the swelling granules (deformable particles) become embedded into a continuous matrix of entangled amylose molecules.[83] Such a complex polymer mixture sets as a viscoelastic gel in which the molecular association (principally involving bonding between chains) are physical rather than chemical. Assuming such a biphasic system, the rheological behavior would depend on the level and characteristics both of the dispersed phase (swelling granules enriched in amylopectin) and the continuous matrix (amylose network), as well as upon their interaction with water. Starch gelation is an important quality determinant in many processed foods, including baked and extruded soups.[17]

Starch gels are in metastable, nonequilibrium states and thus undergo structure transformation (further chain aggregation, recrystallization) during storage. Crystallites eventually begin to form, and this is accompanied by a gradual increase in the rigidity and phase separation between the polymer and solvent (syneresis). If the original starch solution is diluted and cooled slowly, the amylose molecules have sufficient time to align themselves in such a way that several hydrogen bonds can be formed among adjacent parallel chains.[4] This phenomenon is known as retrogradation and is manifested by the formation of precipitates or gels. The principal influence of retrogradation is upon the texture, acceptability, and digestibility of starch-containing foods.[17] Considerable amounts of experimental data are available (calorimetry, X-ray diffraction) to suggest that the main underlying cause of bread staling is retrogradation. However, despite several research efforts directed to the elucidation of this phenomenon, the exact mechanism of the retrogradation, particularly at a molecular level, is not clear. Retrogradation is a complex process and depends on many factors, such as type of starch, starch concentration, cooking and cooling regimes, pH, and the presence of solutes (e.g., lipids, salts, and sugars).[17]

It has been demonstrated that retrogradation consists of two separate processes: (1) gelation of amylose molecules exuded from the granules during gelatinization and (2) recrystallization of amylopectin.[81]

Amylose and amylopectin from gelatinized starch reassociate and increase the rigidity between and within the swollen granules. Amylose usually gels (solidifies) outside the starch granules immediately after gelatinization; amylopectin usually remains inside the swollen granules, where it slowly recrystallizes. It was reported that gelation of amylose is complete within hours, while amylopectin crystallization takes much longer.[83]

Amylose, with its lineal nature, is considered primarily responsible for starch gelation. It can form firm gels in concentrations as low as 1.5%.[4] In contrast with amylose, amylopectin gelation is a much slower process and requires much higher concentrations (uisually ≥15%).[17] In this respect, amylopectin gelation kinetics are different from those of amylose and are believed to reflect network formation via crystallization of the outer short chains of the molecule.[85] The slow gelation-crystallization rate of amylopectin agrees kinetically with the staling events of bread.[17]

An important question which needs to be addressed is whether retrogradation and crystallinization are synonymous or interchangeable, or whether the methods used to measure them give comparable results. Another significant unanswered question is whether starch retrogradation can be defined in terms of one component (amylopectin) only, or whether both amylose and amylopectin must be considered.[4]

3. Other Properties

Applications of starch are dictated by the functional properties starch imparts to the system where it is added. Among the most important properties are swelling power, solubility, and water absorption. Table 11 shows some functional properties of starches from different amaranth species compared with other sources.[1,29,30,71,86] As can be seen, an increase in temperature correlates with increased swelling power, a feature dictated by the gelatinization phenomenon. In general, it appears that amaranth starch presents higher values of swelling power than waxy maize.[54,71] Perhaps this could be a result of amylopectin chain lengths in these starches. However, some amaranth starches containing high amylose level (~12.0%)[71] present extremely high values; a clear example of this phenomenum is *A. paniculatus*.[30]

In the case of solubility, amaranth starch exhibited the highest values reflecting the shorter amylopectin chains present, but *A. hybridus*[71] was found to give the smallest solubility values, which contrasts with its high relative value for water absorption. Indeed, water absorption does not appear to follow any given pattern.

Table 11 **Functionality properties of starches (%)**

Species	Swelling Power (°C) 60	70	90	Solubility (°C) 60	70	90	Water Absorption	Ref.
A. cruentus	2.8	9.2	10.9	0.2	—	39.2	245.5	71
A. hypochondriacus	5.8	9.7	10.6	9.1	14.2	37.4	127.3	1
A. hybridus	2.7	7.7	9.0	1.1	2.2	3.6	227.1	71
A. paniculatus	12.9	15.1	35.1	25.5	45.4	55.5	—	30
A. hypochondriacus	—	2.9	3.3	—	80.6	80.8	—	29
Waxy maize	2.5	5.8	7.3	0.3	3.2	40.9	193.5	86
Wheat	4.8	6.1	11.8	1.7	2.3	8.2	71.8	86
Common beans	2.3	2.9	12.9	6.5	8.4	15.4	234.0	86

B. RELATION STRUCTURE-FUNCTION

Manners[49] reported that polymerization and branching degrees have substantial effects on the physicochemical properties of amylopectin molecules. Therefore, it is essential to specify the particular debranching enzymes used in every single amylopectin chains profile study.

Jane and Chen[36] studied the synergistic effect of amylose (potato, normal, and high-amylose maize VII) and amylopectin (high amylose maize V, waxy maize, and normal rice) on some functional properties like viscosity, gel strength, and light transmittance; however, they did not separate the amylopectin chains. Instead, they classified the materials as long, intermediate, and short branch chain amylopectins by running a gel-permeation chromatography of samples debranched by isoamylase. Later on, they looked at the same functional properties of reconstituted starches made of different combinations and proportions of those amyloses and amylopectins. They found significant differences among reconstituted starches made of amylopectins from high amylose maize, waxy maize, and normal rice, and of amylose from potato and maize (normal and high-amylose samples).[36] They concluded that the biggest synergistic effect on the viscosity was observed when long branch chain amylopectin, from maize starch containing the largest amylose proportion, was mixed with amylose of medium molecular size from normal maize.

Structural studies using amaranth starch were done by Konishi et al.[53] They debranched amaranth starch using isoamylase and then the product was fractionated by loading the sample on a Sephadex G-75 column; a biomodal distribution (Figure 7) was observed. In Figure 7a, the first fraction (I) was an amylose remnent, the second fraction eluted (II) consisted of long B- chains (average chain length or CL = 43), and the last one was a mixture of short B-chains and A-chains (CL = 15). Based on this, the authors concluded that the intermediate fraction observed in Figure 7b may be short chain length amylose and/or longer B- amylopectin chains. Therefore, a III/II ratio is considered to be a tentative determinant for the fine amylopectin structure. Sugimoto et al.[27] reported the study of debranched normal and waxy starches from amaranth (Figures 8a and 8c, respectively) and normal and waxy starches from maize (Figures 8b and 8d, respectively) with isoamylase. The fractions were eluted on Sephadex G-75 column; again, a bimodal distribution was observed. The second fraction (II) in all figures consisted of fragments with CL = 42 for normal and CL = 47 for waxy starches. The last fraction (III) presented a CL = 13 for all starch sources. Therefore, they concluded that both amylopectins showed a nearly identical pattern among all samples.

Other structural studies using amaranth amylopectin have been performed.[87] In this case, debranching was carried out with pullulanase; the mixture was eluted on Sepharose CL-2B on a 2 m long and 1.25 cm diameter column. The fractions obtained showed a bimodal distribution with long chains of DP = 40 (II) and short chains of DP = 15 (III). Among these two fractions, the major percentage was obtained for the short chain type (Figure 9).

VIII. AMARANTH STARCH USES

Starch and its derivatives are extensively used in both consumer and industrial levels. Many scientific and technical articles dealing with properties, manufacture, composition, economics, and uses of starch derivatives have been published, and some few among these publications have paid attention to amaranth starch. However, this particular starch has not been used at a commercial level in spite of its unique

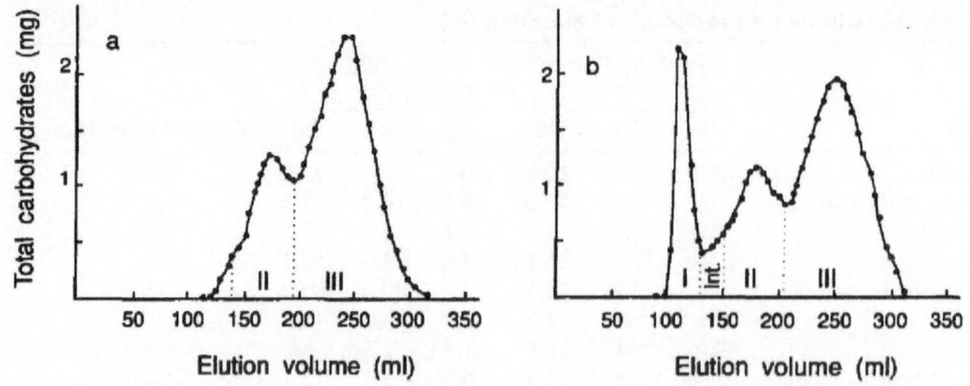

Figure 7 Sephadex G-75 column chromatography of isoamylase-debranched starch of *A. hypochondriacus.* (a) Waxy *A. hypochondriacus*; (b) nonwaxy *A. hypochondriacus*; fraction I = residual amylose; fraction II = amylopectin B chains; fraction III amylopectin A chains; and Int = intermediate fraction.

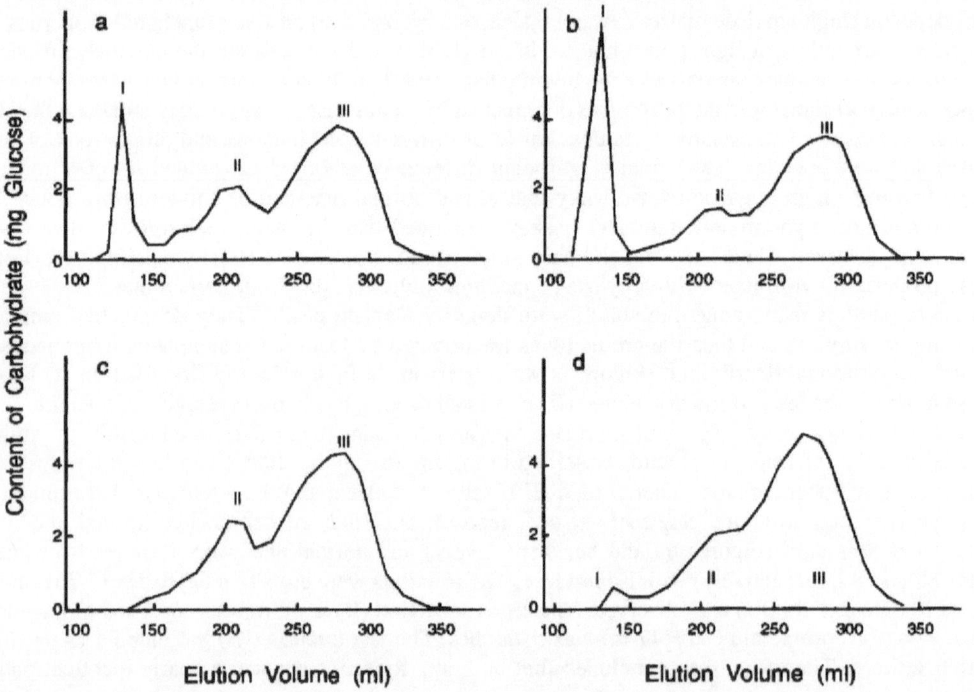

Figure 8 Elution profiles on Sephadex G-75 column of starches debranched with Pseudomonas isoamylase. (a) Normal *A. hypochondriacus*; (b) normal maize; (c) waxy *A. hypochondriacus*; (d) waxy maize.

properties. Starches are materials used worldwide as thickeners, pasting and gelling agents, stabilizers, etc. Among them, maize starch is by far the dominant grain source of carbohydrates as maize syrups, sweeteners, maltodextrins, and principally as a starch source. The functional properties of starch appear to be frequently related to the amylopectin molecular structure. Based on the physicochemical knowledge gathered up to now about amaranth starch, it may be essentially used as salad dressings[88] and pastes.[89] Singhal and Kulkarni[88,89] suggested that amaranth starch can be a potential source for the preparation of pastes and for the incorporation in food products under acidic and canned conditions.

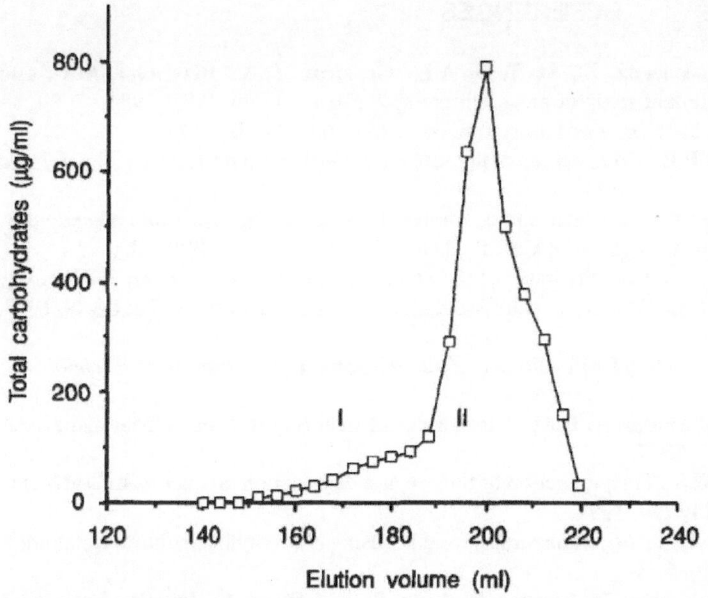

Figure 9 Profile of pullulanase-debranched amylopectin from amaranth. Fraction I = amylopectin B chains and fraction II = amylopectin A chains.

They looked at the viscosity and clarity of the paste, and its stability on a freeze-thaw cycle. The studies performed with *A. paniculatus* rajgeera type showed that this starch is an excellent thickening agent for food products, concluding that it may be used in frozen and canned products. However, the low clarity and poor stability of the starch under acidic conditions limits its applications under these particular conditions. These investigators also reported that amaranth starch resists mechanical shear better than maize starch; therefore, its use as a salad dressing opens a new market for this type of starch. These uses are basically attributed to its low amylose content and its extremely small granule size (1 to 2 µm). These types of starches do not gel, so they have a high viscosity under any temperature. The authors monitored the stability of amaranth starch emulsions for eight weeks, observing that the emulsions did not break as easily as maize starch did. On the other hand, studies on the applicability of *A. paniculatus* starch have shown that waxy starch is not suitable for custard preparation since waxy starch promotes interactions among other ingredients present in the food product.

IX. FUTURE POTENTIAL

Nowadays there is a lively interest in basic and applied starch research, and a primordial goal is the development of new and useful products. However, information on the chemical, physical, and biological properties of the polymer is necessary to understand the essential procedures required to fulfill such a goal.

The functional properties provided by amaranth starch in new products may be the following: thickeners, flavor enhancements, humectants, for prevention of sugar crystallization, sweeteners, and fat replacements. These are only a few of the potential uses of amaranth starch. Its potential resides in the naturally low amylose grain content and/or in the molecular arrangement of its amylopectin.

Since it is now generally recognized that the conformational structure of a molecule is responsible for its functional properties, it is therefore essential to try to understand the structure-function relationship of amaranth starch. To achieve this, more scientific research is necessary in order to elucidate the role of its basic components, amylose and amylopectin, and of other minor constituents.

ACKNOWLEDGMENTS

We gratefully acknowledge the financial support on amaranth starch studies from the Consejo Nacional de Ciencia y Tecnología and from COSNET-SEP.

REFERENCES

1. **Becker, R., Wheeler, E.L., Lorenz, K., Stafford, A.E., Grosjean, O.K., Betschart, A.A., and Saunders, R.M.,** A compositional study of amaranth grain, *J. Food Sci.,* 46, 1175, 1981.
2. **Lorenz, K. and Gross, M.,** Saccharides of amaranth, *Nutr. Rep. Int.,* 29, 721, 1984.
3. **Singhal, R.S. and Kulkarni, P.R.,** Composition of the seeds of some *Amaranthus* species, *J. Sci. Food Agric.,* 42, 325, 1988.
4. **Lineback, D.R. and Rasper, V.F.,** Wheat carbohydrates, in *Wheat: Chemistry and Technology,* Pomeranz, Y., Ed., American Association of Cereal Chemistry, St. Paul, MN, 1988, chap. 6.
5. **Boyer, C.D. and Shannon, J.C.,** Carbohydrates of the kernel, in *Corn: Chemistry and Technology,* Watson, S.A. and Ramstand, P.E., Eds., American Association of Cereal Chemists, St. Paul, MN, 1987, chap. 8.
6. **Becker, R. and Lorenz, K.,** Saccharides in wild rice (*Zizania aquatica*), *Lebensm. Wiss. Technol.,* 14, 134, 1981.
7. **Becker, R., Lorenz, K., and Saunders, R.M.,** Saccharides of maturing triticale, Wheat and rye, *J. Agric. Food Chem.,* 25, 1115, 1977.
8. **López, M.G. and Feather, M.S.,** The production of threose as a degradation product from L-ascorbic acid, *J. Carbohydr. Chem.,* 11, 799, 1992.
9. **Englmaier, P.,** Trifluoroacetylation of carbohydrates for g.l.c. using *N*-Methylbis(trifluoroacetamide), *Carbohydr. Res.,* 144, 177, 1985.
10. **Konig, W.A., Mischnick-Lubbecke, P., Brassat, B., Lutz, S., and Wenz, G.,** Improved gas chromatographic separation of enantiomeric carbohydrates derivatives using a new chiral stationary phase, *Carbohydr. Res.,* 183, 11, 1988.
11. **Carey, E.E., Dickinson, D.B., Wei, L.Y., and Rhodes, A.M.,** Occurrence of sorbitol in *zea mays, Phytochemistry,* 21, 1909, 1982.
12. **Helm, R.F., Conner, A.H., and Young, R.A.,** Quantitative determination of carbohydrates in solution via a reduction/permethylation technique, *J. Carbohydr. Chem.,* 6, 569, 1987.
13. **Decker, P. and Schweer, H.,** Gas-liquid chromatography on OV-225 of tetroses and aldopentoses as their O-Methoxime and O-*n*-Butoxime pertrifluoroacetyl derivatives and C_3-C_6 alditol pertrifluoroacetates, *J. Chromatogr.,* 236, 269, 1982.
14. **McGinnis, G.D. and Biermann, C.J.,** Analysis of monosaccharides as per-*O*-acetylated aldonitrile (PAAN) derivatives by gas-liquid chromatography, in *Analysis of Carbohydrates by GC and MS,* Biermann, C.J. and McGinnis, G.D., Eds., CRC Press, Boca Raton, FL, 1989, chap. 6.
15. **Ahmed, M.U., Dunn, J.A., Walla, M.D., Thorpe, S.R., and Baynes, J.W.,** Oxidative degradation of glucose adducts to protein, *J. Biol. Chem.,* 263, 8816, 1988.
16. **Saunders, R.M. and Becker, R.,** *Amaranthus*: a potential food and feed resource, in *Advances in Cereal Science and Technology,* Pomeranz, Y., Ed., American Association of Cereal Chemists, St. Paul, MN, 1983.
17. **Biliaderis, C.G.,** The structure and interaction of starch with food constituents, *Can. J. Physiol. Pharmacol.,* 69, 60, 1991.
18. **Guilbot, A. and Mercier, C.,** *The Polysaccharides, Starch,* Academic Press, New York, 1985, 209.
19. **Biliaderis, C.G.,** Non-equilibrium phase transitions of aqueous starch, in *Water Relationships in Food,* Levine, H. and Slade, L., Eds., Plenum Press, New York, 1991, 251.
20. **Luallen, T.E.,** Structure, characteristics, and uses of some typical carbohydrate food ingredients, *Cereal Foods World,* 33, 924, 1988.
21. **Lineback, D.R.,** The starch granule organization and properties, *Baker Dig.,* 13, 16, 1984.
22. **Singhal, R.S. and Kulkarni, P.R.,** Review: amaranth — an under-utilized resource, *Int. J. Food Sci. Technol.,* 23, 125, 1988.
23. **Teutonico, R.A. and Knorr, D.,** Amaranth: composition, properties, and applications of a rediscovered food crop, *Food Technol.,* 39 (4), 49, 1985.
24. **Yañez, G.A., Messinger, J.K., Walker, C.E., and Rupnow, J.H.,** *Amaranthus hypochondriacus*: starch isolation and partial characterization, *Cereal Chem.,* 63, 273, 1986.
25. **MacAllister, R.V.,** Nutritive sweeteners made from starch, in *Advances in Carbohydrate Chemistry and Biochemistry,* 36, Tipson, S.R. and Horton, D., Eds., Academic Press, New York, 1979, 15.
26. **Irving, D.W., Betschart, A.A., and Saunders, R.M.,** Morphological studies on *Amaranthus cruentus, J. Food Sci.,* 46, 1170, 1981.

27. **Sugimoto, Y., Yamada, K., Sakamoto, S., and Fuwa, H.**, Some properties of normal- and waxy-type starches of *Amaranthus hypochondriacus* L., *Starch/Stärke*, 33, 112, 1981.
28. **Tomita, Y., Sugimoto, Y., Sakamoto, S., and Fuwa, H.**, Some properties of starches of grain amaranths and several millets, *J. Nutr. Sci. Vitaminol.*, 27, 471, 1981.
29. **Yañez, G.A. and Walker, C.E.**, Effect of tempering parameters on extraction and ash of proso millet flours, and partial characterization of proso starch, *Cereal Chem.*, 63, 164, 1986.
30. **Wankhede, D.B., Gunjal, B.B., Sawate, R.A., Patil, H.B., Bhosale, M.B., Gahilod, A.T., and Walde, S.G.**, Studies on isolation and characterization of starch from rajgeera grains (*Amaranthus paniculatus* Lin.), *Starch/Stärke*, 41, 167, 1989.
31. **Tian, S.T., Rickard, T.E., and Blanshard, J.M.V.**, Physicochemical properties of sweet potato starch, *J. Sci. Food Agric.*, 57, 459, 1991.
32. **Lim, P.Y., Perata, P., Pozueta-Romero, J., Akazawa, T., and Yamaguchi, J.**, Distinct profiles of ADP- and UDP-specific sucrose synthases in developing rice grains, *Biosci. Biotech. Biochem.*, 56, 695, 1992.
33. **Olmedilla, A., Schrauwen, J.A.M., and Wullems, G.J.**, Visualization of starch-synthase expression by *in situ* hybridization during pollen development, *Planta*, 184, 182, 1991.
34. **Subba-Rao, P.V. and Goering, K.J.**, New starches. V. Properties of small starch granules from *Amaranthus retroflexus, Cereal Chem.*, 147, 655, 1979.
35. **Biliaderis, C.G., Page, C.M., Maurice, T.J., and Juliano, B.O.**, Thermal characterization of rice starches: a polymeric approach to phase transitions of granular starch, *J. Agric. Food Chem.*, 34, 6, 1986.
36. **Jane, J.L. and Chen, J.F.**, Effect of amylose molecular size and amylopectin branch chain length on paste properties of starch, *Cereal Chem.*, 69, 60, 1992.
37. **Hood, L.F.**, Current concepts of starch structure, in Food *Carbohydrates*, Lineback, D.R. and Inglett, G.E., Eds., AVI Publishing, Westport, CT, 1982, chap.13.
38. **Zobel, H.F.**, Molecules to granules: a comprehensive starch review, *Starch/Stärke*, 40, 44, 1988.
39. **Morrison, W.R.**, Lipids in cereal starches: a review, *J. Cereal Sci.*, 8, 1, 1988.
40. **Moore, C.O., Tuschoff, J.V., Hasting, C.W., and Schanefelt, R.V.**, Applications of starch in foods, in *Starch: Chemistry and Technology*, Whistler, R.L., Ed., Academic Press, Nerw York, 1984, chap. 19.
41. **Biliaderis, C.G., Giant, D.R., and Vose, J.R.**, Molecular weight distributions of legume starches by gel chromatography, *Cereal Chem.*, 56, 475, 1979.
42. **Thorn, W. and Mohazzeb, S.**, Molecular weights, lengths, and distribution of side-chains in α-D-polyglucanes, *Starch/Stärke*, 42, 373, 1990.
43. **Zobel, H.F.**, Starch crystal transformations and their industrial importance, *Starch/Stärke*, 40, 1, 1988.
44. **Robin, J.P., Mercier, C., Chabonniere, R., and Guilbot, A.**, Litnerized starches. Gel filtration and enzymatic studies of insoluble residue from prolonged acid treatment of potato starch, *Cereal Chem.*, 51, 389, 1974.
45. **MacGregor, A.W. and Morgan, J.E.**, Structure of amylopectins isolated from large and small starch granules of normal and waxy barley, *Cereal Chem.*, 61, 222, 1984.
46. **Hizukuri, S.**, Polynomial distribution of the chain lengths of amylopectins, and its significance, *Carbohydr. Res.*, 147, 342, 1986.
47. **Biliaderis, C.G., Grant, D.R., and Vose, J.R.**, Structural characterization of legume starches. I. Studies on amylose, amylopectin, and beta-limit dextrins, *Cereal Chem.*, 58, 496, 1981.
48. **Kobayashi, S., Schwartz, S.J., and Lineback, D.R.**, Comparison of the structures of amylopectins from different wheat varieties, *Cereal Chem.*, 63, 71, 1986.
49. **Manners, D.J.**, Recent developments in our understanding of amylopectin structure, *Carbohydr. Polymers*, 11, 87, 1989.
50. **Taki, M., Suzuki, K., Taki, A., Hisamatsu, M., and Yamada, T.**, Gel chromatography of residue and extract obtained from starch granules by hot aqueous 1-butanol solution treatment, *Starch/Stärke*, 40, 177, 1988.
51. **Hizukuri, S., Takeda, Y., and Maruta, N.**, Molecular structures of rice starch, *Carbohydr. Res.*, 189, 227, 1989.
52. **MacMasters, M.W., Baird, P.D., Holzapfel, M.M., and Rist, C.E.**, Preparation of starch from *Amaranthus cruentus* seed, *Econ. Bot.*, 9, 300, 1955.

53. **Konishi, Y., Nojima, H., Okuno, K., Asaoka, M., and Fuwa, H.,** Characterization of starch granules from waxy, non-waxy, and hybrid seeds of *Amaranthus hypochondriacus* L., *Agric. Biol. Chem.*, 49, 1965, 1985.
54. **Paredes-López, O., Schevenin, M.L., Hernández-López, D., and Cárabez, A.,** Amaranth starch — isolation and partial characterization, *Starch/Stärke*, 41, 205, 1989.
55. **Adkins, G.K. and Greenwood, C.T.,** The isolation of cereal starches in the laboratory, *Starch/Stärke*, 18, 213, 1966.
56. **Hernández-López, D.,** Aislamiento y caracterización fisicoquímica del almidón de amaranto — efectos hidrotérmicos y de solutos, M.Sci. thesis, CINVESTAV-IPN, Irapuato, México, 1992.
57. **Kintner, P.K.,** *Starch*, Rodale Research Center, Kutztown, PA.
58. **Banks, W. and Greenwood, C.T.,** *Starch and Its Components*, Edinburgh University Press, Edinburgh, 1975.
59. **Sargent, J.G.,** A simple method for the physicochemical characterization of amylose, *Starch/Stärke*, 34, 89, 1982.
60. **Dubois, M., Gilles, K.A., Hamilton, J.K., Rebers, P.A., and Smith, F.,** Colorimetric method for determination of sugars and related substances, *Anal. Chem.*, 28, 350, 1956.
61. **Hizukuri, S. and Maehara, Y.,** Distribution of the binding of A chains to a B chain in amylopectins, in *Biotechnology of Amylodextrin Oligosaccharides*, Friedman, R.B., Ed., ACS Symposium Series, Washington, D.C., 1991, chap. 15.
62. **Hizukuri, S.,** Relationship between the distribution of the chain length of amylopectin and the crystalline structure of starch granules, *Carbohydr. Res.*, 141, 295, 1985.
63. **Barba de la Rosa, A.P., Paredes-López, O., Cárabez-Trejo, A., and Ordorica-Falomir, C.,** Enzymatic hydrolysis of amaranth flour — differential scanning calorimetry and scanning electron microscopy studies, *Starch/Stärke*, 41, 424, 1989.
64. **Mora-Escobedo, R., Paredes-López, O., and Gutiérrez, G.F.,** Effect of germination on the rheological and functional properties of amaranth seeds, *Food Sci. Technol.* (*Lebensm. Wiss. U. Technol.*), 24, 287, 1991.
65. **Kainuma, K., Oda, T., and Suzuki, S.,** Observation on the changes of starch granules during gelatinization with the novel photopastegraphy method. Part I. Design and produce of photopastegraph, *Denpun Kogyo Gakukaishi*, 16, 52, 1968.
66. **Okuno, K. and Sakaguchi, S.,** Glutinous and non-glutinous starches in perisperm in grain amaranths, *Cereal Res. Commun.*, 9, 305, 1981.
67. **Zobel, H.F., Senti, F.R., and Brown, D.S.,** Studies of starch gelatinization by differential thermal analysis, *Cereal Sci. Today*, 10, 1654, 1965.
68. **Donovan, J.W.,** Phase transitions of the starch-water system, *Biopolymers*, 18, 263, 1979.
69. **Hizukuri, S.,** in *Starch Science Handbook*, Nakamura, M. and Suzuki, S., Eds., Asakura Shoten, Tokyo, 1978, 209.
70. **Lorenz, K.,** *Amaranthus hypochondriacus* — characteristics of the starch and baking potential of the flour, *Starch/Stärke*, 33, 149, 1981.
71. **Stone, L.A. and Lorenz, K.,** The starch of *Amaranthus* — physico-chemical properties and functional characteristics, *Starch/Stärke*, 36, 232, 1984.
72. **Paredes-López, O., Cárabez-Trejo, A., Pérez-Herrera, S., and González-Castañeda, J.,** Influence of germination on physico-chemical properties of amaranth flour and starch microscopic structure, *Starch/Stärke*, 40, 290, 1988.
73. **Slade, L. and Levine, H.,** Beyond water activity: recent advances based on an alternative approach to the assessment of food quality and safety, *Crit. Rev. Food Sci. Nutr.*, 115, 30, 1991.
74. **Biliaderis, C.G.,** Structures and phase transitions of starch in food systems, *Food Technol.*, 46(6), 98, 1992.
75. **Hoseney, R.C., Zeleznak, K.J., and Yost, D.A.,** A note on the gelatinization of starch, *Starch/Stärke*, 38, 407, 1986.
76. **Zobel, H.F., Young, S.N., and Rocca, L.A.,** Starch gelatinization: an X-ray diffraction study, *Cereal Chem.*, 65, 443, 1988.
77. **Lund, D.,** Influence of time, temperature, moisture, ingredients, and processing conditions on starch gelatinization, *Crit. Rev. Food Sci. Nutr.*, 20, 249, 1984.
78. **Lai, L.S. and Kokini, J.L.,** Physicochemical changes and rheological properties of starch during extrusion, *Biotechnol. Prog.*, 7, 251, 1991.

79. **Biliaderis, C.G. and Zawistowski, J.**, Viscoelastic behavior of aging starch gels: effects of concentration, temperature, and starch hydrolysates on network properties, *Cereal Chem.*, 67, 240, 1990.
80. **Paredes-López, O. and Hernández-López, D.**, Application of differential scanning calorimetry to amaranth starch gelatinization — influence of water, solutes and annealing, *Starch/Stärke,* 43, 57, 1991.
81. **Krueger, B.R., Knutson, C.A., Inglett, G.E., and Walker, C.E.**, A differential scanning calorimetry study on the effect of annealing on gelatinization behavior of corn starch, *J. Food Sci.*, 52, 715, 1987.
82. **Morris, V.J.**, Starch gelatinization and retrogradation, *Trends Food Sci. Technol.*, 2, 1990.
83. **Ring, S.G.**, Some studies on starch gelation, *Starch/Stärke,* 37, 80, 1985.
84. **Waniska, R.D. and Gomez, M.H.**, Dispersion behavior of starch, *Food Technol.*, 46, 112, 1992.
85. **Ring, S.G., Colonna, P., I'Anson, K.J., Kalichevsky, M.T., Miles, M.J., Morris, V.J., and Orford, P.D.**, The gelation and crystallization of amylopectin, *Carbohydr. Res.*, 162, 277, 1987.
86. **Paredes-López, O., Barba de la Rosa, A.P., Hernández-López, D., and Carabez-Trejo, A.**, Amaranto — Características alimentarias y aprovechamiento agroindustrial, Secretaría General de la Organización de los Estados Americanos, Programa Regional de Desarrollo Científico y Tecnólogico, Washington, D.C., 1090, 31.
87. **Bello-Pérez, L.A., López, M.G., and Paredes-López, O.**, data not published, 1993.
88. **Singhal, R.S. and Kulkarni, P.R.**, Utilization of *Amaranthus paniculatus* (rajgeera) starch in salad dressing, *Starch/Stärke,* 42, 52, 1990.
89. **Singhal, R.S. and Kulkarni, P.R.**, Some properties of *Amaranthus paniculatus* (rajgeera) starch pastes, *Starch/Stärke,* 42, 5, 1990.

Chapter 7

Amaranth Oil: Composition, Processing, and Nutritional Qualities

Robert Becker

CONTENTS

I. INTRODUCTION

Those promoting amaranth as a new cereal grain usually identify the high protein content of the grain as its most important asset. This may be true, but the lipid content of the grain is certainly significant and worthy of increased emphasis.

As will be documented in this chapter, amaranth oil has many attributes. Researchers are now beginning to identify agronomic and environmental factors important to controlling yield and influencing composition. Development of processing technology for isolation of grain components and preparation of new products is also progressing. Amaranth grain has a higher lipid content than most cereal grains. This is very important because the lipids are the main source of the high energy content of amaranth. The fatty acid composition of amaranth grain is very similar to rice bran oil and maize oil and has a nutritious ratio of saturated to unsaturated fatty acids. The oil has a lower digestibility than cottonseed oil, probably due to fairly large amounts of squalene, but studies have not shown any deleterious effects of the oil. Amaranth has been reviewed by a number of authors.[1-5]

II. EFFECTS OF FERTILIZER

Bressani et al.[6] grew four selections of three amaranth species in four replications and fertilized with 0, 30, 60, and 90 kg/ha of 12-24-12 fertilizer (Figure 1). The data showed inconsistent effects on yield and chemical composition (fat and protein content) and no effect on protein quality with fertilizer application. For two *A. cruentus* selections, increased fertilization resulted in a decrease in fat content. For *A. caudatus,* there was an increase, and for *A. hypochondriacus,* there was an increase then a decrease at the highest fertilizer level. In other experiments, fertilizers did not affect species differences.[7] When *A. cruentus* was grown with no nitrogen applied and with 150 lb nitrogen per acre, no significant differences in fat content were observed.[8]

III. MORPHOLOGY, MILLING, AND OIL EXTRACTION

Amaranth seed morphology is unlike common cereal grains. The seed is small, typically approximately 1 mm in diameter, and lenticular shaped. The embryo is campylotropous, i.e., circular with the ends nearly touching and enclosing the perisperm (Figures 2A and 2B). This type of seed structure also occurs in quinoa (*Chenopodium quinoa*) and sugarbeet (*Beta vulgaris*). Microscopically, the seed coat is a single layer of cells closely associated with the perisperm and embryo. The embryonic tissue of *A. cruentus* has physical and histochemical properties typical of matrices containing lipid and lipid complexes. The perisperm contains primarily starch.[9,10]

0-8493-5374-2/94/$0.00+$.50

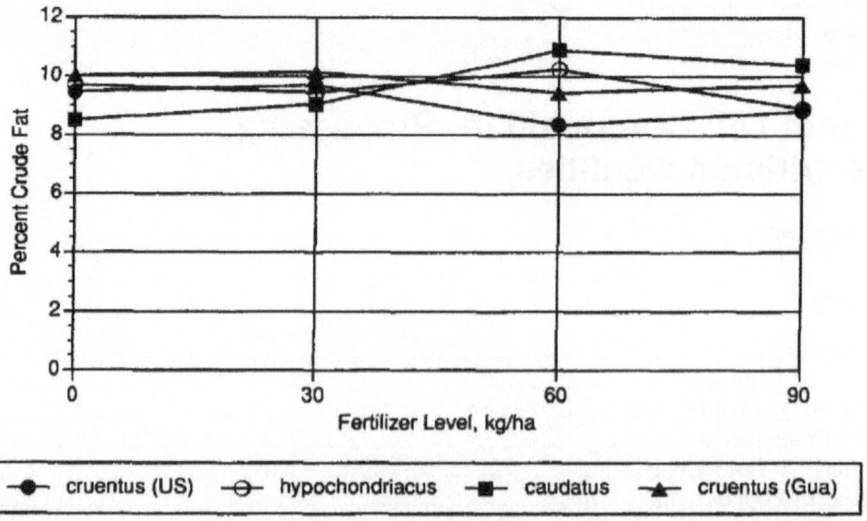

Figure 1 Effect of fertilizer applications on fat content of seeds from three amaranth species, including *A. cruentus* from the U.S. (US) and Guatemala (Gua). (From Bressani, R., Gonzalez, J.M., Elias, L.G., and Melgar, M., *Qual. Plant. Plant Foods Human Nutr.*, 37, 59, 1987. With permission.)

Because of these morphological characteristics, milling techniques could be devised to selectively remove the outer bran and embryo and to isolate perisperm tissue. Successive passes through a barley pearler were used to remove increasing amounts of the embryo and seed coat (bran); analysis demonstrated that this material contained most of the crude fat material in the seed (Figure 3).[11] The maximum amount of crude fat localized in the bran/embryo fraction was about 19%, compared to 7.7% in the starting material and about 2% remaining in the perisperm. The bran and embryo also contains most of the seed protein, as shown by Betschart et al.[11] using the barley pearler. Becker et al.[12] used a stone mill with 5-in diameter stones and obtained similar results.

To extract the oil from amaranth seed, the seed must be milled to a flour; the efficiency of the oil extraction is dependent upon the degree of milling of the seeds. Seed samples milled in a Wiley Mill, hammer mill, attrition mill, and abrasion mill gave somewhat different yields of oil when extracted overnight with a Soxhlet apparatus. The more finely milled flour yielded more oil than coarser flours.[13] Cold press extraction has not been studied.

The extracting solvent is also important. Extraction with hexane, ethyl ether, or petroleum ether can be expected to extract more total lipids than more polar solvents; bound lipids are typically extracted with water-saturated n-butanol.

IV. LIPID COMPOSITION

Typical total lipid contents of the common *Amaranthus* species used for grain production range from about 0.8 to 8%, most of which are higher than conventional cereal grains with the exception of oats, some millets, and soybeans. Weedy amaranths (*A. spinosus*) contains 17% total lipid and a garden ornamental (*A. arthropurpureus*) contains 9.75%.[14] Lorenz and Hwang[15] studied 4 *A. cruentus* samples that ranged from 6.84 to 7.80% total lipids, with free lipids comprising 90 to 93% of the total, and bound lipids the remainder. Triglycerides were the major nonpolar component of the free lipids[14,15] and showed little variations between varieties. Bound lipids have been identified as phospholipids (5%) and glycolipids (2.6%).[14] The low level of polar lipids may account for some of the poor baking quality of amaranth flour, although specific proteins are the primary components of good breadmaking.

The tocopherol and tocotrienol (tocol) contents of *A. cruentus* and *A. hypochondriacus* seed oil have been determined by Lehmann.[32] Lehmann found amaranth seed oil to have a relatively high ratio of tocotrienols to tocopherols, which may have beneficial health effects. When compared to vegetable oils, amaranth oil has a total tocol content similar to or somewhat higher than olive oil. For comparison, the reported[16] total tocol content of several seed oils in micrograms per gram are olive oil (191), sesame oil (528), sunflower (835), soybean oil (1233), safflower oil (609), and maize germ oil (1148). This low tocol

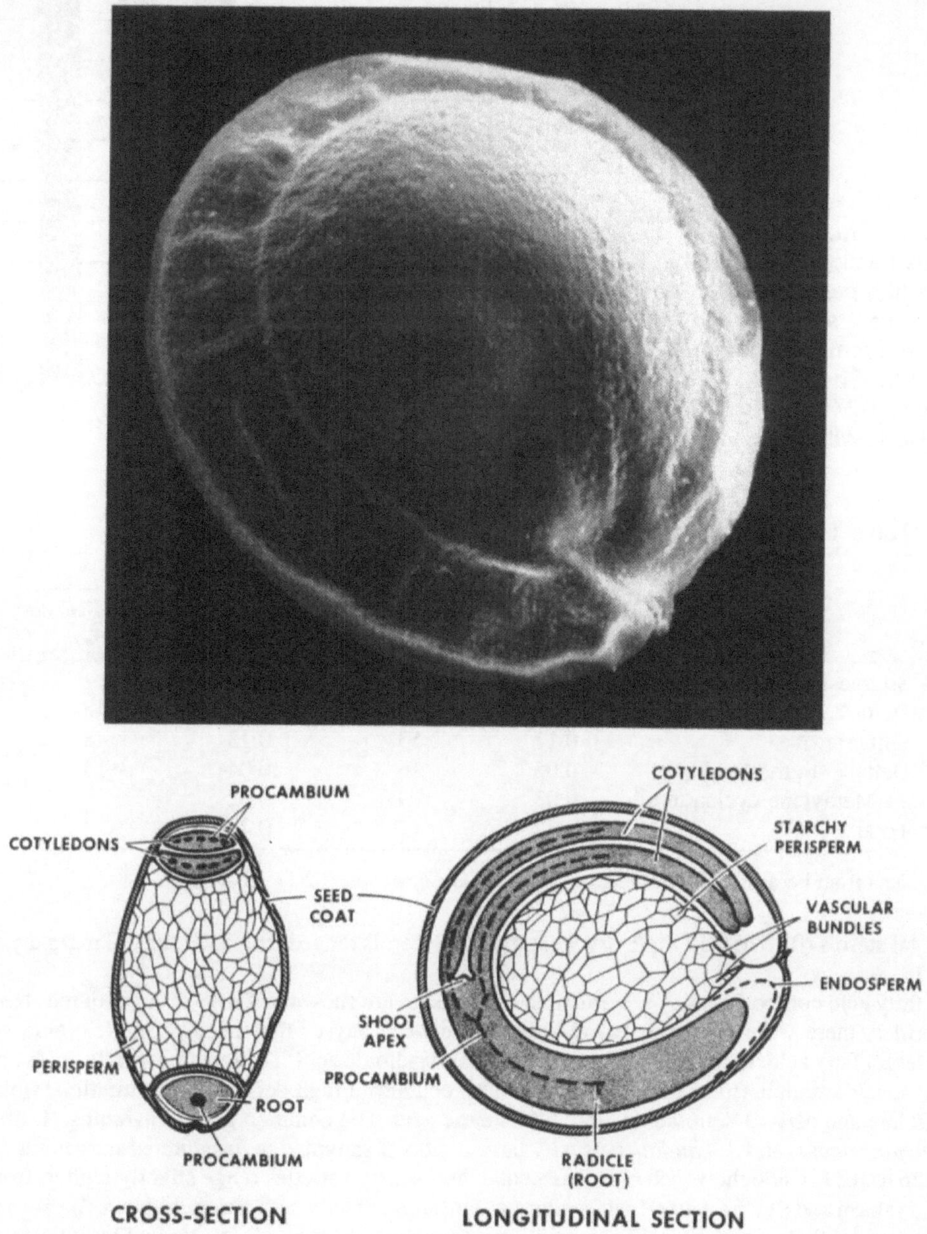

Figure 2 (A) Intact *A. cruentus seed* and (B) illustration of *A. cruentus* seed in cross and longitudinal sections. (From Irving, D.W., Betschart, A.A., and Saunders, R.M., *J. Food Sci.*, 46(4), 1170, 1981. With permission.)

content may explain the reduced oxidative stability observed in both the seed and the oil. Tocopherols possess biological activity as vitamin E compounds, with alpha-tocopherol having the highest activity. The tocotrienols have been reported to inhibit beta-hydroxy-beta-methylglutaryl-coenzyme A reductase, the rate limiting enzyme in cholesterol biosynthesis, and lower serum cholesterol in humans and are implicated in vitamin E recycling of the LDL-cholesterol molecule.[17] Danz and Lupton[18] have attributed the cholesterol lowering effect of amaranth to its dietary fiber content.

The sterol content of six amaranth varieties were determined by Fernando and Bean.[19] They found spinasterol to be the preponderant sterol in both weedy and vegetable amaranths, ranging from 48 to 53%

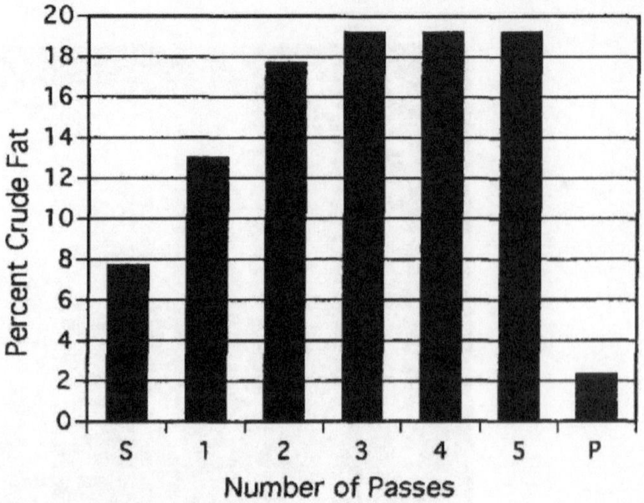

Figure 3 Crude fat content of *A. cruentus* fractions removed by first through fifth pass through a barley pearler: intact seed (S) and perisperm (P). (From Betschart, A.A., Irving, D.W., Shepherd, A.D., and Saunders, R.M., *J. Food Sci.*, 46, 1181, 1981. With permission.)

Table 1 **Sterol composition of amaranth seed**

	A. cruentus		*A. dubius*	
	mg/g	Total Sterols (%)	mg/g	Total Sterols (%)
Stigmasterol	0.03	9	0.03	11
Delta 7 ergosterol	0.04	12	0.05	18
Spinasterol	0.17	53	0.13	48
Delta 7 stigmasterol	0.05	16	0.04	15
24-Methylene cycloartenol	0.03	9	0.02	7
Total	0.32		0.27	

Data from Fernando, T. and Bean, G., *J. Am. Oil Chem. Soc.*, 62(1), 89, 1985.

of the total sterols (0.13 to 0.17 mg/g dry weight). Total sterols ranged from 0.27 to 0.32 mg/g dry weight (Table 1).

The fatty acid compositions of several amaranth species are shown in Table 2 . As with the sterols and triglycerides, there were no consistent differences between any of the species. Linoleic acid was the preponderant fatty acid, with lesser amounts of oleic and palmitic acid. The remaining fatty acids occurred in much lower amounts (trace to 3%). Amaranth oil contains a high degree of unsaturation, with about 70% linoleic and oleic, 1% linolenic, and 20% stearic acid. The common grain amaranths, *A. cruentus, A. hypochondriacus,* and *A. caudatus*, usually have a ratio of saturated to unsaturated fatty acids ranging from 0.26 to 0.31, while the vegetable, ornamental, and weedy varieties range slightly higher, from 0.33 to 0.36. Prakash and Pal[20] reported higher saturated/unsaturated fatty acid ratios, with one *A. cruentus* line having a ratio of 0.50. The fatty acid composition of amaranth oil most closely resembles wheat germ oil of most major cereal grain oils.

Ayorinde et al.[21] analyzed 18 varieties of amaranth seed cultivated in West Africa. Samples were ground to a powder in a mortar and pestle and Soxhlet-extracted 24 h with pentane. Fatty acid methyl esters were quantitated by gas chromatography (GC). They observed a mean percent oil content of 2.80 ± 0.80 and mean squalene content of 3.01 ± 0.81 in the oil, with wide interspecific and intraspecific variations; agronomic variations were inconclusive. Hand grinding of the seeds and pentane extraction may not give representative and reproducible results.

The unsaponifiables of amaranth oil are comprised primarily of squalene, the isoprenoid 2,6,10,15,19,23-hexamethyl-2,6,10,14,18,22-tetracosahexaene, which occurs as about 6% of the oil (Table 3). Total unsaponifiables in *A. cruentus* seed oil were 8.84%.[22] This is a considerably higher amount of squalene than is present in oils from other cereal grains. There have been no reports on methodology for isolation and purification of squalene from the oil, but technology from other commodities is probably transferable. Squalene may be determined by GC and high pressure liquid chromatography (HPLC).[13,23,24] Squalene

Table 2 **Crude fat and fatty acid content of amaranth seed oil**

Selection	Fat, %	C14:0	C16:0	C18:0	C18:1	C18:2	C18:3	C18:3+ C20:0	C20:0	C22:0	C24:0	Ref.
A. caudatus	11–12.5	Tr	17–20	1.9–2.6	25–29	47–51		0.9–3.2				6
A. caudatus (A-713)	12.5	Tr	17.2	1.9	25.2	51.6		3.2				6
A. caudatus (A-982)	12.4	Tr	20.1	2.6	28.7	46.8		1.0				6
A. caudatus (A-1113)	11.6	Tr	18.6	2.5	28.5	47.4		2.0				6
A. caudatus (11 lines)	1.7–5.2	tr–3.1	20.6–25.1	0.5–3.7	27.7–37.0	37.4–39.8	0.5–2.4					20
A. cruentus	9.2–13	Tr–1	13–22	2.7–4.1	20–35	38–62	1–1.1	1.0	0.7–1			6
A. cruentus (82S-1011)	12.9	Tr	16.8	2.8	35.5	43.1		1.0				6
A. cruentus (82S-434)	10.9	Tr	21.6	4.0	34.5	38.3						6
A. cruentus (82S-1034)	9.2	Tr	21.2	4.1	25.8	48.9						6
A. cruentus (78S-40)		1.0	20.0	4.0	22.0	50.0	1.0		1.0		1.0	19
A. cruentus		Tr	13.4	2.7	20.4	62.1	1.1		0.7			13
A. cruentus (4 lines)	0.8–3.9	Tr–0.2	24.3–31.1	0.8–2.3	24.4–36.2	36.1–47.4						20
A. dubius (78S-223)		1.0	25.0	3.0	21.0	47.0	1.0		1.0		1.0	19
A. dubius (2 lines)	3.6–4.2		22.9–24.7	0.1–0.3	23.4–24.6	51.8–52.2						20
A. hybridus	8.5	Tr–1	19–21	2.0–3	22–23	50–53	1.0	1.4	1.0		1.0	6
A. hybridus (82S-1004)	8.5	Tr	19.0	2.9	22.2	53.3		1.4				6
A. hybridus (81S-394)		1.0	21.0	2.0	23.0	50.0	1.0		1.0		1.0	19
A. hybridus (2 lines)	1.7–5.0		25.4–28.1	0.9–1.7	20.9–25.2	41.8–50.1						20
A. hypochondriacus	8.0–11	Tr	18–24	2.3–3.3	20–26	48–58		0.4–1.2				19
A. hypochondriacus (A-718)	10.5	Tr	17.7	2.3	20.3	57.9		0.4				19
A. hypochondriacus (A-720)	10.4	Tr	19.7	2.7	22.4	53.3		1.2				19
A. hypochondriacus (82S-1023)	10.6	Tr	21.3	3.1	26.4	48.3						19
A. hypochondriacus (82S-1024)	9.2	Tr	22.1	3.0	25.1	49.8						19
A. hypochondriacus (82S-674)	8.3	Tr	21.6	2.9	23.0	52.6						19
A. hypochondriacus (82S-SP130)	9.2	Tr	22.9	3.1	24.4	49.6						19
A. hypochondriacus (82S-1008)	7.7	Tr	23.8	3.3	22.9	48.8						19
A. hypochondriacus (12 lines)	2.5–6.8		20.0–27.3	0.3–2.2	22.1–28.8	44.5–55.0						20
A. retroflexus (78S-73)		1.0	21.0	3.0	20.0	51.0	1.0	1.0			1.0	6
A. tricolor		0.1–1	18–23	4.0	21–25	48–51	1.0	1.0	0.7		0.5–1	6
A. tricolor (78S-40)		1.0	23.0	4.0	21.0	48.0	1.0	1.0			1.0	6
A. tricolor (78S-113)		1.0	18.0	4.0	25.0	51.0	1.0	1.0			1.0	6
A. tricolor (unspecified)		0.1	20.0	4.0	24.9	48.8	1.0		0.7		0.5	6
A. tricolor (4 lines)	2.8–5.1	0.3	10.9–24.2	0.1–3.1	22.8–38.3	45.6–51.6						20

Values are percent of oil.

Table 3 **Squalene content of amaranth oil and some commercially important oils**

Fat or Oil	% Oil[a]	% Squalene in Oil	Ref.
A. cruentus	7.0	7.0	13
A. cruentus		4.9	25
Olive	36	0.4	28
Rice Bran	16	0.3	29
Corn	4	0.03	28
Peanut	47	0.03	28
Rapeseed		0.03	28
Sunflower	47	0.01	28
Cottonseed	7	0.01	28
Coconut	35	0.002	28
Dalatias licha (shark spp. liver) 70% squalene in liver oil			30
Centrophorus uyata (Shark spp liver) 90% squalene in liver oil			30

[a] Data from Reference 31.

is used in cosmetics and as a penetrant and lubricant and is commercially obtained from shark and whale liver. Grapefruit in cold storage has been shown to develop a 13-fold increase in squalene in the epicuticular wax, suggesting that squalene has a role in preventing cold injury in the fruit.[24] Squalene, a precursor of cholesterol, may also exhibit feedback inhibition of cholesterol biosynthesis. The high price of amaranth grain probably precludes its use as a single source of squalene, but it may be feasible if the oil is obtained as a by product from the preparation of high value defatted amaranth protein or starch.

V. REFINING AND BLEACHING

Optimum conditions for refining and bleaching the oil have been determined by Lyon and Becker.[13] Finely ground flour was extracted with hexane in a Butt or Soxhlet type extractor for 22 h. The hexane was distilled from the extract and finished under vacuum. The extract was degummed and dewaxed by stirring with 1% water for 15 min at 60°C, allowed to stand overnight at room temperature, and centrifuged at 1250 × *g*. This oil was dewaxed by cooling to 4 to 6°C for 4 to 10 d under nitrogen, then centrifuged at 19,000 × *g*, 4°C.

For alkali refining, 12° Be' (8%) to 20° Be' (14.4%) solutions of NaOH were added to neutrality plus 0.1 to 0.6% excess. After heating to 65°C for 30 min, the solution was cooled to 22°C for 1 to 4 h and centrifuged. Refined oil was bleached by heating under nitrogen with various acid-activated bleaching clays.

The oil can be evaluated using a Lovibond Tintometer or with a spectrophotometer between 350 and 750 nm. The spectrum of refined amaranth oil is shown in Figure 4. Standard methods exist for the measurement of free fatty acids, peroxide values, fatty acid composition, etc. Squalene survives typical refining procedures.[13] Squalene may be determined by HPLC and GC.[13,24]

Some characteristics of amaranth oil are given in Table 4. The amaranth oil described had not been refined, so it should not be expected to correspond closely with refined corn oil. However, the most noticeable difference between amaranth oil and corn oil is the larger amounts of unsaponifiables in amaranth oil (8.84 vs. <2); much of this is presumably due to squalene. Refined amaranth oil contained free fatty acids of about 0.03% and a peroxide value of less than 1.0 meq/kg.[13]

VI. EFFECTS OF POPPING

Amaranth grain will pop when heated, much like popcorn although considerably smaller. Popped amaranth has a nutty, popcorn-like flavor and may be eaten as a snack, as a cold cereal, or formulated with honey or syrup and used as a candy.

Popping causes a darkening of the oil extract. There is an increase in the free fatty acid content from 4.1% in the intact seed to 11.5% in the popped seed oil, accompanied by a 2-fold increase in peroxide value from 5.1 meq/kg to 10.2 meq/kg.[25] An increase in specific gravity and viscosity of the oil from popped seeds indicated that some polymerization had occurred. The reported apparent increase in squalene in puffed seeds may be artifactual.

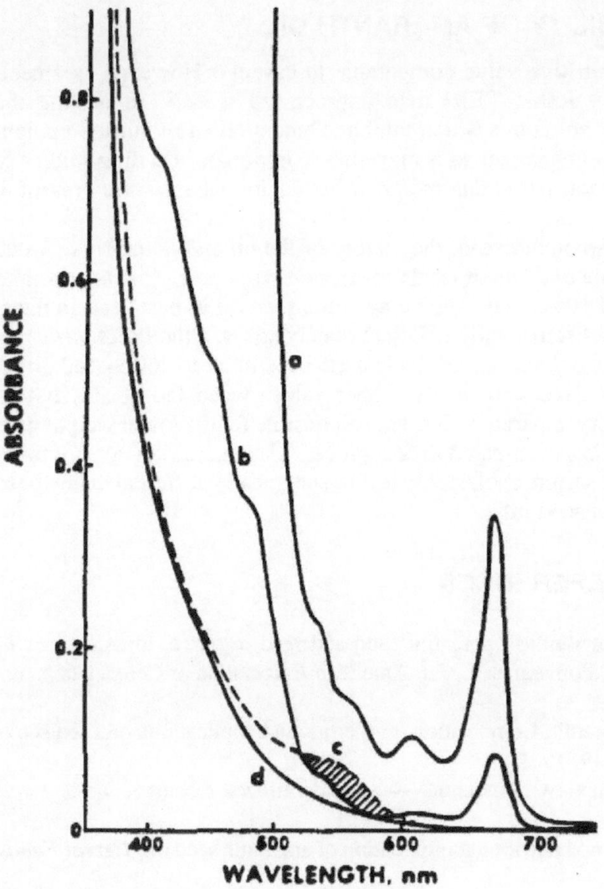

Figure 4 Spectra of amaranth seed oil; (a) degummed crude oil; (b) refined; (c) bleached with 2% Filtrol; (d) bleached, then decolorized with 0.2% Norit-A. (From Lyon, C.K. and Becker, R., *J. Am. Oil Chem. Soc.*, 64(2), 233, 1987. With permission.)

Table 4 Characteristics of amaranth oil

	Oil from Untreated Amaranth Seeds[a]	Oil from Popped Amaranth Seeds[a] Range	Corn Oil[b] AOCS Standard
Iodine value	100.0 87–113[c]	96.8	103–128
Saponification value	217 139–158[c]	237	187–193
Mean molecular weight of fatty acids	246	324	
Refractive index	1.470	1.470	1.470–1.474
Color value	3R + 2.8Y	3R + 2.1Y	
Viscosity (h)	40.1	42.3	
TBA value	1.22	1.59	
Specific gravity	0.9155	0.9186	0.915–0.920
Acid value, mg KOH/g	6.2 15–25[c]		
% Free fatty acids as oleic acid	4.1	11.5	
Peroxide value	5.1	10.2	
Unsaponifiable residue, %	8.8[d]	not over 2.0	

[a] Data from Reference 25; [b] data from Reference 28; [c] data from Reference 27; [d] data from Reference 22

VII. DIGESTIBILITY OF AMARANTH OIL

Raw amaranth grain is thought to have a nutritive value comparable to casein.[26] However, wet heat-processing gave a higher Protein Efficiency Ratio (PER) than unprocessed seeds,[27] suggesting the existence of unknown thermolabile factors or structures which inhibited amaranth's full nutritive potential; bioavailability of nutrients and palatability characteristics may also be important. Feeding studies of defatted and nondefatted amaranth flours demonstrated that the growth reducing factors were present in the raw, fat-free flours.

To fully assess the nutritive potential of amaranth seed, the quality of the oil and its nutritive value and potential toxicity was evaluated.[27] Hexane extracts of crude oil from *A. caudatus, A. cruentus,* and *A. hypochondriacus* were fed to rats at 5 and 10% levels, replacing cottonseed oil in basal casein diets. At both levels, PER and weight gains were not statistically different than controls, although *A. cruentus* had lower digestibility. *A. caudatus* and *A. hypochondriacus* oils had effects similar to cottonseed oil on rat serum cholesterol, while *A. cruentus* oil gave statistically higher values when fed at 5%. It was concluded that, except for the lower digestibility, amaranth oil is not responsible for the lower weight gain and lower protein quality of raw amaranth grain as compared to heat processed amaranth. In none of these cases did feeding amaranth oil result in rat serum cholesterol levels statistically different than those observed in rats fed similar amounts of cottonseed oil.

REFERENCES

1. **Saunders, R. M. and Becker, R.,** Amaranth: a potential food and feed resource, in: *Advances in Cereal Science and Technology,* Vol. 6, Pomeranz, Y., Ed. American Association of Cereal Chemists, St. Paul, MN, 1984.
2. **Teutonico, R. A. and Knorr, D.,** Amaranth: Composition, properties and applications of a rediscovered food crop, *Food Technol.,* 39(4), 49, 1985.
3. **Singhal R. S. and Kulkarni, P. R.,** Review: amaranths — an underutilized resource, *Int. J. Food Technol.,* 23, 125, 1988.
4. **Becker, R.,** Preparation, compositional and nutritional implication of amaranth seed oil, *Cereal Foods World,* 34, 950, 1989.
5. **Breene, W. M.,** Food uses of grain amaranth, *Cereal Foods World,* 36(5) 426, 1991.
6. **Bressani, R., Gonzalez, J. M., Elias, L. G., and Melgar, M.,** Effect of fertilizer application on the yield, protein and fat content, and protein quality of raw and cooked grain of three amaranth species, *Qual. Plant. Plant Foods Human Nutr.,* 37(1), 59, 1987.
7. **Carlsson, R.,** Quantity and quality of *Amaranthus* grain from plants in temperate, cold and hot, and subtropical climates — a review, in *Proc. 2nd Amaranth Conf.,* Rodale Press, Emmaus, PA, 1979, 48.
8. **Becker, R., Wheeler, E. L., Lorenz, K., Stafford, A. E., Grosjean, O. K., Betschart, A. A., and Saunders, R. M.,** A compositional study of amaranth grain, *J. Food. Sci.,* 46(4), 1175, 1981.
9. **Irving, D. W., Betschart, A. A., and Saunders, R. M.,** Morphological studies on *Amaranthus cruentus, J. Food Sci.,* 46(4), 1170, 1981.
10. **Irving, D. W. and Becker, R.,** Seed structure and composition of potential new crops. *Food Microstruct.,* 4 (1), 43, 1985.
11. **Betschart, A. A., Irving, A. A., Shepherd, A. D., and Saunders, R. M.,** *Amaranthus cruentus:* Milling characteristics, distribution of nutrients within seed components, and the effects of temperature on nutritional quality, *J. Food Sci.,* 46(1), 1181, 1981.
12. **Becker, R., Irving, D. W., and Saunders, R. M.,** Production of debranned amaranth flour by stone milling, *Lebensm. Wiss. Technol.,* 19, 372, 1986.
13. **Lyon, C. K. and Becker, R.,** Extraction and refining of oil from amaranth seed, *J. Am. Oil Chem. Soc.,* 64(2), 233, 1987.
14. **Opute, F. I.,** Seed lipids of the grain amaranths, *J. Exp. Bot.,* 30(116), 601, 1979.
15. **Lorenz, K. and Hwang, Y. S.,** Lipids in amaranths, *Nutr. Rep. Int.,* 31(1), 83, 1985.
16. **Speek, A. J., Schrijver, J., and Schreurs, W. H. P.,** Vitamin E composition of some seed oils as determined by high-performance liquid chromatography with fluorometric detection, *J. Food Sci.,* 50, 121, 1985.

17. **Qureshi, A. A., Qureshi, N., Wright, J. J. K., Shen, Z., Kramer, G., Gapor, A., Chong, Y. H., Dewitt, G., Ong, A. S. H., Peterson, D. M., and Bradlow, B. A.,** Lowering of serum cholesterol in hypercholesterolemic humans by tocotrienols (palmvitee), *Am. J. Clin. Nutr.*, 53, 1021S, 1991.
18. **Danz, R. A. and Lupton, J. R.,** Physiological effects of dietary amaranth (*Amaranthus cruentus)* on rats, *Cereal Foods World,* 37(7), 489, 1992.
19. **Fernando, T. and Bean, G.,** A comparison of the fatty acids and sterols of seeds of weedy and vegetable species of *Amaranthus* spp., *J. Am. Oil Chem. Soc.*, 62(1), 89, 1985.
20. **Prakash, D. and Pal, M.,** Seed protein, fat and fatty acid profile of *Amaranthus* species, *J. Sci Food Agric.*, 58, 145, 1992.
21. **Ayorinde, F. O., Ologunde, M. O., Nana, E. Y., Bernand, B. N., Afolabi, O. A., Oke, O. L., and Shepard, R. L.,** Determination of fatty acid composition of *Amaranthus* species, *J. Am. Oil Chem. Soc.*, 66(12), 1812, 1989.
22. **Schwartz, D. P.,** Improved method for quantitating and obtaining the unsaponifiable matter of fats and oils, *J. Am. Oil Chem. Soc.*, 65(2) 246, 1988.
23. **Sulpice, J. C. and Ferexou, J.,** Squalene isolation by HPLC and quantitative comparison by HP and GLC, *Lipids,* 19(8), 631, 1984.
24. **Nordby, H. E. and McDonald, R. E.,** Squalene in grapefruit wax as a possible natural protectant against chilling injury, *Lipids,* 25(12), 807, 1990.
25. **Singhal, R. S. and Kulkarni, P. R.,** Effect of puffing on oil characteristics of *Amaranthus* (Rajgeera) seeds, *J. Am. Oil Chem. Soc.*, 67(12), 952, 1990.
26. **Sánchez-Marroquin, A., Maya, S., and Pérez, P. L.,** Agroindustrial potential of amaranth in Mexico, in *Proc. 2nd Amaranth Conf.* Rodale Press, Emmaus, PA, 1980, 95.
27. **Garcia, L. A., Alfaro, M. A., and Bressani, R.,** Digestibility and nutritional value of crude oil from three amaranth species, *J. Am. Oil Chem. Soc.*, 63(3), 371, 1987.
28. **Swern, D.,** Structure and composition of fats and oils, in *Bailey's Industrial Oil and Fat Products,* Swern, Daniel, Ed., John Wiley & Sons, New York, 1964, 36.
29. **Fitelson, J.,** The occurrence of squalene in natural fats, *J. Am. Oil Chem. Soc.*, 26, 506, 1943.
30. **Heller, J. H., Heller, M. S., Springer, S., and Clark, E.,** Squalene content of various shark livers, *Nature,* 179, 919, 1957.
31. **Watt, B. K. and Merrill, A. L.,** Composition of Foods, Agicultural Handbook No. 8, Agricultural Research Service, U.S. Department Agriculture Washington, D.C. 1963.
32. **Lehmann, J. W.,** personal communication, 1993.

Chapter 8

Popping Amaranth Grain, State of the Art

L. R. Tovar, M. A. Valdivia, and E. Brito

CONTENTS

I. INTRODUCTION

The most conspicuous way of consuming grain amaranth in Mexico is by popping the seed in a hot plate, first cleaning the seed of soil and plant debris that has remained from harvest. Even wet treatment of the seeds with mild soda ash solutions ("tequesquite water") follows the cleaning process by small processors in the Central Highlands of Mexico; this water addition seems to improve the popping capability of the seed. After popping the seeds, honey or saccharose solutions are added; the seeds are agglomerated and form what is called "alegrías" candies which means "many joys" in Spanish; these candies resemble small cakes and are still sold in several cities of Central Mexico. These operations, from harvest to the manufacture of the "alegrías", take considerable labor as well as time, factors that increase the price of the popped amaranth to the consumer. This way of consuming amaranth grain goes back to pre-Columbian times;[1] thousands of years ago, people managed to identify those plants that contained different chemicals like curare, quinine, and the like; therefore it would be reasonable to believe that among the dwellers of the old Tenochtitlan, the Aztec capital, someone discovered that heat exposure to the amaranth seed makes it pop into small white kernels similar to popcorn. In northern India, amaranth is popped and consumed just the same as in Central Mexico; in the former country, *A. hypochondriacus* is planted at 8000 ft above sea level and in the plains.[2,3] Organoleptic changes occur in the grain after popping, sort of a nutty and certainly pleasant flavor. On the other hand, the size of the popped seed is about four to seven times the size of the raw seed; apparent density of the popped grain is decreased, and if the seed is overheated during the popping operation, it is roasted and nutrient losses can be detected.[4]

In Mexico there are two main locations where amaranth grain is industrially popped, Tulyehualco, D.F. and the town of San Miguel del Milagro, Tlaxcala. In these villages, besides using the traditional hot plate, there are two approaches to the popping operation: (1) a fluid bed system in which the seeds are cleaned and then popped and (2) a vibrating hot plate. In the U.S., workers at the Rodale Research Center have been very active disseminating information on amaranth as well as doing some practical developments in the popping of the seed, i.e., a bench scale amaranth popper in 1980 and air-drying of just post-harvested grain.[5]

0-8493-5374-2/94/$0.00+$.50

Most of the amaranth processing and consumption in Mexico and India are related to popping the grain; however, to the authors' knowledge, there is no thorough report available describing the popping procedure for amaranth grain and the equipment required. Betschart et al.[6] popped amaranth grain in a stream of hot air at 220°C for 10 to 15 s; no details were given of the equipment. Walker et al.[7] used a pilot toaster manufactured by Surface Combustion (Toledo, OH) to pop milo, barley, red and white wheats. These workers developed a prototype machine to pop grains with a capacity of 1500 to 2000 lb per hour. Brito and Tovar[8] developed and built a laboratory scale fluid bed amaranth popper that could also dry and clean the seed.

When popping amaranth seeds, some partially pop, others do not pop, and a portion of the latter are even roasted or burned; thus, a criterion of quality of a given batch of amaranth grain is its poppability, in other words, the percentage of popping of a certain amount of seeds. Moisture content of the seed, variety, and the popping procedure followed seem to be factors in the poppability of a given batch of seeds. Sánchez Marroquin[9] used indistinctly the terms "popping" and "toasting". Bressani et al.[10] produced toasted amaranth flour by heat treatment of the raw amaranth seed at 150°C for 60 to 90 s. Tovar and Carpenter[11] "roasted to the point of popping" amaranth seeds at 170° for 90 s. In this paper, amaranth popping means dry-heat expansion of the seed to at least a fourfold volume of the raw grain and no signs whatsoever of burnt areas in the popped seed; these latter were observed under the light microscope.

The purpose of this paper is to review different approaches of popping amaranth at an industrial scale, as well as to shed light on the causes of amaranth popping. Certainly, it is important to put in perspective what different authors meant by popping, toasting, and roasting amaranth seeds and changes in the nutritional content of the seeds due to these different processing conditions. In this line of thought, and since grain amaranth has a high content of lysine and tryptophan, it is relevant to figure out how a corn-based diet, like the one consumed in the rural areas of Mexico and other countries of Central America, could be supplemented with the dry heat treated amaranth seeds.

II. MORPHOLOGY OF THE GRAIN BEFORE AND AFTER DRY HEAT TREATMENT

Irving et al.[12] described in detail the morphology of *A. cruentus*; the seed is lenticular in shape and in Tulyehualco — Federal District Area, in Central Mexico, the color of commercial varieties goes from off-white to light brown, brown, and black. The embryo is nearly circular in shape surrounding the starchy perisperm, and when popping the seed, the embryo is the most sensitive constituent of the grain to dry heat; microscopic examination of partially popped seeds shows that the germ of several of them is charred. Perhaps this is due to the fact that the matrix of the embryo is made of lipid complexes and protein bodies.[13] The perisperm is a storage tissue in which starch appears as amylopectin and protein occurs as very small deposits between starch granules. The seed coat is closely associated with the perisperm, although there is a close attachment of the seed coat to the endosperm, which may serve as a protective structure to the embryo.

A. SCANNING ELECTRON MICROSCOPY OF THE RAW, POPPED, AND ROASTED GRAIN

A. cruentus seeds from Tulyehualco-Federal District Area were popped in a laboratory scale fluid bed system.[8] On the other hand, seeds from this batch were roasted in the process of popping in a hot plate, then the popped, the roasted, and the raw seed (the latter from the same batch) were mounted on aluminum stubs and gold coated. Specimens were observed and photographed in a Zeiss DSM 950 scanning electron microscope. Raw seeds are shown in Figure 1. In contrast, in the popped seed the perisperm expands with a "soap-bubble" structure; the color is bright white, resembling popcorn, and the seed coat appears loose with much less attachment to the perisperm. Furthermore, some areas of the seed coat are broken since the lenticular shape of the seed has been lost; the color of the remaining seed coat is light brown. Unsymmetrical shapes due to the sudden expansion can be observed in a light microscope; however, the structure of the embryo in most of the popped seeds is kept (Figure 2). The popped seed expands to about four to eight times its original volume.[14] The roasted amaranth is shown in Figure 3; browning of the seed coat as well as of the perisperm due to severe heat treatment occurs, the lenticular shape has vanished, the embryo is unrecognizable, and the volume of the roasted seed is slightly increased compared to the raw seed.

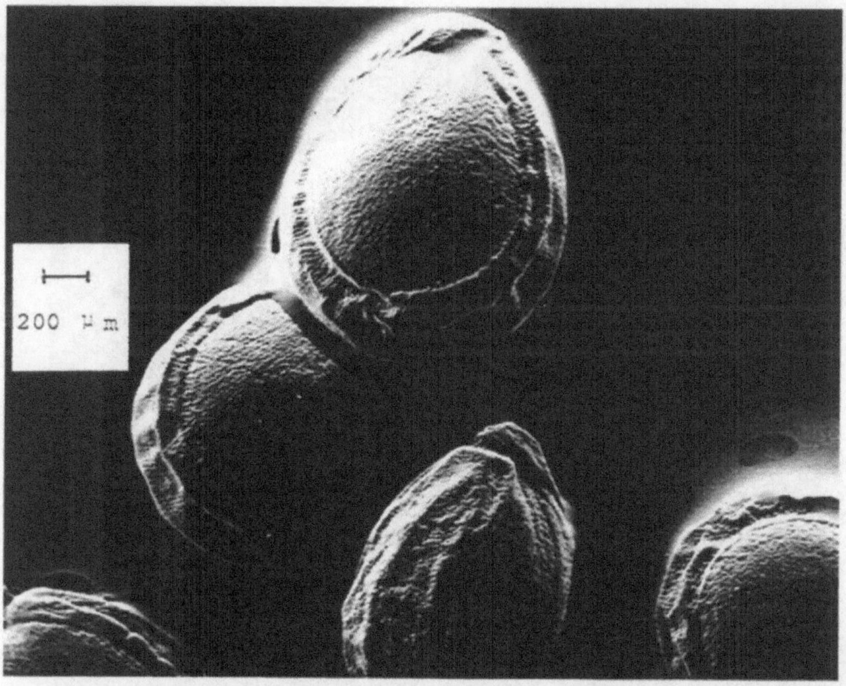

Figure 1 Scanning electron micrograph of *A. hypochondriacus.*

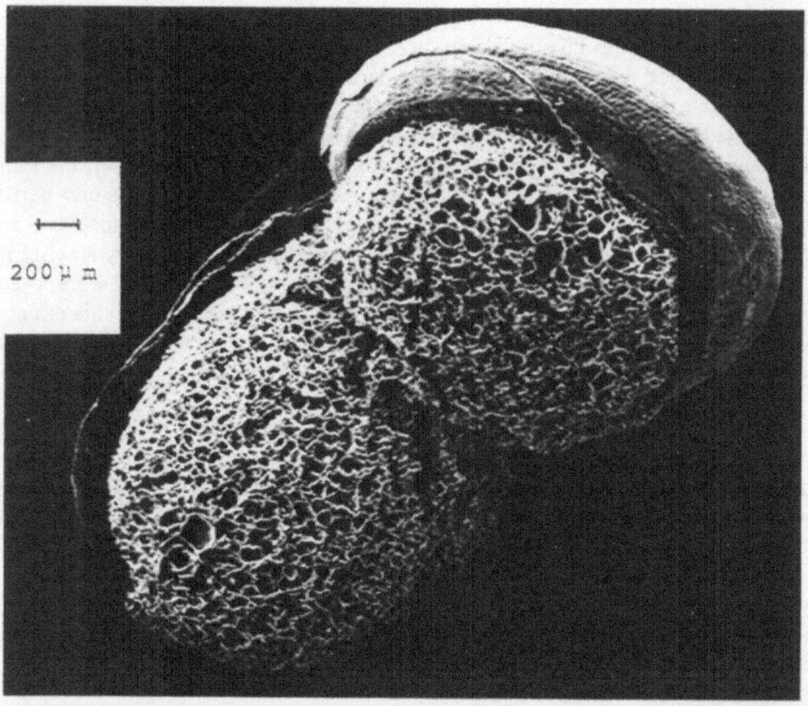

Figure 2 Scanning electron micrograph of *A. hypochondriacus* seed popped by direct contact.

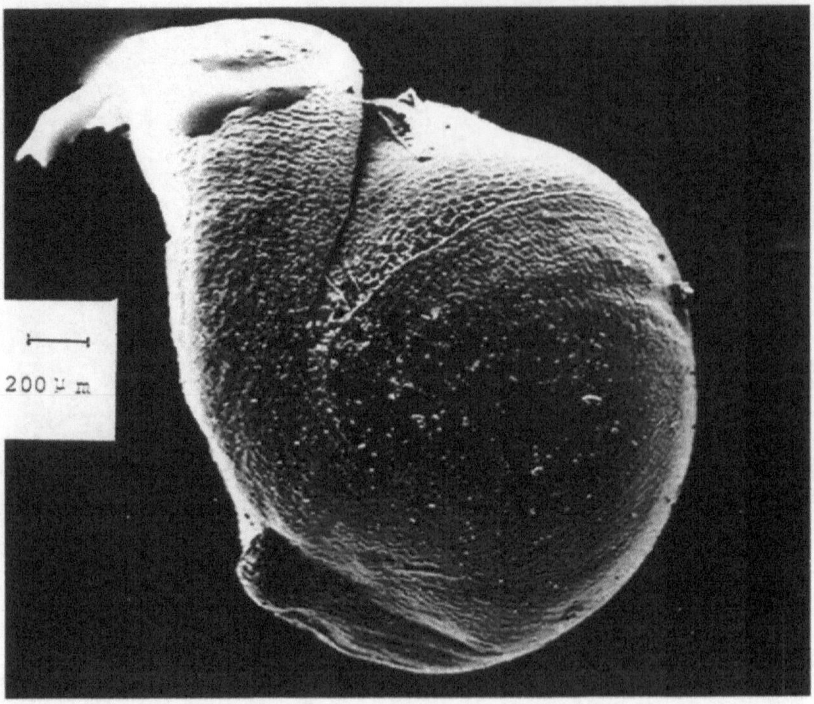

Figure 3 Scanning electron micrograph of *A. hypochondriacus* seed roasted by direct contact.

1. Popping, Toasting, and Roasting Amaranth Seeds

Time exposure of the seed to a given temperature, heat transfer either by conduction or convection, seed variety, and moisture levels of the raw seed are some of the variables to consider for amaranth popping. Amaranth popping could take place by direct contact or in a stream of hot air. The volume per weight unit in the raw seed was 1.05 ± 00.1 ml/g, whereas that for the popped amaranth ranged from 6.25 ± 0.25 to 10.00 ± 0.2 ml/g. Amaranth toasting refers to heat exposure of the seed at about 150°C for 60 to 90 s and usually takes place by direct contact of the seed on a hot surface; no apparent expansion of the seed occurs.[10,16] Amaranth roasting occurs due to overexposure of the seed to severe heat, and extensive charring may be observed; sometimes, in the same seed, some partial popping might happen prior to roasting. In an amaranth popping trial evaluating several species for 3 consecutive years, Stiebritz et al.[17] fractioned each dry-heat exposed amaranth sample by sieving it through a U.S. standard 16-mesh screen. The kernels that passed through this screen were predominantly unpopped and roasted kernels. The best popping species under the conditions of this experiment were the *A. cruentus*, Mexican grain type and the *A. hypochondriacus*, Mercado grain type with a mean percentage loss of 8.4 and 10.7, respectively.

2. Amino Acid Composition of Processed Amaranth Grains

Due to this ambiguous frontier among the popped, partly popped and roasted amaranth by the dry heat process and since lysine is the amino acid that might be most damaged by severe heat treatment, Table 1 summarizes what several workers have reported on the amino acid content of the heat-treated amaranth seed. After lysine, arginine and sulfur amino acids are the next most damaged. In the roasted seed, racemization of methionine and aspartic and glutamic acid was reported.[4] In a recent study,[18] D-amino acids have been detected in roasted coffee as well. This suggests that the popping process should be carefully monitored in order to avoid these losses which inhibit the supplementary role of grain amaranths in diets such as the one based on corn as the staple food.

3. Corn Tortilla Supplemented with Popped Amaranth

The amino acid deficiencies of tortillas and the supplementary value of popped amaranth have been evaluated by several workers.[10,11,19] When heat damage of the amaranth seed occurs during the popping procedure, tryptophan is not damaged, thus roasted amaranth supplements tortilla at least in its second

Table 1 **Amino acid losses of dry-heat treated grain amaranth by direct contact relative to the raw seed[a]**

Amino Acid	*A. hypochondriacus* (4) Popped & Roasted	*A. cruentus* (6) Popped	*A. caudatus* (16) Popped	*A. caudatus* (16) Toasted
Lysine	55.9	14.1	11.8	22.1
Arginine	28.7	4.4	6.1	12.5
Methionine + Cysteine	22.5	7.1	5.3	10.3

Note: Figures shown reflect percentages. Reference number in parentheses.

most limiting amino acid.[11] It seems that in the amino acid analyses of the popped amaranth reported by several authors,[4,10,16] the samples analyzed might have had, besides popped amaranth, partly popped and unpopped seeds, and even roasted grain; this suggests that when popped amaranth is produced at industrial scale, the heat damaged seeds should be separated and used for other purposes, not necessarily as a source of lysine and other essential amino acids present in the raw seed. Vargas-Lopez et al.[20] have lime-treated amaranth seeds, resembling the traditional procedure for making corn tortillas; they concluded that lime-treated amaranth could be made into flour for tortilla preparation either by itself or for enriching lime-treated corn flour. From the nutritional standpoint, this approach might be better compared with the popping procedure, due to the fact that no lysine losses seem to occur when cereals such as corn or wheat are treated with hot lime aqueous solutions;[21] however, the handling of the amaranth seed, as a supplementary source of lysine and tryptophan, at industrial scale in corn flour plants could be somewhat troublesome, mainly because of its small size, a fact that makes amaranth seed losses possible in the different stages of the lime-treated corn flour production, as well as possible difficulties in grinding the lime-treated amaranth seeds in the mills actually in operation at such industrial plants.

III. POSSIBLE PHYSICOCHEMICAL REASONS FOR THE POPPING OF THE AMARANTH GRAIN WHEN EXPOSED TO HEAT

Amylose content and starch granule size influence starch gelatinization; when popping amaranth, partial gelatinization takes place first and then a sudden change of state of the water within the seed occurs. The latter goes from liquid to steam in a very short period of time, i.e., 3 to 6 s depending on the heat transfer mode. This endogenous steam partially cooks different tissues in the seed, but mainly the perisperm, as well as expanding starch granules within the cells, usually breaking open the seed coat. On the other hand, dry heat in the popping operation contributes to denature storage proteins in the seed. It would appear that the dense protein present in the embryo not only provides some of the structural barrier so that steam pressure is built up for popping, but may even inhibit starch granule hydration in some of the neighboring cells in the endosperm. In Table 2, properties of starch of *A. hypochondriacus*, compared with other common starches, are shown. Granule size, amylose content, solubility, and swelling power, or the combination of these parameters, might play a role in the popping of these amaranth species. The starch granules of *A. hypochondriacus* are uniform in size, round, and polygonal (Figure 4). They are as small as the starch granules seen in taro or dasheen root from Polynesia.[22] Small-granule starches, as the one found in *A. hypochondriacus*, are relatively rare and have commercial potential value as dusting starches, for example, in cosmetics and candy dusting. No information, to the authors' knowledge, on the properties of starch modified in the popping process for amaranth grains is available.

A. MOISTURE CONTENT OF THE SEED AND ITS EFFECTS ON POPPING

The moisture content of several species of grain *Amaranthus* ranges from 6 to 11%.[23] The interactions of glutinous starch granules of the amaranth grain with water, for example, the water binding capacity of starch from *A. hypochondriacus*, is much higher than the one from wheat,[13] and the removal of moisture by flash evaporation due to a sudden and drastic increase in temperature seems to be a major factor that influences the process of popping. It is well established for milo and other cereals that some percentage of the grains to be popped only partly pop; the other fraction does not pop.[7] In the latter article, the percentage of fully popped milo increased at moisture levels of 18%, slightly above normal levels, and the starch digestibility of the fully popped milo was significantly higher than those for the partly popped, not expanded, and raw seeds. Metzger et al.[24] found that the maximum popping volume for air popping of popcorn was attained at a moisture content of 14%.

Table 2 **Some physicochemical properties of starch from different sources**

	A. hypochondriacus	Wheat	Waxy Corn	Tapioca
Granule size (m)	1	10	15	20
Amylose (%)	1–7	25	1	17
Solubility at 95°C (%)	40.2	8.2	23.0	48
Swelling power at 95°C	3.9	13.5	64.0	71
Gelatinization range (°C)	52–56	62–68	63–72	52–64
Granule shape[a]	R,P	O, R	R,P	T, R, O

[a] R = round; P = polygonal; O = oval; T = truncated.

Adapted from Saunders, R. M. and Becker, R., *Advances in Cereal Science and Technology*, Vol. 6, Pomeranz, Y., Ed., American Association of Cereal Chemists, St. Paul, MN, 1984, chap. 6 and Pomeranz, Y., *Functional Properties of Food Components*, Academic Press, New York, 1985, chap. 2.

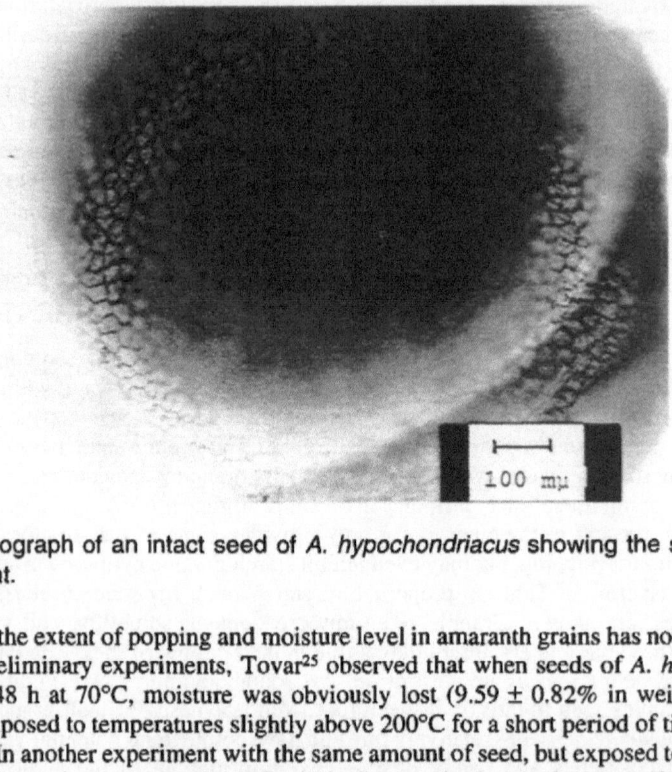

Figure 4 Light micrograph of an intact seed of *A. hypochondriacus* showing the starch granules through the seed coat.

On the other hand, the extent of popping and moisture level in amaranth grains has not been described in the literature. In preliminary experiments, Tovar[25] observed that when seeds of *A. hypochondriacus* were left in an oven 48 h at 70°C, moisture was obviously lost (9.59 ± 0.82% in weight). When this amaranth grain was exposed to temperatures slightly above 200°C for a short period of time, no popping of the grain occurred. In another experiment with the same amount of seed, but exposed to an atmosphere of 75% relative humidity and room temperature for 48 h, the sample absorbed water and the popping was not different from a control sample which was stored at 40% relative humidity and 22°C. The sample size in each treatment was 3.00 ± 0.05 g of clean *A. hypochondriacus* from the state of Morelos, Mexico; the popping method was by direct contact. In both treatments the popping volume was 6.7 ml/g; furthermore, the popping process dried the grain, and in Table 3 numerical results of triplicate runs of each treatment are summarized. These observations suggest that there seems to be an optimal moisture content of the amaranth grain and the extent of popping.

IV. AMARANTH POPPERS: HOT AIR OR DIRECT CONTACT

If a gas is introduced into the bottom of a container of solid particles of irregular shapes and sizes (such as amaranths grain, chaff and even soil from the field) and the upward gas velocity is gradually increased, after a time the particles will lift and begin to float. At a certain gas flow, an upward drag force just cancels

Table 3 **Percentages of weight variation and poppability of *A. hypochondriacus* under different moisture content**[a]

	Sample	
	% Moisture at 75% R. H.	Moisture at 40% R.H.
Moisture gain	9.5 ± 1.0	–
Moisture loss after popping	21.4 ± 3.2	12.3 ± 2.8
Popped material	87.0 ± 4.8	94.0 ± 6.4
Partly popped	4.7 ± 0.6	3.4 ± 0.7
Unpopped	8.1 ± 0.9	2.5 ± 0.3

[a] Mean ± SD.

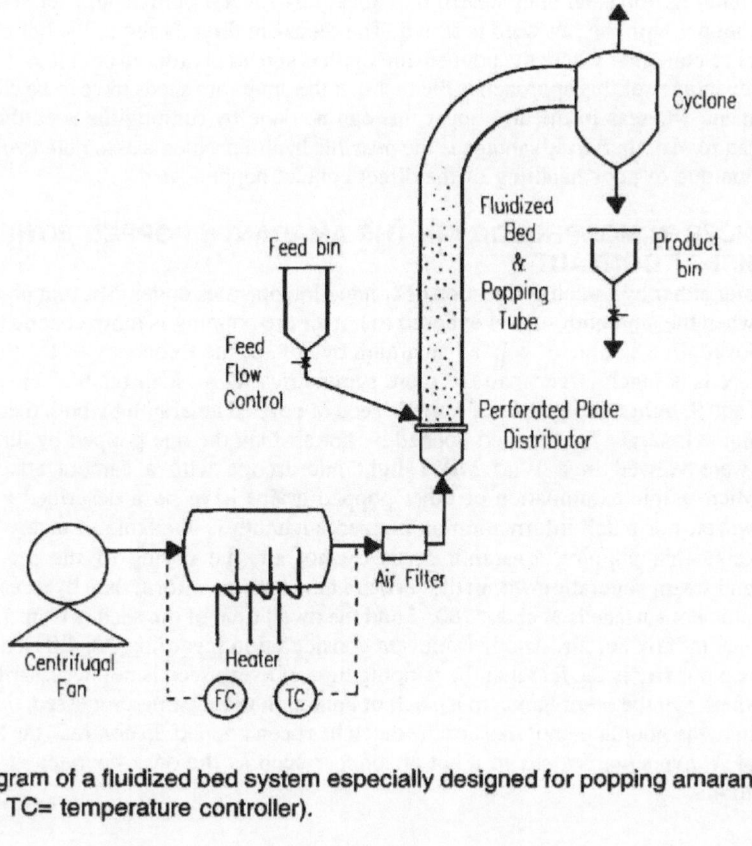

Figure 5 Diagram of a fluidized bed system especially designed for popping amaranth seeds (FC= flow controller; TC= temperature controller).

the downward pull of gravity; particles become free to move relative to one another, and the lightest or less dense of them would leave the container. Thus, a piece of equipment based on this principle would clean and pop amaranth grains on a continuous basis. The popping of the seeds occurs when the flowing gas is hot air, i.e., from 220 to 250°C; then the popped amaranth increases in volume and decreases in density relative to the raw seed therefore separating itself from the just-fed raw seed or the late–to–pop seed. Figure 5 schematizes the above description.[15] The fluidization column was a Pyrex glass cylinder, 0.10 m in diameter and 0.86 m long, with a stainless steel perforated plate distributor at the bottom. The fluidizing air was drawn from a centrifugal fan, and the air flow was measured by a digital anemometer and regulated by interchangeable plates located at the suction of the fan. The air was heated with LP gas to the popping temperature and filtered before entering the fluidized bed. Air temperature was controlled at the outlet of the heater; when there was a drop in temperature the heater was turned on, and when the temperature was above the set point, the heater was turned off. A flame detector sent a signal to close an LP gas valve when the flame was off in the burner. The popped amaranth was collected in the cyclone bin and further screened to separate the popped seeds from the unpopped ones, as well as the seed coat residues. Improvements to this concept have been made in industrial scale applications.[25,26] They were as follows:

1. The input of the raw seed into the bottom of the column was closely regulated so the output of the popped seed made a continuous steady state operation.
2. The air flow through the popper, gas consumption in the combustion chamber, and temperature of the air in several points of the column were adjusted; thus, the quality of the popped grain (tenderness and crispness) was much better than the one processed in a hot plate or the direct contact approach.

All construction materials of these poppers can stand temperatures up to 500°C. In the Tulyehualco-Federal District Area in Central Mexico, the hot plate for popping amaranth is still used. An operator with a broom has to move the seeds, otherwise they will be roasted; when popped, they are pushed away from the hot plate to a mat at the ground level and collected for further processing. A modification of this approach is a vibrating iron steel plate heated underneath by several burners and set in an angle of 10 to 15° in which a hopper with the raw seed is stored. The seeds are directly fed to the hot plate, and the end of it discharges in a container where the popped amaranth is spread in order to cool it to room temperature. One of the disadvantages of this approach is the fact that the amaranth seeds have to be cleaned in another piece of equipment, whereas in the air popper this can be done by running the seed through it with air at room temperature. Another disadvantage is the possible health problems associated with microbiological contamination due to poor handling of the direct contact popped seed.

A. DIFFERENCES IN MORPHOLOGY IN THE AMARANTH POPPED EITHER BY HOT AIR OR DIRECT CONTACT

Since heat transfer either by forced convection or conduction operates under different physical principles, it appears that when the amaranth seed is exposed to hot air the popping is more chaotic than when direct contact is employed. In a sample of popped amaranth by hot air, the frequency of the three or even four "soap-bubble" seeds is much larger than the more symmetrical two "soap-bubble" ones (Figures 6 and 7). In Figures 8 and 9, light micrographs of a single seed of popped amaranth by both methods are shown; the popping volume is larger for the seed popped by hot air than the one popped by direct contact. The popped seeds were viewed in a Wild MPS1 light microscope with a camera attached (Heerbrug, Switzerland). Microscopic examination of other popped grains have been described previously in the literature;[27] however, not much information on popped amaranth is available to date.

We hypothesize that popping amaranth seeds by hot air, the drying of the seed, partial starch gelatinization, and steam generation within the seed are much more uniform than by direct contact; in this latter approach, the hot surface is at about 200°C and the dwell time of the seed is from 15 to 30 s, longer than popping amaranth by hot air. Another relevant consideration regarding the differences between the two modes of heat transfer is the fact that the popping time of every seed is not necessarily the same with respect to the others. On the other hand, in a batch of amaranth grain being processed, the first seeds that pop must remain in the hot plate until most of the batch has been popped. In contrast, the average popping time of seeds of *A. hypochondriacus* in a hot air popper such as the ones operated at industrial scale, ranges from 2 to 4 s.

V. FINAL REMARKS

The conversion of an artisan making "alegrias" in small towns of a less developed country into an industrial worker producing popped amaranth and even other popped grains, keeping standards of quality for an increasing demand and competing with other producers, makes the subject relevant in countries such as Mexico, India, Peru, Thailand, and even the U.S. where the seed is being planted and used for human consumption. The popcorn industry in the United States was based on several studies that were published back in the 1920s.[28,29] Popcorn is purchased by weight and sold by bulk volume, which are criteria that could also apply for popped amaranth.

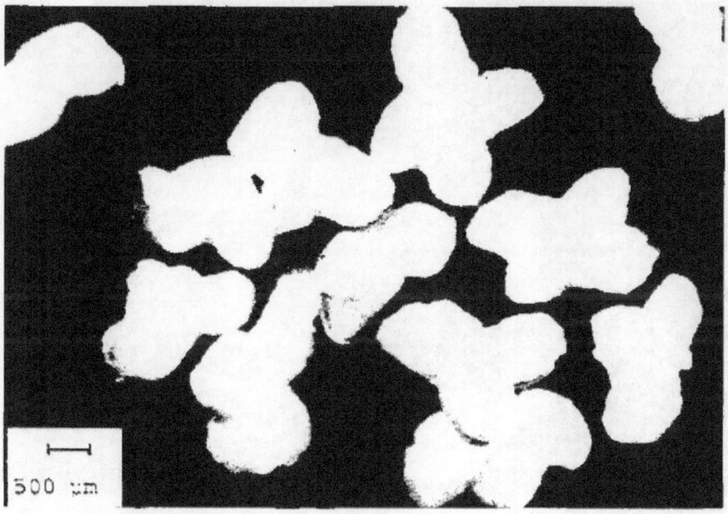

Figure 6 Hot-air popped amaranth seeds, most of them show the three or four "soap-bubble" configuration.

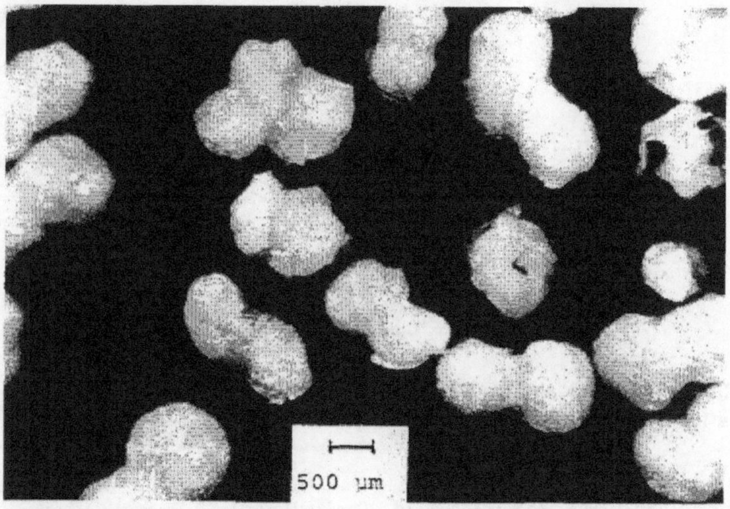

Figure 7 Hot-plate popped amaranth seeds; three on the right were partially popped, but most of the seeds show the two "soap-bubble" configuration.

A. SOME QUESTIONS TO BE ANSWERED REGARDING THE POPPING PROCEDURE

Tenderness and crispness and their possible correlation with popping volume should be established for *A. hypochondriacus*. Moisture content of the amaranth seed and maximum popping volume for both hot air and direct contact popping should be evaluated. When amaranth is popped, the seed coat is partially lost; therefore, losses of dietary fiber, minerals, and perhaps some water-soluble vitamins do occur. This is another area of research that should be considered.

No data is available on the mechanisms of flavor and odor development during the dry heat treatment of amaranth seed; studies on this subject could eventually support quality control criteria for industrial processing of popped amaranth.

Studies on the partial gelatinization of starch during amaranth popping are of interest since this phenomenon occurs at high temperature and low moisture; information on starch gelatinization kinetics at several popping conditions could provide a better understanding of the poppability of a given amaranth

Figure 8 A hot-air popped amaranth seed showing the three "soap-bubble" configuration. Detachment of the seed coat from the endosperm can be observed, the embryo structure is preserved.

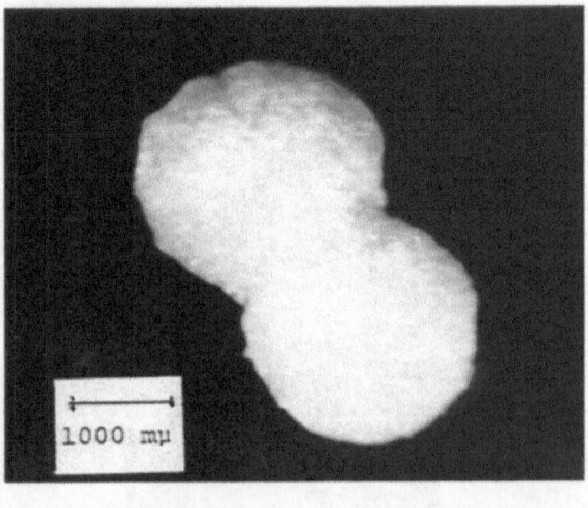

Figure 9 A hot-plate popped amaranth seed showing the two "soap-bubble" configuration.

species as well as of other grains. Regarding lipid composition of *A. cruentus*, squalene is about 5% of the oil,[23] and the heat effect due to popping in this compound is not known.

B. TRENDS AND POSSIBLE NEW DEVELOPMENTS

The fluidized bed system or the hot air amaranth popper could be used for coating fine or intermediate-size particles if the proper modifications are made to it. In one arrangement, the coating solution is sprayed downward onto the substrate as it is fluidized by air from below. Another approach is when the coating solution is applied from the bottom at the same direction as the flow of the popped seeds through the column. In the former case, one has to be aware of the heat labile compounds used for the coating, and in the latter, the coating should be made with the already popped seed and air at room temperature. Whittaker and Ologunde[30] showed that when grain amaranth was fortified with ferrous fumarate and fed to rats, the hemoglobin concentration of the supplemented group improved compared to a control group. Popped amaranth could be fortified with ferrous fumarate, ascorbic acid, and other nutrients including binding agents, flavors, and sweeteners to reduce iron deficiency anemia among infants and young children. Another approach in this direction is to encapsulate the amaranth seed before popping, either for confectionary purposes or nutrient enrichment or both.[31,32]

ACKNOWLEDGMENTS

We thank the amaranth processors in Central Mexico for the many tours given to us in their factories and for supplying popped seeds. Technical assistance on part of the photomicroscopy of A. Carabez and J. Ortega, from the Histology Department of the Faculty of Medicine, UNAM is gratefully acknowledged.

REFERENCES

1. National Research Council, Amaranth: Modern Prospects for an Ancient Crop, National Academy Press, Washington, D.C., 1984, 7.
2. **Sauer, J. D.,** The history of grain amaranth and their use and cultivation around the world, in *Proc. 1st Amaranth Sem.,* Bailey, N.N. and Snyder, S.G., Eds., Rodale Press, Maxatawny, PA, 1977, 9.
3. **Pant, K. C.,** The effect of amino acid or protein supplementation on the nutritional quality of grain amaranth, *J. Food Sci. Technol.* (India), 23, 298, 1986.
4. **Tovar, L. R., Brito, E., Takahashi, T., Miyazawa, T., Soriano, J., and Fujimoto, K.,** Dry heat popping of amaranth seed might damage some of its essential amino acids, *Plant Foods Hum. Nutr.,* 39, 299, 1989.
5. **Kauffman, C. S., Bailey, N. N., Volak, B. T., Weber, L. E., and Volk, N. R.,** *Amaranth Grain Production Guide,* Rodale Research Center, Rodale Press, Emmaus, MA, 1984, 24.
6. **Betschart, A. A., Irving, D. W., Shepherd, A. D., and Saunders, R. M.,** *Amaranthus cruentus*: milling characteristics, distribution of nutrients within seed components, and effects of temperature on nutritional quality, *J. Food Sci.,* 46, 1181, 1981.
7. **Walker, H. G., Bet Lai, W. C., Rockwell, W. C., and Kohler, G. O.,** Preparation and evaluation of popped grains for feed use, *Cereal Chem.,* 47, 513, 1970.
8. **Brito, E. and Tovar, L. R.,** Procesamiento de amaranto en un lecho fluidizado, in Investigaciones recientes sobre amaranto, 1a. Edición, Reyna, T. Ed., Instituto de Geographia, Universidad Nacional Autonoma de México, México D.F., 1988, 87.
9. **Sánchez Marroquin, A.,** Dos cultivos olvidados de importancia agroindustrial: el amaranto y la quinua, *Arch. Latinoam. Nutr.,* 33, 11, 1983.
10. **Bressani, R., Sánchez Marroquin, A., and Morales, E.,** Chemical composition of grain amaranth cultivars and effects of processing on their nutritional quality, *Food Rev. Int.,* 8, 23, 1992.
11. **Tovar, L. R. and Carpenter, K. J.,** The effects of alkali-cooking of corn and supplementation with amaranth seed on its deficiencies in lysine and tryptophan, *Arch. Latinoam. Nutr.,* 32, 961, 1982.
12. **Irving, D. W., Betschart, A. A., and Saunders, R. M.,** Morphological studies on *Amaranthus cruentus, J. Food Sci. 46, 1170, 1981.*
13. **Saunders, R. M. and Becker, R.,** *Amaranthus*: a potential food and feed resource, in *Advances in Cereal Science and Technology,* Vol. 6, Pomeranz, Y., Ed., American Association of Cereal Chemists, St. Paul, MN, 1984, chap. 6.
14. **Vázquez, C. M. G., Espitia, R. E., and Márquez, S. A. R.,** Efecto de localidad, método de reventado y genotipo en el potencial de reventado y la calidad proteinica del amaranto, in *Memorias Coloquio Nacional del Amaranto,* Gobierno del Estado de Querétaro, México, 1987, 242.
15. **Brito, E., Rodríguez, G., Tovar, L. R., and Valdivia, M. A.,** Desarrollo tecnológico para el procesamiento contínuo de amaranto y su evaluación químico biológica, presented at the National Food Science and Technology Bout, México D.F., 1988, 13.
16. **Pedersen, B., Hallgren, L., Hansen, I., and Eggum, B. O.,** The nutritive value of amaranth grain (*Amaranthus caudatus*). I. Protein and minerals of raw and processed grain, *Plant Foods Hum. Nutr.,* 36, 309, 1987.
17. **Stiebritz, S., Harrington, M., and Gilbert, L. C.,** Amaranth: food research and development summary report 1980-1983, Rodale Press, Emmaus, PA, 1985, 11.
18. **Palla, G., Marchelli, R., Dossena, A., and Casnati, G.,** Occurrence of D-amino acids in food. Detection by capillary gas chromatography with L-phenylalaninamides as chiral selectors, *J. Chromatogr.,* 475, 45, 1989.
19. **Morales, E., Lembcke, J., and Graham, G. G.,** Nutritive value for young children of grain amaranth and maize-amaranth mixtures: effect of processing, *J. Nutr.,* 118, 78, 1988.
20. **Vargas-Lopez, J. M., Paredes-López, O., and Espitia, E.,** Evaluation of lime heat treatment on some physicochemical properties of amaranth flour by response surface methodology, *Cereal Chem.,* 67, 417, 1990.

21. **Tovar, L. R.,** The effects of treatment with alkali on the nutritional characteristics of proteins, *Diss. Abstr. Int.,* 43, 2503B, 1983.
22. **Pomeranz, Y.,** Functional Properties of Food Components, Academic Press, New York, 1985, chap. 2.
23. **Teutonico, R. A. and Knorr, D.,** Amaranth: composition, properties, and applications of a rediscovered food crop, *Food Technol.,* 39, 44, 1985.
24. **Metzger, D. D., Hsu, K. H., Ziegler, K. E., and Bern, C. J.,** Effect of moisture content on popcorn popping volume for oil and hot-air popping, *Cereal Chem.,* 66, 247, 1989.
25. **Tovar, L. R.,** unpublished data, 1993.
26. **Solis, E.,** unpublished data, 1993.
27. **Reeve, R. M. and Walker, H. G.,** The microscopic structure of popped cereals, *Cereal Chem.,* 3, 227, 1969.
28. **Carr, R. H. and Ripley, E. F.,** What puts "pop" in popcorn, *Proc. Indiana Acad. Sci.,* p. 261, 1920.
29. **Weatherwax, P.,** The popping of corn, *Proc. Indiana Acad. Sci.,* p. 147, 1921.
30. **Whittaker, P. and Ologunde, M. O.,** Study of iron bioavailability in a native Nigerian grain amaranth cereal for young children, using a rat model, *Cereal Chem.,* 67, 505, 1990.
31. **Lew, C. W. and Barlow, D. E.,** U.S. Patent Appl. 277, 296, 1988; U.S. Patent 4,880,646, 1989.
32. **Cooley, J. and Douglas, D.,** U.S. Patent Appl. 177,762, 1988; U.S. Patent 4,888,186, 1989.

Chapter 9

Food uses and Amaranth Product Research: A Comprehensive Review

Kent A. Schnetzler and William M. Breene

CONTENTS

I. AMARANTH: AN OVERVIEW

Of the more than 60 different species of the genus *Amaranthus* that grow throughout the world[1] most are considered opportunistic weeds[2] and are associated with disturbance of soils by cultivation. Indeed, *A. lividus* is a principal weed in Japan, Jordan, Greece, and Austria and, in the U.S., has become a serious weed problem in vegetable production in Ohio.[3] Farmers and gardeners in North America know *A. retroflexus* by its familiar name, redroot pigweed.

0-8493-5374-2/94/$0.00+$.50

A. ORIGINS AND HISTORY

Few plants have cultural or historical origins comparable to the Amaranthaceae family. Historical records show that amaranth was consumed both as a vegetable and as a cereal grain, with grain production reaching a peak during the Mayan and Aztec periods in Central America. Mayan civilization was probably among the first to use amaranth as a high-yielding crop, with the Aztec and Inca civilization of Peru later recognizing its importance. Anthropological studies show the high regard these civilizations held for amaranth, so much so that it was incorporated into their religious ceremonies and beliefs.[4] Writings indicate that before the Spanish Conquest, the Aztecs produced 15 to 20 thousand tons of the grain per year, which placed the crop third after maize and beans.[5]

When Cortes invaded Mexico in 1519 and began the conquest of the Aztec empire, ceremonies involving statuary partially made from amaranth seeds were viewed as perversions of the Catholic Eucharist and were banned from Aztec culture. To destroy the religion and expunge its pagan rituals, Cortes banned all amaranth cultivation, ordered fields burned or trampled, and cut off the hands of people found growing or possessing the grain.[6]

Vietmeyer[6] noted that almost overnight one of the most important crops of the Americas fell into disuse and obscurity, while maize and beans went on to become two of the major crops that feed the world. Because of Cortes, amaranth grain species are little known and have since lain mostly dormant as a food crop. Production fell sharply between 1577 and 1890 with little record of grain production thereafter. During this period, however, the grain was introduced to Africa and Asia, and many ornamental forms were introduced into European gardens.[7]

B. AN ANCIENT CROP REDISCOVERED

Saunders and Becker[4] wrote that amaranth has the dubious distinction of being one of the few plant food species in history eliminated from popular cultivation due to a legislative fiat. That amaranth was so important to the Aztec and other New World diets makes it a promising unconventional crop to study. It is, however, only one of many underutilized food crops indigenous to Third World areas, but largely neglected by researchers and policymakers.[8] In 1975, the National Academy of Sciences, in an effort to broaden the world's food base, conducted a study and published a document entitled *Underexploited Tropical Plants with Promising Economic Value.* Amaranth was named one of these 36 potential food crops.[9]

Amaranth illustrates the fact that the food crops of today are mostly accidents of history. While most people believe that the world eats wheat, rice, maize, potatoes, and other staples because they are the best food plants, in fact, it is because such plants were best suited for cultivation by early man. Many crops exist only because of coincidence and serendipity.[10] In the last century, even with sophisticated knowledge of breeding and genetics, only rutabaga and minor fruits and nuts such as boysenberry, cranberry, pecan, kiwifruit, and macadamia have been domesticated.[6]

Nature contains about 80,000 species of edible plants, and of the 3000 or so species of plants used throughout human existence, only about 150 have reached commercial production levels. Apart from edible fruits and vegetables, most of the world receives the majority of its protein and calories from approximately 21 species. These include eight cereal grains, eight food legumes, two oil-containing food legumes, and three starch food crops — notably wheat, rice, maize, and millets; beans, peanuts, and soybeans; potato, sweet potato, and cassava; sugarcane, sugar beet and bananas.[8,11] Sattaur[12] estimated that just 30 species satisfy 95% of our nutritional needs, while the majority of humanity actually lives on no more than 12. Such a narrowing of the food base could lead to serious problems if continuous efforts are not made against disease, insects, and environmental conditions.[11]

To diversify the food base, it is important not to overlook lesser-known indigenous crops with promise to become global resources. A logical extension of this would focus on food crops that are known to have had important roles in past civilizations and are today still produced and used to some extent.[11] Marx[13] noted that some observers are coming to the view that it is a mistake to replace traditional forms of agriculture. It would appear to be better to maintain and improve food sources already present in less developed countries rather than replace them with Western crops. From a nutritional viewpoint, it is difficult to justify introduction and expansion of new food sources which require both cultural and ecological adaptation, especially if excellent native sources already exist.[14] Only a century ago, the soybean, sunflower, and peanut were considered crops unworthy of research effort; today, they are important world crops.[8] Of all the crops overlooked, amaranth appears especially promising because of its importance to Aztec, Mayan, and Incan civilizations. Such a food source which promises enough protein, fat, and carbohydrate and is replete with essential amino acids and fatty acids would be invaluable, especially considering that the choice of foods is comparatively selective.[15] Much work remains for the commercialization of amaranth and little or no precedents for the reintroduction of an ancient staple crop exist. However, with intensive research and agronomic interest, amaranth too could rise to prominence.

C. THE DUAL FOOD USAGE OF THE AMARANTH

Amaranth is consumed as a cooked vegetable, or green, in many of the lesser developed countries of Central and South America, Africa, and Asia. In West Africa, *Amaranthus* species are used solely as ingredients for vegetable soup, while the seeds are discarded.[16] However, in the more developed countries, the emphasis has been on the amaranth seed, or grain. Therefore, it is common practice to categorize the various amaranth species as vegetable or grain types, although there are no clear dividing lines. The general practice is to refer to light-seeded amaranths as "grain" types (although the dark seeds are also edible).[17] Macler et al.[18] noted that this dual usage, along with advantages of high edibility, food value, and potentially low waste biomass, might suggest novel use of *Amaranthus* in space. Amaranth grain has characteristics and properties similar to those of the monocotyledonous cereal grains. True cereals contain an abundant starchy endosperm, with small amounts of lipids and proteins found in the embryo. Amaranths, too, are starchy grains, but the starch is stored in the perisperm. Since the embryo occupies a great proportion of the grain, amaranths are a good source of both lipids and protein.[19] However, since it is dicotyledonous and not a true cereal, it is technically referred to as a pseudocereal;[20] the edible seeds of other dicots such as buckwheat and quinoa (*Chenopodium quinoa*) also share this term.[17]

Oke,[21] Saunders and Becker,[4] Teutonico and Knorr,[22] Bressani,[11] and Singhal and Kulkarni[23] have published comprehensive reviews of amaranth. Breene[20] examined food uses of the grain while Mallory et al.[24] have compiled over 300 citations from 1915 to 1988 into a bibliography entitled *Human Nutrition and Food Applications of Grain Amaranth. Amaranth-Food Properties and Agroindustrial Potential* (in Spanish), a 1990 book with more than 250 references, is available from Unidad Irapuato, CIEA-IPN, Apdo. Postal 629, 36500 Irapuato, Gto., Mexico. Teutonico and Knorr[22] identified *A. tricolor* as the main vegetable type, but pointed out that *A. caudatus, A. gracilis, A. graecizans,* and *A. spinosus* are also consumed as vegetables. They noted that the leaves of all *Amaranthus* species are probably edible, with even the grain types being acceptable if they are harvested in a young, tender state. This practice of dual usage for leaf and grain production is generally observed among minor landholders who grow the crop in a garden setting.[17]

Among the grain types, *A. hypochondriacus, A. cruentus,* and *A. caudatus* have been studied most extensively for their food uses, with some attention given to *A. hybridus, A. hybridus* × *A. hypochondriacus* crosses, and *A. edulis.*

Authors have been unaware of the taxonomic synonymy among amaranth species and this has led to confusion in the amaranth literature. Synonymous, taxonomically accepted species include *A. hypochondriacus* syn. *leucocarpus, leucosperma,* and *flavus; A. blitum* syn. *lividus* and *oleraceus; A. caudatus* syn. *edulis* and *mantegazzianus*; *A. tricolor* syn. *gangeticus* and *mangostanus; A. cruentus* syn. *paniculatus;* and *A. viridis* syn. *ascendens* and *gracilis.*[4] Wilkin,[25] however, felt certain that *A. cruentus* was not synonymous with *A. paniculatus* and suggested that *A. quitensis* should not be treated as distinct from *A. hybridus.*

Although mainly starch, amaranth contains a substantial quantity of protein and more fat than most cereals. Many workers have reported the protein and fat contents of amaranth, which average 15 to 16% and 7 to 8%, respectively. Mucciarelli et al.[26] reported that the protein content (N × 5.85) of *A. quitensis* was a remarkably high 21.7%. Becker et al.[27] performed one of the more comprehensive compositional studies. They examined grain from 2 different samples of *A. cruentus,* 1 each of *A. hypochondriacus, A. edulis,* and an *A. cruentus* × *A. hypochondriacus* cross, a crop × weed hybrid and 4 samples not identified as to species, for their contents (dry basis) of protein (15.3 to 18.2%), fat (5.6 to 8.1%), crude fiber (3.2 to 5.8%), and other components. Singhal and Kulkarni[28] examined the grain of 5 species from a total of 11 sources; they reported some of the widest ranges of protein (11% in *A. spinosus* to 23.3% in *A. teunifolious*) and fat contents (5.2% in *A. polygamus* to 19.4% in *A. teunifolious*), presumably on a dry basis.

Protein content in foods is routinely calculated by multiplying nitrogen content as determined by the Kjeldahl procedure by a factor based on the nitrogen content of the protein. The factor 6.25 is often used, on the assumption that protein contains 16% nitrogen and that all of the nitrogen resides in the protein. Many studies have used the factor 5.85, based on determination of the "true protein" content of

A. cruentus.[4,27] Correa et al.[29] determined Kjeldahl nitrogen to protein conversion factors for *A. hypochondriacus* (6.00), *A. cruentus* (6.02), *A. gangeticus* (5.92), *A. mantegazzianus* (5.92), and *A. anclancalius* (6.00).

A sample containing 16% protein as calculated using the factor 6.25 would contain only 15% protein using the factor 5.85. Furthermore, composition can be reported on the dry basis or the as is (wet) basis; sometimes the distinction is not made in published reports. Too often, the nitrogen conversion factor is not given. If the 16% protein sample (above) had been analyzed on the dry basis, the 15% value would be further reduced (on an as is basis) to 13.5% for grain containing 10% moisture. Hence, one must interpret reported protein content based on the conversion factor used and on whether values are expressed on an as is or dry basis. The absence of this information can result in a fairly substantial error.

At any rate, amaranth grain is as good as or better, quantitatively, than the cereals as a protein source. Amaranth protein has been shown in numerous studies to be superior in quality to that of the cereals. It is high in the amino acid lysine which is characteristically low in cereal proteins; therefore, amaranth is well suited for blending with cereals.

D. TRADITIONAL AMARANTH FOODS

Amaranth was a staple crop of the Aztecs and was also cultivated by the Mayas, Incas, and other pre-Columbian peoples. The Aztecs' practice of making *zoale* by combining popped amaranth with honey or molasses and human blood in the shapes of idols and consuming it as a religious ritual was proscribed by the Spanish conquistadores.[7] Teutonico and Knorr,[22] Early,[30] Bressani,[11] and Sauer[31] have listed a number of amaranth-based foods that are consumed today in various parts of the world (Table 1). It has long been known that amaranth seeds expand or "pop" when heated; *alegria*, which translates to "joy", is a Spanish name for a popular Mexican food made by combining popped amaranth with molasses or syrup. *Bollos* and *laddoos* are other confectionery type amaranth foods. Roasting or parching of seeds has traditionally been employed to enhance flavor and may also improve nutritive value by inactivating antinutritional factors or by altering such components as proteins to make them more digestible. Roasting or parching are preparatory steps in making *atole* and *sattoo*. *Chicha* is an alcoholic beverage or beer.

E. AMARANTH AS A COMMODITY GRAIN

Lehmann et al.[32] estimated U.S. production of amaranth grain at 1.3 to 1.8 million pounds in 1988, and while no commodity market exists, several companies handle large quantities of the crop. The small amaranth crop produced today is marketed directly to consumers by mail-order and to small- and medium-sized food processors for marketing to health food stores and some supermarkets. Consumer marketing of amaranth has stressed its nutritive qualities; therefore, it is not surprising that it is perceived as a "health food" and reaches consumers in large measure through health food stores and co-op stores. Amaranth products are not cost-competitive with cereal products as regular supermarket items, but health-conscious consumers are willing to pay a premium price for them. Thus, while such products have been quite successful in the health food market, few have yet to reach mainstream markets.

Lehmann et al.[32] listed more than 170 companies interested in amaranth. Several large food companies are reportedly conducting research on incorporating amaranth into various products (breads, hot and cold cereals, and pancake mixes), but seem unwilling to risk much money on producing and marketing such products. As a result, U.S. production is limited and prices remain high.

The 1988 U.S. domestic wholesale price was about $0.85/lb, although some sources reported prices as low as $0.55/lb. Even at the lower price, amaranth is still more expensive than competing grains, wheat and rice, which sold for $0.06/lb and $0.12/lb, respectively.[33] It has been estimated that the price of amaranth would have to fall to approximately $0.10/lb in order to become a commodity.[34] The price the U.S. farmers received for the grain in 1989 was $0.45 to 0.70/lb; the range in price reflected differences in quality, quantity shipped, organic vs. conventional production, and services provided by the grower.[35] In Mexico, farmers not wishing to sell their crop to buyers find it difficult to transport and market the grain; there, the lack of a market is the most limiting factor in the development of grain amaranth.[36]

Amaranth has been grown by midwestern U.S. farmers on a commercial scale since 1981 with approximately 2500 acres cultivated in 1989,[37] and Freese[38] identified amaranth as 1 of 20 popular farm diversifications. Thus, the technical base to produce large amounts of the grain exists; unfortunately, the production side is ahead of the marketing side.[39] The delicate nature of this small market is illustrated by the fact that a farmer entering the market with amaranth grain from several hundred acres could cause a massive surplus and drastically lower prices. Markets must first be identified, and better, contracted

Table 1 **Some traditional amaranth foods around the world**

Food	Ref.	Country or Region	Description of the Food
Alegria	22	Mexico	Confection of popped seeds with molasses or syrup
Turrones	30	Peru	
Niguas	31	Guatemala	
Atole	22	Mexico	Beverage or gruel of roasted, ground seeds, syrup, water
Bollos	22	Peru	Confection of popped, ground seeds, and syrup
Laddoos	22	India	Confection
Sattoo	22	Nepal	Gruel of parched, ground seeds, and water
Chapatti-like pancakes	22	Himalayas	Made from amaranth flour
Pinole	30	Mexico	Popped, milled flour
Mash'ka	30	Peru	
Alboroto	11,31	Guatemala	Confection of sorghum and amaranth, similar to *alegria*
Chicha	30	Peru	Beer made from fermented seeds

buyers arranged. Perhaps if one or two major food companies were to begin using the crop, increased demand would create production incentives, wholesale prices would decrease, and a positive net result would be obtained.

Kauffman[37] noted that commercial production in Mexico has increased in recent years, likely a response to large-quantity purchases of the grain by U.S. health food companies; unfortunately, secrecy concerning scale of production or Mexican sources makes it nearly impossible to determine the amount or source of amaranth grain imported into the U.S. In Peru, the domestic market for amaranth has suffered and commercial production has decreased. However, some European countries are reportedly seeking the quality Peruvian grain. Interest in the crop continues to grow in Kenya and both domestic and international markets show potential; resources are severely lacking, though. In China, the focus has been on amaranth as a forage crop, and it is possible that more land cultivated with both grain and vegetable amaranth exists there than anywhere else.

The U.S. production of amaranth has significantly increased in the last 3 years, with the number of acres planted increasing from 2500 to 4500. By comparison, however, more than 57.6 million acres of wheat were harvested in the U.S. in 1991. Because the crop is small, the U.S. Department of Agriculture does not keep statistics on amaranth; it has been estimated that Nebraska planted approximately 2000 acres in 1991, Colorado 300, Minnesota 200, and Montana 75 to 100, with the rest scattered throughout the Midwest.[40]

F. AMARANTH AS A FOOD INGREDIENT

Because amaranth is a substantially higher priced grain than the cereals, it is commonly included as an ingredient in cereal-based foods to boost their nutritive value. These foods are often touted as, for example, "amaranth" breads, cereals, cookies, crackers, granola bars, or whatever when they only contain relatively small percentages (e.g., 5 to 15%) of amaranth ingredient. Until conventional cereal production volumes are achieved, 100% amaranth products are unlikely to appear. Where amaranth is incorporated at less than 50% of a formulation, its nutty or grainy taste is usually undetected by the consumer. In blind tests, the consumer often prefers the amaranth-containing product.[41] The health food aspects are often emphasized in merchandising these foods with such slogans as "mystical grain of the Aztecs" or "golden grain of the Aztecs".[32] From a few dozen small- to medium-sized food companies, many amaranth-based products including flours, pancake mixes, popped amaranth, muffin and bread mixes, hot- and cold-cereals, multigrain cereals, amaranth breads, rolls, pastas, cookies, graham crackers, pilafs, tortillas, granolas, and tempeh are produced and marketed (Figure 1).[42] Schaeffer[34] analyzed amaranth foods in the U.S. marketplace and reported that of 33 products manufactured by 11 processors, 14 were cereals, 7 grains, 4 flours, 2 crackers, 1 was a pancake miz, 1 cookies, 1 canned hot dish, 1 spaghetti, and 1 tempeh.

Figure 1 Amaranth products in the U.S. marketplace. (From the Rodale Institute Research Center, Kutztown, PA. With permission.)

In 1992, the Amaranth Institute petitioned the U.S. Food and Drug Administration (FDA) for amaranth grain to be recognized as GRAS (Generally Regarded as Safe). A legal notice filed in the U.S. Federal Register documented a petition (Docket No. 91G-0452, petition GRASP 1G0372) to consider amaranth grain as a direct human food ingredient. The FDA responded (May, 1993) that as a traditional food, GRAS status for amaranth is not required.

Some of the more important cereal- and vegetable-based amaranth-containing foods will be discussed in this chapter. In addition, recent and current research efforts to develop new or "second generation" amaranth foods and to fractionate amaranth into its components such as starch, protein, and oil for use in foods and other consumer products will be addressed.

II. CONSUMER FOOD PRODUCTS CURRENTLY AVAILABLE

A. WHOLE AMARANTH PRODUCTS

Whole grain amaranth is sold in small packaged units (about 1 lb) and in bulk in co-op and specialty stores. Amaranth flour is packaged and sold at retail in several forms, as whole amaranth flour and as high- or low-bran products. Popped or puffed amaranth, and flours made from it, are also commercial products. Toasted flours and flakes are additional forms. A number of suppliers offer these forms of amaranth products in quantity for use as ingredients in formulated food products.

B. BAKING MIXES

Amaranth flour, flakes, or fractionated flour are included in specialty mixes for making pancakes, muffins, and other baked items. These are generally health food oriented products of the multigrain (e.g., 5- to 12-grain types).

C. MALTED (PRESPROUTED) FLOURS

The practice of sprouting seeds has important effects on the chemical composition, nutritive value, and acceptability characteristics of foods for human consumption.[43] Although sprouts have been consumed for centuries in the Orient, the practice of sprouting various cereal and legume seeds has only recently become popular in the western world.[44] Chavan and Kadam[45] have reviewed chemical changes and uses of sprouted cereals in foods.

In some cases, amaranth has been allowed to germinate, or sprout, before being dried and ground into flour. Paredes-López and Mora-Escobedo[46] showed that sprouting had a favorable influence on the nutritive properties of *A. hypochondriacus* seeds; it increased the true protein content by 25 to 30%, probably due to a decrease in carbohydrate and fat. They noted that presprouting rendered the flours more red; Hunter a and b values increased and L values decreased. This was attributed to the biosynthesis of the betacyanine pigment amaranthin. Paredes-López et al.[47] observed a decrease in Brabender viscograph peak viscosity of presprouted flour from 0 to 12 h germination, then an increase at 24 h and finally, a decrease at 36 h. Respective peak viscosities were 1380, 210, 2000, and 500 Brabender units (BU); peak pasting temperatures were 71, 90, 90 and 85°C, respectively. They suggested presprouting as a means of modifying functional properties of the flours.

Mendoza and Bressani[48] "hydrolyzed" amaranth flour (*A. cruentus* and *A. caudatus*) by reacting it for 1 h at 50°C with 1/10 its weight of presprouted, dried amaranth flour in an aqueous slurry. Balasubramanian and Sadasivam[49] reported a 31% increase in lysine content in *A. hypochondriacus* after 24 h of sprouting and subsequently drying and milling into flour. Such so-called malted flours can be used in the same product applications as regular flours.

Seeds of *A. hypochondriacus, A. cruentus,* and *A. caudatus* were germinated at 32°C for 0, 24, 48, and 72 h. Protein, crude fiber, and ash contents (dry weight basis) did not change, whereas lipid and phytic acid content decreased with increasing germination time. All vitamin contents increased during the process, particularly riboflavin and ascorbic acid, as did total and reducing sugars and damaged starch. However, germination did not increase the protein value or biological utilization of the grain.[43] Mora-Escobedo et al.[44] germinated seeds of *A. hypochondriacus* at 32°C for 24 and 48 h. Brabender viscograph patterns were affected by germination and water absorption values were similar to those of high-protein amaranth flours produced by enzymatic action on raw flour;[50] oil absorption (important in mouthfeel and flavor retention) increased with the longest germination time and was better than some plant protein isolates. Flour from seeds germinated at 48 h were completely dispersible and all were slightly pink, suggesting that amaranthin pigments were synthesized during germination.

D. BREAKFAST CEREALS

Amaranth is a natural for inclusion in breakfast cereals. It can be found in both the hot types and the cold ready-to eat (RTE) types, as well as in granola and granola bars. In Europe, it has been incorporated into muesli-type cereals. Again, it is usually present as one of several grains, including the conventional cereals and other somewhat more exotic grains such as buckwheat, triticale, quinoa, flax, etc. The cold cereal forms include flakes and various forms of puffed, popped, and extruded products. Sprouted amaranth is an ingredient in some dry breakfast cereals. Contreros et al.,[51] cited by Lehmann,[4] found that breakfast cereals containing 15% amaranth were well received by both health food and mainstream supermarket consumers. In Peru, a breakfast food made from amaranth flakes (to be used instead of imported oatmeal) found favor among local consumers.[52]

E. SNACKS

Snacks that crunch and crackle and make delightful sounds when eaten are a must for picnics, parties, and various manifestations of the "happy hour". Such foods are perceived by many as "junk foods". Inclusion of some amaranth, with its superior nutritive value, could partially alleviate this junk food image and, at least partially, alleviate the guilt feelings that some people experience as a result of enjoying these treats.

Amaranth-containing snack foods can take the form of extruded "corn curl" types or "nacho" type chips. Popped amaranth is also used as an ingredient in snacks, sometimes along with popped popcorn, in caramel corn or cracker jack-type sweet snacks. A popular weight loss program has included such a snack as one of its food items; the popped amaranth added high quality protein to the product. Brito[53] described a granola made from amaranth, oats, corn fiber, sesame and cashew seeds, and sugar that had good acceptance.

F. BAKED GOODS

There is a specialty market for multigrain breads, muffins, crackers, and cookies. It will probably remain a specialty market rather than a mass market for quite some time. Lorenz[54] found that up to 10 or 15% of the wheat flour in a formulation could be replaced by amaranth flour with little detrimental effect on bread quality; a 10% replacement lowered loaf volume by 7 to 10% and resulted in a darker colored bread. However, a taste panel preferred the "nutty" flavor of these breads over those made using only wheat flour. Although wheat flour is an essential ingredient in leavened breads, larger quantities of amaranth or other gluten-free grains might be used in unleavened, or flat, breads.[55]

Sánchez-Marroquín et al.[56] prepared tortillas using a high-protein fraction (29 to 30% protein) or a starchy fraction (7 to 9% protein) amaranth flour to replace nixtamalized corn flour in the recipe. The fractions were obtained by grinding and air-classifying seeds of *A. cruentus* and *A. hypochondriacus*. They found that up to 15 or 20% replacement of corn flour with either fraction boosted the nutritional quality of the tortillas without appreciably altering sensory quality. The high-starch fraction seemed most promising, since at the 20% replacement level it increased nutritional quality with the least effect on overall product quality. Whole flour and milling fractions of *A. cruentus* were used in 90:10, 80:20, and 50:50 mixtures with nixtamalized corn flour to prepare tortillas and arepas.[57] The use of 80:20 mixtures was recommended for preparation of tortillas, as the spared corn could be freed for other uses (especially considering the scarcity of corn and other cereals in developing countries).

Teutonico and Knorr[22] reported that crackers maintained their characteristic quality with up to a 20% replacement of wheat flour by amaranth flour. One would expect similar results with cookies. A variety of bakery products, muffins, wafers, *bolillos* (French bread), cookies, sweet bread, and *mantecadas* (sweet, fatty bread) were prepared from mixtures of *A. cruentus*-wheat flour by Sanchez-Marroquin et al.[59] Enriched cookies and *bolillos* with Net Protein Retention (NPR) values of 3.6 and 4.3, respectively, were recommended for improving the diets of poorer people. Vargas-Lopez et al.[59] used response surface methodology to evaluate the effects of temperature, alkali concentration, and cooking time on physico-chemical properties of *A. hypochondriacus* flour for tortillas. The nixtamalized flour possessed the basic functional properties needed for tortilla production.

G. MISCELLANEOUS FOOD PRODUCTIONS

Amaranth grain can be found in several other commercial products on which little or no published research is available. A canned "amaranth with garden vegetables main dish" is a sort of stew. It might more properly be called "garden vegetables with amaranth" since the latter is nearer the bottom than the top in the list of ingredients.

Also available are pasta products which can be in any of the popular forms and shapes such as spaghetti, macaroni, lasagna, etc. The structural ingredient in pasta products in semolina (durum wheat flour); amaranth flour probably constitutes 5 to 15% of the finished products. There is a fairly reliable rule of thumb in food formulation that major ingredients can be replaced with substitutes of similar composition at up to a 15% level before gross differences in product characteristics begin to appear. Based on increased protein content, improved amino acid balance, and good acceptability, Sanchez-Marroquin et al.[60] selected a 90:10 and 80:20 semolina:amaranth flour mix for producing spaghetti and elbow macaroni, respectively.

There are probably other products being offered for sale in local areas. For example, amaranth sprouts, as such or as a medley of sprouted grains, is one possibility. Saunders and Becker[4] and Teutonico and Knorr[22] listed some examples based on cookbooks and home recipes. These included breadings, salad dressings and sauces, pilafs, crepes, tostados, dumplings, fritos, puddings, and marzipan. A recent patent[61] presented recipes for bread, imitation corn bread, cake, muffins, pancakes, waffles, French toast, cookies, doughnuts, pretzels, hush puppies, dumplings, batter for frying, crepes, pie crust, tortillas, imitation peanut (and other nut) butter, imitation mayonnaise, amaranth milk, noodles, crackers, and puddings. The stated purpose of this patent was to provide nutritious and tasty hypoallergenic foods for persons with allergies to such food sources as wheat, corn and other cereals, legumes, milk, eggs, nuts, and yeast. Jones[62] published grain-free recipes that incorporated amaranth flour.

III. BASIC AND PRODUCT DEVELOPMENT RESEARCH

So far, the consumer products market for amaranth has been relatively small and confined to health foods and other specialty food items that can command the prices required in order to show a profit. Indeed, if some amaranth product were to suddenly become so popular that raw product demand would double or triple over the present production, a severe strain would be placed upon the entire system. It does appear, however, that amaranth production and utilization are both expanding; meanwhile considerable research is being conducted to lay the groundwork for the development of additional amaranth products for consumer markets. Product development and food use are important research areas as they give meaning to agricultural production efforts and create the driving force behind a dynamic food chain.[11]

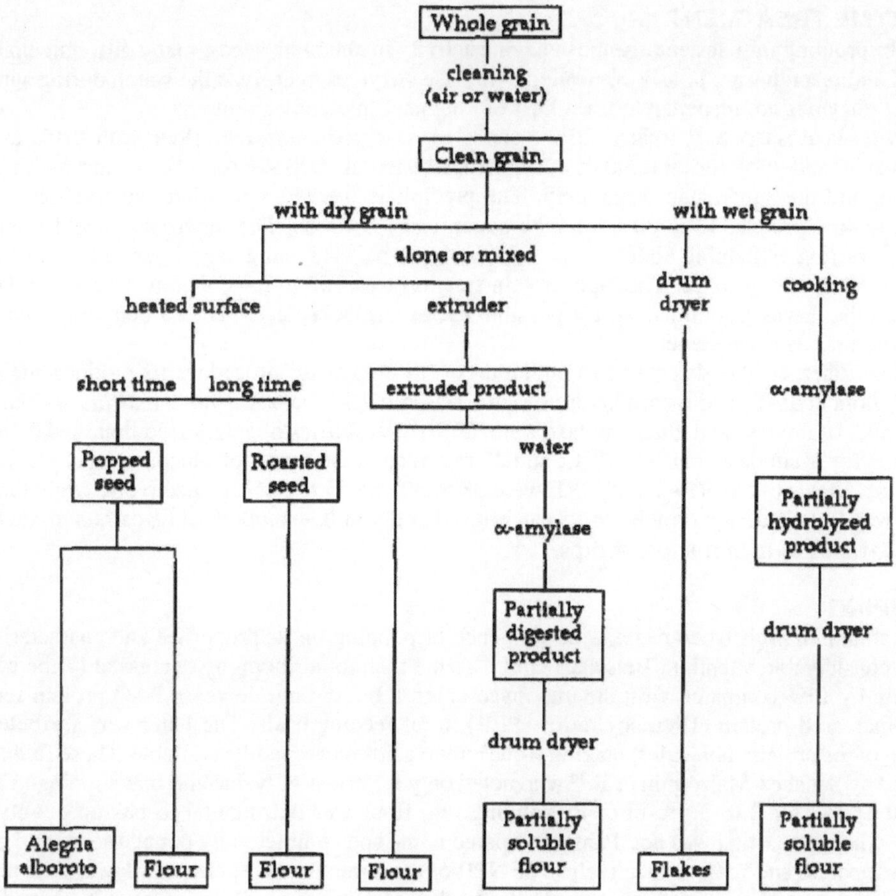

Figure 2 Processing of whole grain amaranth. (From Bressani, R., *Food Nutr. Bull.*, 10(2), 49, 1988. With permission.)

Bressani[11] illustrated some of the many efforts and technologies being applied to the whole grain (Figure 2). After cleaning with air or water, the dry grain can be popped, and if moistened to 20 to 25% water, it can be flaked. Alone or mixed with other products, the dry grain can be extruded to produce flours. After being suspended in water, these extruded flours could be enzymatically treated and dried to give a partially soluble product. Such a product could also be produced by a 10 min cooking, followed by starch hydrolysis and final drying; this would also be an excellent product for feeding children, as it could be suspended in a 15 to 20% solids solution.

A. AMARANTH MILLING

Because of its unique morphological structure and small seed size, amaranth must be milled differently from the cereal grains.[63,64] Betschart et al.[65] ran *A. cruentus* grain through a Strong-Scott barley pearler in five passes to obtain a "seedcoat-embryo" flour fraction and a "perisperm" flour fraction. Becker et al.[66] used a Morehouse stone mill to make whole grain flour or protein-rich or starch-rich fractions by adjusting the moisture content of *A. cruentus* grain and/or the gap between the grinding stones. Sanchez-Marroquin et al.[67,68] employed the barley pearler approach, as well as grinding/air-classifying, to obtain a 1-R (protein-rich) and a 2-R (carbohydrate-rich) flour fraction. Fractionation of amaranth flour into starch and protein components was studied by Wiesner,[69] who compared some conventional wheat mills, the Quadrumat Junior mill, and the Buhler-Miag Experimental mill with a modified Strong-Scott barley pearler. The latter was determined to be most satisfactory as it yielded protein isolates of 85 to 89% protein. Saunders[70] studied the effectiveness of a tangential abrasive dehulling device and produced high-protein fractions from *A. cruentus* and *A. hypochondriacus* with the small-scale equipment.

B. ENZYME TREATMENT

Although sprouting initiates endogenous enzyme activity in amaranth seeds, many different enzymes are involved and reactions are poorly controlled. Amylase enzymes hydrolyze the starch during sprouting,[49] and other enzymes are responsible for a host of chemical interconversions.

Barba de la Rosa et al.[71] treated 20% slurries of *A. hypochondriacus* flour with 0.1% α-amylase enzymes at 70 and 90°C for up to 90 min, centrifuged them at 9000 × *g* for 30 min, and freeze dried the precipitate and the supernatant separately. The precipitate yielded a high-protein fraction containing about 30 to 40% protein, 13 to 20% fat, 20% sugar, and 8% starch. The supernatant yielded a carbohydrate-rich fraction containing about 1% protein, with the balance being carbohydrate having a dextrose equivalent (D.E.) of 14 to 22. The high-protein fraction was said to have a composition similar to that of whole milk; it was recommended for possible use as a milk replacer. The carbohydrate-rich fraction showed promise as a sweetener.

Paredes-López et al.[50] determined conditions of an enzymatic procedure to produce high-protein amaranth flour (HPAF) and a carbohydrate-rich fraction (CRF) from *A. hypochondriacus* flour. Conditions for both α-amylase and glucoamylase were: 20% (w/v) slurries of gelatinized flour and 0.10% (v/w) of enzyme; for α-amylase, pH 6.5, 70°C, and 30 min incubation time; for glucoamylase, pH 4.5, 60°C, and 60 min. Yields of the HPAF and CRF were 38 to 39 and 55 to 56%, respectively, while the D.E. of the CRF was 17 to 21. High protein flours can be used as dry milk extenders while carbohydrate fractions can be converted to high-fructose syrups.

C. POPPING

Because amaranth pops when heated, the influence of popping on its properties and characteristics has received considerable attention. Betschart et al.[65] found that hot air popping decreased lysine content in the protein by 15% compared with the unpopped control, but did not decrease FAO protein scores and actually increased protein efficiency ratios (PER) in rat feeding trials. The latter was attributed to the unfolding of the protein molecules, making some amino acids more readily available. These findings were confirmed by Sanchez-Marroquin et al.[72] who noted only a 7% loss of lysine and that popping in a Rodale popper at 180°C for 2 to 3 sec prior to grinding into flour was detrimental to baking quality (bread, cookies), whereas toasting was not. Pant[73] compared home and commercially popped seeds and observed lysine losses of 36 and 25%, respectively. The NPR of the home-popped grain was lower than that of the commercially popped grain (which was similar to that of raw grain). Bressani et al.[74] popped (175 to 195°C for 15 to 25 s), flaked (200°C rotating drums for 1 to 3 s), and toasted (150°C for 60 to 90 s) seeds of *A. caudatus*. Available lysine increased in popped (5.7 g per 16 g N) and flaked (6.5 g per 16 g N) seeds, but decreased in toasted (4.3 g per 16 g N) seeds when compared to unheated sample (5.3 g per 16 g N). Protein quality was highest in the popped sample (NPR 3.19), followed by the flaked (NPR 2.78), the toasted (NPR 2.24), and the raw (NPR 1.73) samples. Insoluble fiber increased while soluble fiber decreased in processed samples.

Pedersen et al.[55] reported that popping *A. caudatus* seeds on a 200°C aluminum hotplate for 15 to 20 s decreased lysine content of the protein by 17% and produced a slight increase and decrease, respectively, in dietary fiber and starch contents. Tovar et al.[75] referred to a continuous fluid bed popping system that they had developed. The seeds were subjected first to room temperature air (counter-current flow) to remove debris and then to 160°C air (co-current flow) to pop them and whisk them away in the air stream. They suggested, but did not provide evidence, that their system, when perfected, should improve lysine retention because of the lower temperature and shorter duration of heating compared with traditional popping. Interestingly, Imeri et al.[76] noted that autoclaving *A. caudatus* seeds produced a slight increase in available lysine, possibly by causing unfolding of protein molecules.

Functional properties of flours are known to change after processing. Singhal and Kulkarni[77] puffed (popped) *A. paniculatus* and observed that water- and fat-holding capacities were higher (102 g per 100 g to 384.0 g per 100 g and 91 g per 100 g to 190 g per 100 g flour, respectively) for puffed flour. These increases were attributed to decreases in bulk density and have important implications in water, lipid, and flavor binding. Gelation was relatively unaffected while puffed paste viscosities decreased. Unpuffed amaranth flour may be better for applications in fluid foods, soups, batters, and beverages. Foaming ability was adversely affected, again suggesting that unpuffed flours would be better applied in angel food cakes, sponge cakes, meringue, souffles, whipped toppings, icings, and fudges.

Singhal and Kulkarni[78] noted that while amaranth pops to a nutty-flavored, popcorn-like product, poppy seeds fail to do so. Popping at 185 to 195°C readily distinguished expensive poppy seeds from an adulterant, *A. paniculatus* seeds.

A recent patent[79] described a method for popping amaranth seeds without rupturing the vitamin E-containing embryo.

D. EXTRUSION

Extrusion processing is used extensively in the food and animal feed industries to simultaneously heat raw products and formulations and to fashion them into final product forms with a wide range of shapes and textures. The material is fed into a barrel where it is subjected to shear, friction, and heating as it is forced under pressure through a die or dies at the output end of the barrel by one or two rotating screws. Extrusion cooking is a versatile, efficient process and is used commercially to produce a great volume and variety of food products.[80] Although identified as a promising method of amaranth processing, the body of literature that exists is small, and most studies have utilized small, single-screw equipment. As compared with single-screw extruders, modern twin-screw extruders are more intricate, offer better control of internal shear and residence time of heat-sensitive materials, and can be used to process low-moisture products, thus reducing or eliminating further drying requirements.[81] Unfortunately, twin-screw extruders cost 60 to 100% more than the equivalent single-screw machine, and their electrical costs are 20 to 50% higher.[82]

Sánchez-Marroquín et al.[68] used an extruder to heat blends of amaranth flour, wheat flour, and oat flour in order to inactivate any antinutritional factors that may have been present. They did not provide any details as to the kind of extruder, temperature/time regimes, or other processing parameters. A Wenger X-5 single-screw extruder was used to prepare snack variations from 50:50 corn and amaranth blends (plus other minor ingredients). The nutritional qualities of such products merited their recommendation as complements in school-feeding programs.[57] Del Valle et al.[83] used a Brady single-screw extruder to cook a blend of grain ingredients (whole amaranth, pearled oats, dehulled, cracked soybeans) for an infant formula at 160°C for 90 s. Koeppe et al.[84] processed an 80:20 blend of defatted amaranth flour:maize gluten meal at 150°C in a Brabender Model 2802 single-screw extruder and reported no decrease in available lysine due to processing.

A Brady single-screw extruder (Model 2160) was employed by Mendoza and Bressani[48] to process whole seeds of *A. cruentus* (about 150 or 160°C) and *A. caudatus* (1540°C). The extruded material was cooled, ground in a hammermill, and used as an ingredient in a beverage. The small size of the seeds and their approximately 8% oil content allowed for extrusion without grinding or adding additional water. Tillman and Waldroup[85] also extruded whole *A. cruentus* seeds using a Wenger Model X-25 single-screw extruder with the addition of some water in the form of steam. The extruded pellets were ground and fed to poultry.

Lehmann et al.[32] and Schnetzler[81] reported on the extrusion of whole amaranth seeds (*A. hypochondriacus* and *A. cruentus*) and blends of amaranth flour with other ingredients using a Buhler-Miag DNDG-62 twin-screw extruder. By selecting the proper screw configuration, along with a range of barrel temperatures (66, 93, 121°C) and screw speeds (200, 315, 430 rpm), extruded products with varying degrees of expansion were obtained. Bulk densities of the raw, unprocessed *A. hypochondriacus* and *A. cruentus* seeds were 846 and 804 g/l, respectively. Maximum expansion (123 to 124 g/l) was achieved for both species at a barrel temperature/screw speed of 121°C/430 rpm (no water added) and minimum expansion (544 to 578 g/l) at 66°C/200 rpm (small amount of steam added in the preconditioner). Prototype crouton, snack, breakfast cereal, and textured vegetable protein (TVP) products containing amaranth flour and other ingredients (wheat flour, rice flour, soy flour, sugar, salt, flavorings, etc.) were also produced with the Buhler-Miag twin-screw machine. Vargas-Lopez et al.[86] concluded that *A. hypochondriacus* flour extrusion-cooked under alkaline conditions could be used for preparation of tortillas and tortillas-based food products (Brabender 20 DN single-screw extruder, barrel temperature 50, 60, and 70°C, feed moisture 20, 25, and 30%, screw speed 30, 45, and 60 rpm).

Extrusion processing (Wenger X-20 twin-screw extruder, product temperature 160°C) and boiling (10 min) improved protein quality of *A. hypochondriacus* K343 as measured by feed intake, weight gain rate, and feed efficiency of growing rats.[87]

E. DRINKS AND BEVERAGES

Amaranth has been evaluated as an ingredient in drinks and beverages. Sánchez-Marroquín et al.[72] concocted an "amaranth milk" described as "a 10-g suspension of amaranth flour and powdered milk in 100 mL of water." The type of milk (whole or skim) and the proportions of amaranth and milk were not given. Mendoza and Bressani[48] prepared a drink consisting of partially hydrolyzed (malted) *A. cruentus* flour (45.64%), sugar (45.64%), and dry milk powder (8.72%). A 14-member taste panel found it to have good color and taste, but some thought it was too viscous. The solids content was not given; however, viscosity could be reduced by dilution. Balasubramanian and Sadasivam[49] formulated a similar beverage from *A. hypochondriacus* by sprouting the seeds for 24 h, dry roasting and grinding them and combining the flour with sugar and milk. Barba de la Rosa et al.[71] suggested that the protein-rich fraction obtained in their α-amylase treatment/centrifugation procedure could serve as a substitute for whole milk. In Peru, entrepreneurs have developed and sold breakfast drink mixes (to be combined with milk or water), "Kiwigen" and "Kiwi-Instan", made from toasted amaranth, quinoa flour, cocoa, and sugar,[52] and whose names apparently reflect a common Peruvian name for *A. caudatus, kiwicha.*

F. INFANT/WEANING FOODS

Protein-energy malnutrition (PEM) is an important nutritional deficiency that usually occurs during the critical transitional phase of weaning in infants. Prevention involves the timely introduction of weaning foods of proper quality, quantity, and proportion.[88] Such a product would have to: (1) provide all nutrients needed by infants, (2) be relatively inexpensive, (3) be acceptable to the target group,[89] and (4) have sufficient caloric density, thus reducing bulk. The combination of a cereal with an oilseed meal and one or more legumes to produce a weaning food with acceptable organoleptic and nutritional properties has many precedents;[90] thus, the similarities of amaranth grain and cereals (namely a bland taste, ease of cooking, and lack of significant antinutritional factors), combined with the superior nutritional value relative to cereals, would seem to make amaranth an attractive ingredient in infant foods and weaning formulas. Harper and Jansen[91] have discussed and extensively reviewed the production of weaning foods in developing countries by low-cost extrusion technology.

Sánchez-Marroquín et al.[68] prepared infant formulas by combining whole *A. cruentus* flour or its high-protein or high-carbohydrate fraction with wheat flour or oat flour and extruding the blend. They recommended the 50:50 or 60:40 amaranth:oat flour blends because of their satisfactory characteristics of protein content (17 to 23%), fat content (7 to 12%), crude fiber content (1.9 to 3.7%), energy value (380 kcal per 100 g), linoleic acid content (39 to 47% of total fat), protein chemical score (82 to 85), corrected PER (2.3 to 2.5), protein digestibility (89 to 90%), and net protein retention, or NPR (3.4 to 3.5). Ketiku and Olusanya[92] examined nutrient composition of Nigerian weaning foods incorporating an unidentified vegetable amaranth.

Imeri et al.[93] developed a weaning food from *A. caudatus*. The seeds were washed in water or sodium bicarbonate solution to remove astringent flavor notes, cooked for 10 min at atmospheric pressure, and dried on a double-drum dryer. The powder was reconstituted in water or milk as a plain or strawberry flavored product. Del Valle et al.[83] extrusion-cooked (Brady single-screw machine, 160°C for 90 s) a blend of *A. cruentus* seeds, pearled oats, and dehulled, cracked soybeans and combined the extruded product with other ingredients to produce a fine powder consisting of 28.8% amaranth, 28.7% oats, 16.1% soybeans, 18.3% sucrose, 5.9% vegetable oil, and 2.2% minerals/vitamins/flavorings. The product was prepared and fed as a gruel or *atole* to 1 to 3 year-olds weighing 15 to 26 lb with satisfactory results. Weaning mixes composed of puffed amaranth (*A. paniculatus*), jowar (*Sorghum vulgare*), and bajra (pearl millet, *Pennisetum typhoideum*) were inexpensive, nutritious, and remained acceptable after a 1-month storage.[88]

Rathod and Udipi[94] developed weaning mixtures using amaranth grain (malted, roasted, and puffed) in combination with a cereal (malted wheat and milled, puffed, and flaked rice), a green leafy vegetable (drumstick), and skim milk. All of the low-cost, nutrient-rich weaning mixes formulated were acceptable for consumption. Weaning mixtures containing roasted amaranth were acceptable to panelists and children; processing decreased levels of antinutrients.[95] Del Valle et al.[89] compared two infant formulas containing amaranth (*A. hypochondriacus*) with a soy-oats formula. No significant differences were found among the formulas with regards to either nitrogen absorption or retention. The authors concluded that the amaranth and soy-oats formulas were nutritionally similar and could help alleviate malnutrition in certain areas of Mexico.

Besides PEM, vitamin A deficiency and iron-deficiency anemia are nutritional problems in preschool children. Amaranth, because of its uniquely dual food uses, could supply greens rich in needed provitamin A, as well as a cereal-like grain that can supplement cereal-pulse diets. Diets consisting of cereal and vegetable products often contain inhibitory substances and this has led to a high incidence of iron-deficiency anemia. Whittaker and Ologunde[96] observed that fortification of a readily available, inexpensive staple like amaranth would be an effective strategy to enhance long-term iron status. Although amaranth grain has been shown to contain both phytate and tannins, a cereal composed of grain amaranth would be useful in intervention strategies to reduce both PEM and iron-deficiency anemia in infants and young children.[96,97]

G. FORTIFICATION OF CEREALS

In addition to the blends described elsewhere in this chapter (e.g., extruded foods, infant formulas, beverages), amaranth has been combined with cereal grains to supplement their nutritive value. Bressani et al.[98] illustrated this protein complementary effect with maize and wheat flour. Besides nutritional increases, the technology of composite flours (CF), through which a partial substitution of flours produced from locally grown cereals, pseudocereals, roots, tuber, and legumes for wheat flour is accomplished, presents both an option for reducing the effects of growing imports by nonwheat-producing developing countries and an approach for improving the nutrition of populations through selection and proportioning of flour components.[99] Fellers and Bean[100] reviewed composite flours and discussed attempted programs in Brazil, Senegal, Bolivia, Colombia, and the Sudan. The reasons for failure included, among others, a lack of low-cost, high-quality nonwheat flours, starches, or proteins and reduced baking quality or tolerances. The authors felt that while the outlook for *national* CF programs is bleak, there may well be opportunities for specialty CF products that satisfy consumer demand for preference, variety, nutrition, and low cost.

Although Imeri et al.[76] and Bressani et al.[101] found evidence to suggest that threonine was first-limiting and Bressani et al.[102] found methionine first-limiting in Guatemalan grain, many reports[4,27,29,103,104] have indicated that leucine is the first-limiting amino acid. As such, the protein of amaranth is best used in combination with other grains such as wheat,[58,68,105] sorghum,[106] and corn.[56,57,104,107-110]

Vietmeyer[6] reported that the combination of amaranth flour and wheat flour nearly reaches the perfect score of 100 on the nutritionist's scale of protein quality, as amino acids deficient in one are abundant in the other. However, Bressani et al.[101] indicated that threonine is the amino acid that limits protein quality. This would be of importance in efforts to use amaranth flour in mixtures with cereal flours, as rice and wheat are deficient in the same amino acid, threonine, after lysine. Thus, the lysine contributed by amaranth to the total lysine in a mixture with wheat would make threonine first-limiting with only minor improvement in protein quality. Clearly, these studies must continue because of the interest in mixing amaranth with other cereal grains. Bressani et al.[111] noted that although lysine and tryptophan levels are higher in amaranth than in common cereal grains, the amounts are not particularly high. Therefore, breeding and selection programs should take lysine content into consideration as this is one of the positive attributes of the grain. Maeda[112] found that sun-dried leaves of *A. tricolor* could be used to supplement cereal diets low in lysine.

Pedersen et al.[106] combined *A. caudatus* flour with maize, wheat, or sorghum flours at 0, 10, 25, or 50% of each blend. They determined PER values for both the raw and the cooked (slurried in 4.5 parts water, boiled for 30 min, freeze-dried, ground in a hammermill) blends. Substitution of cereals with amaranth improved PER as expected, but cooking had no practical effects. Pant[105] found that a wheat plus amaranth (6:3 protein ratio) combination had a significantly higher NPR (3.48) than wheat alone (2.68), but lower than amaranth alone (3.82) (casein control, 5.14). As little as 10% of amaranth flour raised the PER of semolina from 1.58 to 2.70 (casein, 3.47). Further substitution at the 20% level raised it only slightly, to 2.72.[60] This was explained only if the deficiency of the next-limiting amino acid in amaranth (after leucine) is such that it limits utilization of other essential amino acids and leaves enough lysine to cover some of its deficiency in wheat, at least to the level of threonine, which is likely second-limiting in both proteins.[109]

Tovar and Carpenter[107] observed that popped seeds of *A. hypochondriacus* failed to supplement the value of corn proteins when lysine was made the first-limiting amino acid by the addition of tryptophan. It appeared that available lysine in the popped seeds was reduced even more than the 30% measured by a dye-binding method. Whole flour and milling fractions of amaranth seeds were used in 90:10, 80:20, and 50:50 mixtures with corn flour to prepare tortillas and arepas.[57] All 3 corn-amaranth mixtures had sufficiently good amino acid profiles, while the 50:50 mixture approached the ideal protein (with regard to amino acid content) as established by the FAO/WHO in 1973.

Pacheco de Delahaye and Portillo[108] prepared arepas from 90:10, 80:20, and 70:30 mixtures of white corn and *A. cruentus* flour. After a 3-month storage, arepas prepared from the 90:10 and 80:20 mixtures were acceptable from a sensory standpoint. Lysine contents of the arepas prepared from the 10% amaranth mixture (1.6 g lysine/100 g protein) and the 20% mixture (2.0 g lysine/100 g protein) were higher than that of the commercial corn flour products (0.7 g lysine per 100 g protein). Morales et al.[109] added *A. caudatus* toasted flour to maize meal to provide 20 or 30% of the protein in malnourished children's diets. With as little as 12.7% by weight (20% of protein) of amaranth flour, the corn meal appeared to be able to satisfy protein and fat needs of young children if it provided approximately 90% of diet energy. Young, malnourished children consumed a diet in which maize was supplemented with toasted *A. caudatus* flour. The apparent lack of protein complementation observed may have been due to the high nitrogen absorption and retention from the Peruvian soft-endosperm maize and to protein damage in amaranth during toasting.[110]

H. VEGETABLE AMARANTH

There has been considerably less product development research on the vegetable than the grain forms of amaranth. *Amaranthus* species are grown in portions of subtropical and tropical Asia, Africa, and Central America as green vegetables; in western cultures, greens are not as highly esteemed and per capita consumption is low. Makus[113] listed the principal species and common names of amaranths consumed as greens (Table 2). While there is interest in growing amaranth in the U.S. as a midsummer greens crop, little data exist on which entries would be most acceptable to American consumers.[114] Edible amaranths offer a nutritious alternative in summer when cool-season green, leafy vegetables are not available and seem suited to the soils and climate of the southeastern U.S.[115] Since vegetable amaranth is often grown and consumed locally, information on international trade is nonexistent.[113] Oke,[21] Saunders and Becker,[4] Teutonico and Knorr,[22] and Singhal and Kulkarni[23] have reviewed some of the chemical and nutritive properties of vegetable amaranths. Mallory et al.[116] have published an extensive bibliography entitled *Vegetable Amaranth and Leaf Protein Concentrates*. Teutonico and Knorr[22] reviewed studies that found the flavor of cooked vegetable amaranth to be equal to or better than that of spinach. The greens contain oxalates and nitrates, which can be toxic, but in concentrations similar to those found in other leafy vegetables; therefore, consumption of amaranth greens should not present a health hazard to reasonably healthy individuals. The greens are especially high in calcium and contain more fiber, niacin, and vitamin C than spinach on a fresh-weight basis.[114]

There appears to be little, if any, commercialization of vegetable amaranth products. Makus[113] reported vegetable amaranth consumption in the U.S. in excess of 150 metric tons per year. The greens are consumed as a home garden or local market item in many lesser developed countries and can be prepared fresh as a salad or steamed, boiled, stir-fried, sauteed, baked, or pickled. Other possibilities include such products as soups and stews, baby food, casseroles, pasta, pie crust, quiche, souffles, etc.[4]

Abbott and Campbell[114] evaluated sensory characteristics of 20 steamed amaranths. Several *A. tricolor* entries had nonsignificantly lower scores than spinach while *A. cruentus* entries were least acceptable. Alfaro et al.[117] determined yields (kilogram per hectare), protein and crude fiber contents (dry basis), and eating quality of *A. hypochondriacus* greens at 25, 45, and 60 d after planting. Protein content was highest (29.5%) at 25 d while fiber content (11.1%) and yield (19.7 kg/ha) were lowest. At 60 d, protein content was lowest (14.4%), while fiber content (17.0%) and yield (510.7 kg/ha) were highest. Although total protein yield was highest at 60 d, harvesting at 45 d (22.7% protein, 14.3% fiber, 154.3 kg/ha) was considered best from an overall quantity/quality standpoint. In another study, a pale-seeded variant of *A. hypochondriacus* yielded 60.9 ton/ha of greens that corresponded to 20 ton/ha dry matter. Before flowering, the young stems and leaves contained 11 to 13% and 20 to 26% crude protein, respectively, which decreased to 4 to 5% and 14 to 15%, repectively, after flowering.[118] Soil N enrichment can be expected to increase the nutritional quality of amaranth greens.[115]

Vitamin A deficiency remains a serious health problem in many developing countries. The provitamins of plant foods provide a majority of dietary vitamin A and at a lower cost than animal food sources. Recently, physiological functions other than simply provitamin activity have been attributed to the carotenoids; however, little effort has been made toward determination of complete carotenoid composition of foods.[119] Van der Pol et al.[120] investigated the *trans-cis* isomerization of carotenes in leafy *A. cruentus* after traditional Indonesian cooking methods. Prakash and Pal[121] reported carotenoid levels from 90 to 200 mg/kg and 60 to 200 mg/kg and leaf protein from 14 to 30 g/kg and 15 to 43 g/kg in vegetable and grain amaranths, respectively. Total carotenoid content, mean beta-carotene content, and vitamin A value of Brazilian, leafy *A. viridis* were 347 to 468 μg/g, 110 μg/g, and 18.4 retinol equivalents (RE)/g, respectively.[119] Post-processing iron levels of Nigerian *A. hybridus* was investigated by Latunde-Dada.[122]

I. ANIMAL FEEDS

Although grain amaranth might be considered a rather expensive animal feed, there have been numerous studies of its effects on weight gain and general health of animals consuming it, from both a human food and an animal feed point of view. Saunders and Becker[4] reviewed feeding studies involving voles, rats, chicks, rabbits, and pigs fed grain amaranth. Results were mixed, but some showed poor growth of animals attributed to a variety of substances, including protease enzyme inhibitors (e.g., antitrypsins), lectins (hemagglutinins), oxalates, phytates, saponins, tannins, etc. A reason for the mixed results is that

Table 2 **Species, common names, and areas of consumption of amaranth greens**

Species	Where Grown	Common Names
A. blitum (lividus)	Eurasia	Norpa, chaulai (India) Vleeta (Greece) Horsetooth amaranth (Taiwan)
A. cruentus (paniculatus)	West Africa	
A. dubius	West Africa Carribean Indonesia (Java) Central and South America India	Claroen
A. hybridus	Central and North America	Quintonil (Mexico) Phak khom (Thailand Bayam (Malaysia, Indonesia) Urai (Philippines)
A. palmeri	North America	
A. tricolor (gangeticus)	Southeast Asia West Africa	Hinn choy (China)

Adapted from Reference 113.

there are several species of amaranth and no doubt considerable variation within species; plant composition is influenced by environmental factors and different animal species respond differently to growth-inhibiting compounds. There can also be within-species differences in animal responses due to natural variation and age. Clearly, additional research is needed to better define feeding potential, antinutritional problems, and the most appropriate amaranth type for feed and forage use.[35] Mallory et al.[123] have compiled 135 citations from 1919 to 1988 into a bibliography entitled Amaranth *Bibliography: Forage and Grain for Livestock.*

Tillman and Waldroup[124] reported that broilers could utilize up to 40% *A. cruentus* in their diets if the grain were processed by autoclaving or extrusion. In recent studies, Tillman and Waldroup[85,125] fed extruded *A. cruentus* to broilers at up to 50% of the ration and concluded that its average amino acid availability was similar to that of other grains. Body weights, feed utilization, and carcass yields were not adversely affected by up to 40% inclusion of extruded *A. cruentus* in broiler diets.[126] Acar et al.[127] found that autoclaving *A. cruentus* whole or defatted flour, perisperm fraction, or bran fraction increased the neutral detergent fiber (NDF) content of each and improved the nutritive value of these forms when fed to growing chickens. They attributed NDF increases to the formation during heating of a carbohydrate-protein complex that was poorly soluble in the neutral detergent solution. Laovoravit et al.[128] found that autoclaved *A. cruentus* gave better chick growth than raw amaranth when fed at 30% of the diet. Extruded *A. cruentus* was incorporated into layer diets without adversely affecting egg-production characteristics.[129] *A. cruentus* and *A. hypochondriacus* appeared to be useful feed ingredients for poultry.[130] However, some form of thermal processing was needed for maximum nutritive value; there also appeared to be a species difference in toxicity of the raw grain.

Pedersen et al.[55] noted an increase in dietary fiber due to heating of *A. caudatus*. Vohra et al.[131] reported that whole grain *A. cruentus* diets containing 2% sand were digested very efficiently by Japanese quails, but not by chickens. Alfaro et al.[132] included leaf/stalk meal of *A. hypochondriacus* at up to 60% of rabbit diets; at levels above 15%, growth was retarded. Steam treatment improved nutritive quality. Dairy cows showed no significant production declines when fed with 37% of amaranth crop residues in roughage,[133] and lambs fed forage (25 or 50% of diet) from *A. cruentus* demonstrated equal weight gain and feed utilization as those fed alfalfa forage.[134]

J. ANTINUTRITIONAL ASPECTS OF AMARANTH

While amaranths have long been consumed as cereals, vegetables, and forages, there are conflicting reports as to the presence of significant antinutritional or toxic factors that would limit their food application. For example, Singhal and Kulkarni[23] felt that such factors were unlikely to present any significant hazard, while Pond et al.[135] reported an unidentified factor(s) in *A. cruentus* that intefered with growth of rats.[136] Afolabi et al.[137] attributed weight loss in rats and the resulting negative PER (-2.4) to a high tannin content (0.25%) in *A. caudatus*.

Proposed antinutritional or toxic factors in grain, vegetable, and feral amaranths include: saponins and phenolics (primarily tannins or chlorogenic acid), phytic acid, protease inhibitors, oxalates, nitrates, polyphenols, phytohemagglutinins (lectins), alkaloids, aflatoxins, aluminum accumulation, and antinutritional fatty acids and sterols.

Saunders and Becker[4] noted that factors such as oxalates and nitrates are generally associated with forage and vegetable applications. Potential problems due to tannins, lectins, phytates, and protease inhibitors,[32,136] as well as nitrates, saponins, oxalates, and phenolics,[4,22,23] in amaranth have been reviewed.

Makus[138] evaluated eight accessions of *A. tricolor* and reported that a comparison spinach hybrid typically contained three times more soluble oxalate. Oxalate concentration in *A. gangeticus* averaged 91.0 g/kg (dry weight basis), a level higher than those previously reported in other plants to produce adverse health effects.[139] It is somewhat ironic that the predominant weed form of amaranth (*A. retroflexus*), commonly called pigweed, contains oxalate as high as 120 to 300 g/kg (dry weight basis) and has been shown to be toxic to pigs when consumed as a forage.[4,140] Singhal and Kulkarni[28] reported oxalate levels ranging from 6.09 to 9.66 g/kg in *A. paniculatus*, *A. polygamus*, *A. gracilis*, *A. spinosus*, and *A. teunifolious*. Alfaro et al.[117] determined the most adequate physiological stage of maturity to harvest leaves of *A. hypochondriacus* for human consumption. Oxalate content averaged 4.4 to 4.6% for the first and third harvest, 25 and 60 d from seedling emergence, respectively.

Prakash and Pal[121] evaluated the foliage of 61 accessions of 10 amaranth species (*A. tricolor* ssp. *tristis*, *A. graecizans*, *A. lividus*, *A. hypochondriacus*, *A. cruentus*, *A. caudatus*, *A. dubius*, *A. spinosus*, *A. spinosus* × *A. dubius*, and *A. retroflexus*) comprising both grain and vegetable types. They reported nitrate levels from 4.1 to 9.2 and 1.8 to 8.8 g/kg, respectively, and oxalate levels from 3.0 to 16.5 and 5.1 to 19.2 g/kg, respectively. Oxalate always showed an increase, but nitrates did not significantly change with maturity. Levels of leaf nitrate never exceeded 239 mmol/kg (dry weight basis) after soils were enriched with 0, 50, or 100 kg/ha N and were comparable to those found in spinach, beet, and chard.[115]

Attempts to reduce oxalate content in amaranth using a biotechnological approach might involve selection of low-oxalate cell clones and regeneration into plants[141] while *in vitro* systems for the improvement of both grain and vegetable amaranth have already been developed.[142] Teutonico and Knorr[143] developed a nondestructive method for screening cultured cells of *A. tricolor* to isolate low-oxalate variants. When regenerated, such plants might offer improved bioavailability of calcium, zinc, and iron. Van der Krol et al.[144] envisioned engineering oxalate content in plants by using "anti-sense" genes to reduce expression of a metabolic pathway and its corresponding product.

Of a group of conventional and nonconventional green leafy vegetables (*A. flavus*, colocasia, drumstick, fenugreek, neem, and pumpkin), amaranth had the highest amount of nitrate (56.9 mmole/g), oxalate (7.9%), and antitryptic units (8.1 Trypsin Units Inhibited/mg protein).[145] Correa et al.[146] investigated the antinutritional content of five amaranth species: *A. anclancalius*, *A. gangeticus*, *A. hypochondriacus*, *A. cruentus*, and *mantegazzianus*. Trypsin inhibitor activity ranged from 3.0 to 5.5 TUI/mg and from 0.9 to 2.3 TUI/mg for light- and dark-seeded species, respectively. Polyphenol content ranged from 0.2 to 0.6%, a value comparable to low-tannin sorghum cultivars. Both saponin content and phytohemagglutinin activity were low.

The tannin content (0.15%) of *A. hybridus* seemed to be sufficiently low enough not to have any nutritional impact,[147] while Pedersen et al.[55] reported small amounts (0.3%) of tannins in *A. caudatus*. Tannic acid ranged from 0.21 to 0.64% in 33 diverse amaranth samples.[102] Whittaker and Ologunde[96] reported phytate (7.92 mg/g) and tannin (0.22 mg catechin eq/100 g) levels in raw Nigerian *A. caudatus* grain which dropped to 3.01 mg/g and 0.06 mg CE/100 g, respectively, after processing at 100°C for 20 min. Phytic acid was reduced after 72 h germination time for seeds of *A. caudatus* (0.44 to 0.14%, dry weight basis) and *A. cruentus* (0.32 to <0.04%, dry weight basis).[43] Singhal and Kulkarni[28] reported that the seeds of 5 vegetable amaranth species contained from 4.3 to 6.2 g/kg phytic acid while Brito[53] found tannins (0.067%) and phytates (0.87 g/100 g) at low levels in Venezuelan samples of *A. cruentus* and *A. hypochondriacus*.

Lectins have been isolated and purified from *A. cruentus*[148,149] and *A. leucocarpus*.[150] The proteins, which have the ability to agglutinate red blood cells, had subunit molecular weights of 32,500 to 45,000 Da. The PER of *A. leucocarpus* flour increased from 1.95 to 2.15 as a result of extracting the hemagglutinin.[151]

Pedersen et al.[55] reported low levels, <1.0 trypsin unit inhibited (TUI), of heat-labile trypsin inhibitors in *A. caudatus*; on the other hand, Imeri et al.[76] reported no inhibitor activity, but a highly heat-labile hemagglutinin in *A. caudatus*. No inhibitor activity was found in Venezuelan *A. creuntus* and *A. hypochondriacus*.[53] Bressani et al.[102] found a trypsin inhibitor activity range of 0 to 11 TUI/ml (average, 2) in 33 diverse amaranths from Guatemala, Peru, and Mexico. Koeppe et al.[152] found a low level (0.52 TUI/mg) of trypsin inhibitor in *A. hypochondriacus* grain which they separated into 13 bands by electrophoresis, 3 of which were also active against chymotrypsin. The inhibitor was also very heat stable, retaining 20% of its activity after 7 h at 100°C. Trypsin inhibitor activity in a saline extract of *A. leucocarpus* was extremely low.[151]

Fernando and Bean[153] compared aflatoxin production by *Aspergillus flavus* and *A. parisiticus* on *Amaranthus cruentus*, *A. hypochondriacus*, and *A. hybridus* with that on polished brown rice, corn, peanuts, soybeans, and winged bean. Cooking the amaranth seeds prior to innoculation increased the levels of aflatoxins although such levels were much lower in amaranth than in rice, corn, or winged bean.

Makus[138] reported that the leaf blades of *A. tricolor* contained more than three times as much aluminum as spinach grown at the same location. When grown at pH 6.4, leaf blade aluminum averaged 562 mg/kg, well below the 1000 mg/kg consideration for an aluminum accumulator. The highest leaf aluminum levels were found with soils at pH 6.6 that received low phosphorus supplementation.[154]

Fernando and Bean[2,155] noted the absence of behenic acid (C22:0), a potential antinutritional fatty acid and any other fatty acids considered unsafe for human consumption in *A. tricolor*, *A. retroflexus*, *A. hybridus*, *A. dubius*, and *A. caudatus*.

Thermal processing in the form of autoclaving,[76,124,127,128,130,131] extrusion,[85,87,124,126,129] moist heat,[87,132,156] atmospheric cooking,[96,157,158] toasting, popping, and flaking,[55] and drum drying[158,159] appears to be the best method of reducing the effects of heat-labile or water-soluble antinutrients in amaranth. Pond et al.[156] reported that appropriate heat processing may be important to maximize amaranth's nutritional value. For example, moist heat (10 min in boiling water), but not dry heat (2 h at 100°C) or soaking in water (16 h at room temperature), destroyed or removed the inhibitor(s). Candidates for these moist-heat-labile factors include lectins,[148,149,151] trypsin inhibitors,[55,152] tannins,[76,96] and saponins and phenolics.[160] Whether moist heating of amaranth would release a water-soluble phytate is not entirely known, although popping reduced phytate levels 10 to 20%[55] and cooking at 100°C (but not at 60 or 40°C) for 20 min reduced phytate content.[96]

Imeri et al.[103] and Bressani et al.[74,159] have dismissed the common trypsin inhibitors, lectins, and tannins as explanations for the significant improvement in nutritive value of thermally processed grain, as analyses by many authors have shown only small amounts of these factors. More likely, it could be caused by an increase in the bioavailability of nutrients, particularly amino acids.[74]

Lehmann[136] noted that thermal processing (baking, milling, popping, and roasting) of amaranth has been practiced throughout history; the same, however, can be said for many other indigenous foods. It is tempting to speculate that these methods may have inactivated thermolabile antinutrients.[161] Perhaps these early "processors" recognized the beneficial effect on nutritive value manifested in better growth, vigor, and performance. Or perhaps heating simply may have broken the tough seed coat and made amaranth-based foods more palatable.[136] What is clear, however, is that early farmers selected pale seeds over dark seeds, thereby decreasing the tannins and possibly other antinutrients. Recently, however, Chinese researchers have utilized the tannin-rich, dark seeds to produce award-winning soy sauces with dark colors and good fragrances.[41] Thermal processing was likely a cultural practice that, by fortunate coincidence, reduced or eliminated some antinutrients. Clearly, studies to further explain the increase in nutritive value after appropriate thermal processing must be conducted. Bressani[11] suggested that other kinds of antiphysiological substances and/or changes in amino acid availability, or even a combination, may be responsible while Lehmann[136] recommended studies to assess the impact of heat, moisture, and duration of treatments on antinutrients.

IV. POTENTIAL OF AMARANTH COMPONENTS IN CONSUMER PRODUCTS

In addition to protein-rich and carbohydrate-rich flour fractions, etc., there is the possibility of breaking the amaranth grain or plant down into its component parts. The production of starch, oil, and other products from corn by wet-milling, for example, is a major industry. Some research has already been done on the potential uses of amaranth starch, protein, oil, and minor constituents in consumer products. Potential applications of the components of grain amaranth are numerous and cover several forms.[41]

A. AMARANTH STARCH

Starch is the major component of amaranth grain, constituting about 50 to 60% of the total dry weight.[4,22] The amylose (straight-chain) form ranges from 0 to 22% of total starch, the balance being the branched-chain form, amylopectin.[162,163] Diameters of the starch granules in commercial starches range from 3 to 8 μm for rice starch to about 100 μm for potato starch.[162] By comparison, amaranth starch granules are extremely small at 1 to 3 μm in diameter.[28,64] Starches have been isolated and characterized from several amaranth species.[164-166] The main sources of commercial starch in the world are maize, wheat, potatoes, tapioca, and sorghum; it would appear that the lack of industrial utilization of amaranth starch is due to inadequate knowledge of its physicochemical and functional properties.

Possible product uses of amaranth starch, capitalizing on its small granule size, could include food thickeners, dusting powders for cosmetic as well as food uses, nonallergenic aerosols, laundry starch, and, possibly, biodegradable plastics.[162] Coating food products with minute starch granules is an emerging technology. Lehmann[41] noted that pulverization of corn or wheat starches does not produce the granule shape or coating properties of amaranth. Tomita et al.[163] reported that the unique properties of *A. caudatus* starch, namely, its pasting properties and high susceptibility to amylases, may indicate potential uses as food, feed, and fuel. Yanez et al.[167] suggested its use in gravies, sauces, and soups because of its relatively low gelatinization temperature and good freeze/thaw stability (no separation evident after four freeze/thaw cycles). However, gelatinization temperature can vary over a fairly wide range, reflecting diversity across the germplasm. There can also be differences in reported gelatinization temperatures depending on the particular method used for its measurement. Paredes-López and Hernandez-Lopez[168] used differential scanning calorimetry (DSC) to study the effects of water, solutes, and annealing on the phase transitions of *A. hypochondriacus* starch. Stone and Lorenz[169] concluded that amaranth starch behaved very poorly in breads and cakes from the standpoint of color and appearance, texture, and volume, probably because of its low amylose content.

Singhal and Kulkarni[170] studied properties of *A. paniculatus* starch and compared them to those of corn starch. The amaranth starch was found to have a higher paste viscosity, lower paste clarity, and higher freeze/thaw stability (four cycles required to achieve 30% water separation). While both starches thinned when subjected to shear, the amaranth starch was also sensitive to acidic conditions. In follow-up studies, Singhal and Kulkarni[171,172] investigated utilization of *A. paniculatus* starch in salad dressings and custard preparations, respectively. Oil content in regular mayonnaise (77 to 82%) could be reduced to 30 to 50% in the starch-containing dressings. The waxy nature (50 to 60% amylopectin) and relatively better ability to withstand shear than corn starch made *A. paniculatus* starch better for use in salad dressings. The higher paste viscosity[170] of amaranth starch indicated potential use in products where viscosity is an important consideration, i.e., custard preparations. However, interactions with other ingredients made it unsuitable for this particular application. As with any food product development, interactions among components must be considered. Singhal and Kulkarni[173] cross-linked *A. paniculatus* starch with phosphorous oxychloride to give it increased acid and shear stability and suggested its use in salad dressings and canned foods. However, freeze/thaw stability was adversely affected.

B. AMARANTH PROTEIN

The second most abundant component of amaranth grain is protein. Lehmann et al.[32] or Lehmann[161] and Bressani[5] have provided comprehensive reviews of the proteins of amaranth. Amaranth protein has routinely scored higher nutritionally than that of other seeds, e.g., wheat, barley, soybeans, corn, by FAO/WHO protein scoring, PER, etc.

Proteins are used as functional ingredients in formulated foods. Important functional attributes include water and fat absorption and binding, viscosity, gelation, emulsification, whipping and foam formation, and flavor binding, among others. Correa et al.[29] extracted the protein from five different amaranth species and fractionated it. It consisted, on average, of albumin (65%), globulin (17%), prolamin (ll%), and glutelin (7%). Paredes-López et al.[174] characterized *A. hypochondriacus* protein as consisting of 46 to 49% albumin plus globulin, 3% prolamin, and 30 to 33% glutelin.

Barba de la Rosa et al.[175] devised fractionation procedures for the main protein fractions in *A. hypochondriacus* and characterized them with respect to molecular weight and amino acid composition. Globulins were extracted from *A. hypochondriacus* and studied by ultracentrifugation and chromatographic techniques; sedimentation constants of 10 S and 12.7 S and a molecular weight of 166,000 were reported. The major globulin fraction appeared to be 10 S.[176] Gorinstein et al.[177] demonstrated different extraction methods for alcohol-soluble proteins of *A. cruentus*, *A. flavus*, *A. caudatus*, and *A. hypochondriacus* and separated the fraction by electrophoresis into 80 to 85% polypeptides of 10 to 14 kDa and 7% of 20 kDa, the rest being minor fractions. Comparison of electrophoretic patterns of the alcohol-soluble fractions and total proteins of amaranth with prolamins and total proteins extracted from maize, rice, oats, and sorghum did not show any relationship. In a closely related study, Gorinstein et al.[178] examined electrophoretic patterns and amino acid compositions of fractions from the same four amaranth species. In agreement with Konishi et al.,[179] this study indicated that prolamin is not a storage protein in amaranth seed as it is in corn.

Konishi et al.[179] purified a globulin extracted from *A. hypochondriacus* seed and Konishi and Yoshimoto[180] examined its emulsifying properties. Based on their measurements of its emulsion activity (EA), emulsion stability (ES), and surface hydroscopicity, along with its relatively high heat stability, they concluded that it could be useful as an emulsifier. Konishi et al.[181] extracted two albumin fractions from *A. hypochondriacus* and *A. cruentus*, while Mora-Escobedo et al.[182] characterized albumins and globulins from *A. hypochondriacus*. Marcone and Yada[183] also studied globulin from *A. hypochondriacus*. Lehmann et al.[32] quoted work by Wiesner who found *A. cruentus* and *A. hypochondriacus* protein isolates to have poor emulsifying and foaming properties, limited water absorption, and good gel-forming properties. It must be remembered that individual protein fractions are very likely to behave differently from mixtures and that the manner of extraction can alter functional properties, i.e., denatured proteins behave quite differently than undenatured ones.

There is controversy as to which fraction, i.e., albumins, globulins, or glutelins, is the main protein storage component in amaranth. Perhaps this is due to differences in the protein fractionation methods employed.[184] Using a sequential solvent extraction method, Soriano-Santos et al.[184] studied protein extraction efficiency in *A. cruentus* seeds and the influence of ionic strength and pH on solubility and separation of albumins from globulins. Bressani and Garcia-Vela[185] fractionated protein from *A. caudatus, A. hypochondriacus,* and *A. cruentus* using the Landry and Moreaux fractionation scheme. Differences in reported globulin/albumin (G/A) ratios in amaranth grain could reflect differences in species, cultivation conditions, sample preparations, or extraction procedures. Konishi et al.[186] discussed pitfalls in the determination of the G/A ratio and proposed an improved fractionation method.

Calderon de la Barca et al.[151] found about 2% of the protein of *A. cruentus* to exist as lectin. Recent work at the University of Michigan[32,136,187,188] suggested that a lectin from amaranth dubbed "amaranthin" binds to colon cancer cells and premalignant tumors and may have potential as a screening device for these highly undesirable abnormalities. The term amaranthin was a poor choice, in this case, because it has been applied for quite some time to the betacyanine pigment responsible for the red color of various parts of amaranth plants.

It appears that there is a need for more research on the nature and functionality of amaranth proteins.

C. AMARANTH OIL

Amaranth seeds are probably too low in oil content (7 to 8%) and too high in market price to compete with most other commercial edible oils. Lyon and Becker[189] did a pilot plant-scale hexane extraction, refining, and bleaching of *A. cruentus* oil to produce a light-colored product that appeared suitable for use as an edible oil. However, Becker[33] concluded that unless a specialty starch or protein could also be produced from the defatted meal, a substantial decrease in raw product cost will probably be required before amaranth oil goes commercial.

Amaranth oil is not particularly unique, being similar in composition to corn and cottonseed oils.[189] Garcia et al.[190] reported that crude amaranth oil had a lower digestibility than cottonseed oil, but that it was not responsible for growth depression when the seed was fed raw. Despite the diversity of the germplasm, the fatty acid composition does not show the extent of variability that can be seen in, for example, sunflowers and rapeseed. However, Ayorinde et al.[16] reported wide variations that appeared to be agronomically related in the fatty acid composition of 18 West African varieties of amaranth. Published fatty acid profiles, in terms of percentages that each of the principal ones contributes to the total, show the following ranges: linoleic, 25 to 62; oleic, 19 to 35; palmitic, 12 to 25; stearic, 2 to 8.6 and linolenic, 0.3 to 2.2.[2,4,33,111,155,189,191,192] The saturated/unsaturated fatty acid ratio ranged from 0.29 to 0.43. The preponderance of unsaturated fatty acids is a plus from the standpoint of nutrition, and the low content of linolenic acid should ensure good stability against autoxidation. Acid values indicated high contents of free fatty acids in amaranth oil, while low saponification numbers, compared to other vegetable oils, suggested high contents of unsaponifiable fat; possibly, this was due to substances such as squalene or wax.[190]

Prakash and Pal[193] published a comprehensive survey of seed protein, fat, and fatty acid profiles of 41 lines of amaranths comprising 9 species of both grain and vegetable types. Protein content of the defatted meal ranged from 103 to 183 g/kg and fat content of the seeds from 8 to 68 g/kg. Palmitic, oleic, and linoleic acids were the major fatty acids found with the saturated/unsaturated fatty acid ratio ranging from 0.12 to 0.50.

Vegetable oils that are rich in lauric, palmitic, stearic, and oleic acids are used to partially or totally replace cocoa butter in chocolate-flavored products. Epoxy fatty acid-rich oils are used as stabilizers in plastic formulations and in the preparation of long-chain compounds, while cyclopropenoid fatty acid-containing oils have received attention for their effects on animals and co-carcinogenic properties.[194] Oil from *A. paniculatus* contained palmitic (19.4%), stearic (3.9%), oleic (21.9%), linoleic (43.9%), epoxy (vernolic, 7.8%), and cyclopropenoid (malvalic, 1.5% and sterculic, 1.6%) fatty acids and was considered a minor source of these epoxy and cyclopropenoid fatty acids.[194]

Amaranth oil has been reported to contain larger amounts (7 to 8%) of squalene (2,6,10,15,19,23-hexamethyl-2,6,-10,14,18,22-tetracosahexaene) than other common vegetable oils;[33] Ayorinde et al.[16] reported smaller amounts (2 to 5%) in West African species. For comparative purposes, squalene contents in olive, wheat germ, and rice bran oils are in the 0.1 to 0.7% range. Becker et al.[27] found levels from 4.6 to 6.7% in samples of *A. cruentus* and *A. edulis* oil while Lyon and Becker[189] reported levels from 5 to 8% in *A. cruentus*. Squalene content was reported to be 4.65, 2.42, and 3.97%, respectively, in oils of *A. paniculatus*, *A. polygamous*, and *A. gracilis*.[195] Singhal and Kulkarni[192] observed that puffing (popping) of *A. cruentus* seeds decreased the percent unsaturation from 75.5 to 62.3% (similar to coconut oil), the greatest effect being on linoleic acid (decreased from 46.8 to 27.0%). Squalene content increased by 15.5 from 4.9% in unpuffed seed oil; the only plausible explanation was that puffing possibly destroyed some components of the oil so that it became relatively richer in squalene. This phenomenon might be used to isolate squalene on a commercial scale. Singhal and Kulkarni[78] suggested quantitative determination of squalene as an index of adulteration of spice poppy seeds with seeds of *A. paniculatus*. Dailey et al.[196] isolated a tiny, but pure squalene fraction from leaves of *A. palmeri*.

Squalene is an important ingredient in skin cosmetics and penetrants and in lubricants for computer disks.[32,197] Presently, this expensive (to $53,000 per ton) isoprene derivative is industrially synthesized or obtained from shark livers, with Japan, and to a much lesser extent, Norway controlling the world market; however, it can be easily extracted from amaranth grain by simple vacuum distillation.[39] Recently, the Japanese cosmetic industry has switched to synthetic sources of squalene because of unreliable supplies and qualities of squalene from marine oils. In 1990, the U.S. market alone for squalane (the hydrogenated form of squalene) was expected to exceed 30 metric tons per year.[32] Recent estimates place the world market at 1400 metric tons per year (Japan 800, U.S. 200 to 300, Europe 100 to 200, and Korea 150 to 200).

Squalene is also a known obligatory biological precursor of sterols. Currently, it is hypothesized that squalene can reduce serum cholesterol by inhibiting hepatic 3-hydroxy-3-methylglutaryl coenzyme A (HMG-CoA) reductase, the key regulatory enzyme for cholesterol biosynthesis. However, data from a chick-feeding study suggested that the squalene in amaranth had no effect on cholesterol synthesis in the liver.[128] Two cyclopropylamine squalene derivatives were found to be effective inhibitors of cholesterol synthesis.[198]

The heat stability of fats and oils is thought to depend both on their fatty acid profiles and the composition of the unsaponifiable fraction; therefore, squalene could be a potential oxidation inhibitor. Addition of squalene (0.4%) retarded the formation of polar compounds in heated (170°C) rapeseed oil and model lipid systems.[199] If squalene does indeed have antioxidant properties, then perhaps amaranth oil might possess greater oxidative stability compared to other oils.

D. AMARANTH FIBER

Amaranth contains high levels of nonfermentable fibers which suggests that the grain may lower serum cholesterol and LDL levels without increasing the risk of colon cancer. Recently, such health-contributing attributes have gained importance in cereal marketing and processing, e.g., oats and high-fiber cereals.[32] Kruse[200] noted that the current trend toward increased fiber and bran consumption is evidenced by an 800% increase in the demand for oat bran in 1988 and an expected increase of 17% in oat consumption during the next few years. Przybyla[201] listed amaranth as a potential bran fiber source and suggested applications in bakery products and cereals. While volatile, the marketplace for high-fiber, cholesterol-lowering, nutritionally-sound foods is apparently worthy enough to risk large capital ventures[32] and may present opportunities for cereal and snack food formulations.

Pedersen et al.[202] noted that crude fiber values, representing primarily lignin and most of the cellulose, of amaranth have been reported. Values generally ranged from 3 to 8%. However, dietary fiber includes lignin and cellulose, as well as the hemicelluloses, pectic substances, gums, and certain other carbohydrates. Schnetzler[81] determined fiber content of *A. hypochondriacus* K343 and *A. cruentus* R149 seeds using the enzymatic-gravimetric method of Prosky et al.[203] and reported total dietary fiber values (TDF) of 12.9 and 15.0% (dry weight basis), respectively. Bressani et al.[102] found that in grain samples from Guatemala, crude fiber content was 7.1% on average, as compared with 3.9 and 3.0% for samples from Peru and Mexico, respectively. Singhal and Kulkarni[28] examined seeds of 5 Indian amaranth species and found crude fiber values ranging from 4.9 to 12.9%; considering the inadequacy of the crude fiber method, it would be of great interest to obtain TDF values for these high-crude fiber species. Dietary fiber contents of Malaysian white (*A. viridis*) and red spinach (*A. gangeticus*) was reported to be 3.6 and 4.1 g per 100 g, respectively.[204]

Dietary fiber can be separated into insoluble (IDF) and soluble (SDF) fractions. IDF includes cellulose, some hemicelluloses, and lignin while SDF includes gums, pectins, and other hemicelluloses. The TDF fraction of *A. hypochondriacus* K343 consisted of 75.3% IDF and 24.7% SDF, while that of *A. cruentus* R149 was higher in insoluble fiber and, consequently, lower in soluble fiber (85.7% IDF and 14.3% SDF).[81]

Pedersen et al.[202] reported the levels of TDF in pale- and dark-seeded varieties of *A. caudatus*. The pale seeds contained about 8% TDF and the black seeds about 16%. The SDF fraction made up between 30 and 44% of the TDF in the pale-seeded varieties, but only 18% in the black seeds. Processing seemed to have only minor effects on the levels of TDF. The TDF content of pale-seeded varieties of *A. cruentus* was reported to be 7.93% on average, while a black-seeded variety contained 16.4% TDF.[55] Bressani et al.[74] reported TDF values of 13.5% for *A. caudatus*. The TDF fraction was composed of 51.8% IDF and 48.2% SDF.

Schnetzler[81] studied the effect of amaranth type (*A. hypochondriacus* or *A. cruentus*), barrel temperature (66, 93, or 122°C), screw speed (200, 315, 430 rpm), and moisture content (none or steam-preconditioned) on dietary fiber content of twin-screw extruded whole amaranth. Only "temperature" and "type x temperature" were found to be significant for soluble fiber. Perhaps this represented thermal decomposition and redistribution of insoluble to soluble fiber, but it was dependent on amaranth type. Fiber content of *A. hypochondriacus* K343 was not greatly affected by the range and combination of variables studied; few treatments were significantly different in TDF, IDF, or SDF content. The dietary fiber content of *A. cruentus* R149 appeared to be much more responsive to the range and combination of processing conditions. TDF content was significantly reduced (by 13 to 31%) for 12 of the 18 treatments, while IDF content was reduced (by 17 to 42%) for every treatment. SDF content increased (by 52 to 118%) for 14 of the treatments as compared to the unprocessed grain.

The physiological effects of dietary amaranth (*A. cruentus*) on rats was studied by Danz and Lupton.[205] Amaranth diets resulted in lower serum cholesterol values than those of fiber-free controls and lower liver cholesterol values than those of cellulose (a poorly fermented fiber). In the colon, dietary amaranth was fermented poorly and neither decreased colonic pH nor increased cecal surface area or intestinal length. The authors suggested that amaranth behaved like soluble fibers in lowering serum cholesterol, but similar to insoluble fibers in its actions in the colon. Since cholesterol-lowering properties of oat bran and pectin are assumed to be in the SDF fraction and amaranth is generally low in SDF, it might appear that a fiber fraction other than SDF possesses such hypocholesterolemic properties.

E. AMARANTH PIGMENT

Amaranth is a common name for one of the seven original coal tar dyes certified by the Food and Drug Act of 1906 as safe for use in foods, drugs and cosmetics. It was designated as FD&C Red No. 2 under the Food, Drug and Cosmetic Act of 1938. In 1976, it was banned for use in foods, drugs and cosmetics under the Delaney Clause. Possibly, this dye got its name from the red-violet pigmentation of amaranth plants.

Lehmann et al.[32] and Lehmann[206] reviewed the status and prospects for use of the betacyanine pigment amaranthine in consumer products. Another betacyanine pigment, betanine, which is responsible for the red color of beets, has been commercialized as a natural food color. Huang and Von Elbe[207] extracted amaranthine from the leaves of *A. tricolor*, described it chemically, and found it to be comparable in most respects to betanine. The practice of these authors to designate the pigment as "amaranthine" rather than "amaranthin" may have been adopted to minimize confusing it with amaranthin, the lectin. Amaranthine, if properly extracted, would have the advantage for food usage of being a "natural" coloring agent.

F. MEDICINAL PRODUCTS

Amaranth seeds have been shown to be relatively high in tocopherols (vitamin E) and tocotrienols.[32] The tocotrienols have been shown to have antitumor activity,[208] to possess *in vitro* antioxidative activity,[209] and to suppress the synthesis of cholesterol in chicks via the same mechanism as that proposed for squalene.[210] Thus, they may have potential as maintenance drugs for lowering blood serum cholesterol levels in hyperlipoproteinemic individuals. A recent patent application[211] outlined a novel process for the extraction of medicinally useful vitamin E isomers from amaranth and other sources. Qureshi et al.[212] reported a smaller cholesterol-lowering effect of tocotrienols in humans, possibly due to lack of dietary control; Tan et al.[213] observed a hypocholesterolemic effect in humans with palm oil-vitamin E-concentrate and attributed it to high concentrations of tocotrienols.

REFERENCES

1. **Lorenz, K. and Gross, M.,** Saccharides of amaranth, *Nutr. Rep. Int.*, 29(3), 721, 1984.
2. **Fernando, T. and Bean, G.,** Fatty acids and sterols of *Amaranthus tricolor* L., *Food Chem.*, 15, 233, 1984.
3. **Teitz, A.Y., Gorski, S.F., and McDonald, M.B.,** The dormancy of livid amaranth (*Amaranthus lividus* L.), *Seed Sci. Technol.*, 18, 781, 1990.
4. **Saunders, R.M. and Becker, R.,** *Amaranthus*: a potential food and feed resource. in *Advances in Cereal Science and Technology*, Vol. 6, Pomeranz, Y., Ed., American Association of Cereal Chemists St. Paul, MN, 1984.
5. **Bressani, R.,** The proteins of grain amaranth, *Food Rev. Int.*, 5(1), 13, 1989.
6. **Vietmeyer, N.D.,** The revival of the amaranth, *Ceres*, 15(5), 43, 1982.
7. **Hemmendinger, A. and Laidig, G.L.,** VIII. Grain amaranth, *Amaranthus* spp., in *Feasibility of Introducing New Crops: Production-Marketing-Consumption (PMC) Systems*, Knox, E.G. and Theisen, A.A., Eds. A report prepared by Soil and Land Use Technology, Inc., Columbia, MD. Rodale Press, Emmaus, PA, 1981.
8. National Research Council, Amaranth: Modern Prospects for an Ancient Crop, National Academy Press, Washington, D.C., 1984.
9. National Academy of Sciences, Underexploited Tropical Plants with Promising Economic Value, National Academy of Sciences, Washington, D.C., 1975.
10. **Vietmeyer, N.D.,** Lesser-known plants of potential use in agriculture and forestry, *Science*, 232, 1379, 1986.
11. **Bressani, R.,** Amaranth: the nutritive value and potential uses of the grain and by-products, *Food Nutr. Bull.*, 10(2), 49, 1988.
12. **Sattaur, O.,** Botanical entrepreneurship, *Ceres*, 55(1), 17, 1991.
13. **Marx, J.L.,** Amaranth: a comeback for the food of the Aztecs?, *Science*, 198, 40, 1977.
14. **Gross, R., Koch, F., Malaga, A.F., de Miranda, A.F., Schoeneberger, H., and Trugo, L.C.,** Chemical composition and protein quality of some local Andean food sources, *Food Chem.*, 34, 25, 1989.
15. **Opute, F.I.,** Seed lipids of the grain amaranths, *J. Exp. Bot.*, 30(116), 601, 1979.
16. **Ayorinde, F.O., Ologunde, M.O., Nana, E.Y., Bernard, B.N., Afolabi, O.A., Oke, O.L., and Shepard, R.L.,** Determination of fatty acid composition of *Amaranthus* species, *J. Am. Oil Chem. Soc.*, 66(12), 1812, 1989.
17. **Kauffman, C.S.,** Realizing the potential of grain amaranth, *Food Rev. Int.*, 8(1), 5, 1992.
18. **Macler, B., Bamberg, E., Moffatt, E., Bui, H., Minor, H. and Nishioka, L.,** Effects of salinity and nitrogen on growth, productivity and food value of *Amaranthus* in controlled culture, in Proc. 4th Amaranth Symp., Amaranth: Perspectives on production, processing and marketing. Minneapolis, MN, 1990, 129.
19. **Cortella, A.R. and Pochettino, M.L.,** South American grain chenopods and amaranths: a comparative morphology of starch, *Starch/Stärke*, 42(7), 251, 1990.
20. **Breene, W.M.,** Food uses of grain amaranth, *Cereal Foods World*, 36(5), 426, 1991.
21. **Oke, O.L.,** Amaranth, in: *Handbook of Tropical Foods*, Chan, Jr., H.T., Ed., Marcel Dekker, New York, 1.
22. **Teutonico, R.A. and Knorr, D.,** Amaranth: Composition, properties and applications of a rediscovered food crop, *Food Technol.*, 39(4), 49, 1985.

23. **Singhal, R.S. and Kulkarni, P.R.,** Review: Amaranths — An underutilized resource, *Int. J. Food Sci. Tech.,* 23, 125, 1988.
24. **Mallory, E., Blair, M., Weber, L. and Purrman, B.,** *Amaranth Bibliography: Human Nutrition and Food Applications of Grain Amaranth,* Rodale Press, Emmaus, PA, 1989.
25. **Wilkin, P.,** The status of *Amaranthus bouchonii* Thellung within *Amaranthus* L. section *Amaranthus*: New evidence from studies of morphology and isozymes, *Bot. J. Linn., Soc.,* 108, 253, 1992.
26. **Mucciarelli, S.I.L. de, Arellano, M.A.L. de, Cid, J.A., Luques, N.A.G. de and Fernandez, S.,** Chemical composition and nutritive value of the protein of *Amaranthus quitensis, Arch. Latinoam. Nutr.,* 40(1), 69, 1990 (in Spanish).
27. **Becker, R., Wheeler, E.L., Lorenz, K., Stafford, A.E., Grosjean, O.K., Betschart, A.A., and Saunders, R.M.,** A compositional study of amaranth grain, *J. Food Sci.,* 46(4), 1175, 1981.
28. **Singhal, R.S. and Kulkarni, P.R.,** Composition of the seeds of some *Amaranthus* species, *J. Sci. Food Agric.,* 42, 325, 1988.
29. **Correa, A.D., Jokl, L. and Carlsson, R.,** Amino acid composition of some *Amaranthus sp.* grain proteins and of its fractions, *Arch. Latinoam. Nutr.,* 36(3), 466, 1986.
30. **Early, D.K.,** Amaranth production in Mexico and Peru, in *Advances in New Crops,* Proc. 1st Natl. Symp. New Crops Research, Development, Economics. Janick, J. and Simon, J.E., Eds., Timber Press, Portland, OR, 140, 1988.
31. **Sauer, J.D.,** The grain amaranths: A survey of their history and classification, *Ann. Mo. Bot. Gard.,* 37, 561, 1950.
32. **Lehmann, J.W., Breene, W.M., Guanella, T., Putnam, D.H., and Rose, G.,** Minnesota Amaranth Feasibility Study. Parts I and II. A Final Report to the Minnesota State Legislature and Department of Agriculture, Institute for the Development of Amaranth Products, Bricelyn, MN.
33. **Becker, R.,** Preparation, compositional and nutritional implications of amaranth seed oil, *Cereal Foods World,* 34(11), 950, 1989.
34. **Schaeffer, J.R.,** Considerations and constraints to the marketing of amaranth grain as a new commodity, Proc. 4th National Amaranth Symp., Amaranth: Perspectives on Production, Processing and Marketing, Minneapolis, MN, 1990, 97.
35. **Weber, L.E., Applegate, W.W., Baltensperger, D.D., Irwin, M.D., Lehmann, J.W., and Putnam, D.H.,** *Amaranth Grain Production Guide,* Rodale Press, Emmaus, PA, 1990
36. **Espitia, E.,** Amaranth germplasm development and agronomic studies in Mexico, *Food Rev. Int.,* 8(1), 71, 1992.
37. **Kauffman, C.S.,** The status of grain amaranth for the 1990s, *Food Rev. Int.,* 8(1), 165, 1992.
38. **Freese, B.,** They've ADAPTed, *Successful Farming,* 11, 14, 1990.
39. **Tucker, J.B.,** Amaranth: the once and future crop, *Bioscience,* 36, 9, 1986.
40. Anon., Amaranth reaches new levels, *The Forum (Fargo, ND),* p. A6, March 15, 1992.
41. **Lehmann, J.W.,** Grain amaranth. Developing 21st century products from the legendary sister crop of maize, *Int. Food Ingred,* July, 3, 26, 1988.
42. **Weber, L.E., Hubbard, E.S., Nelson, L.A., Putnam, D.H., and Lehmann, J.W.,** Amaranth Grain Production Guide, Rodale Press, Emmaus, PA, 1988.
43. **Colmenares de Ruiz, A.S. and Bressani, R.,** Effect of germination on the chemical composition and nutritive value of amaranth grain, *Cereal Chem.,* 67(6), 519, 1990.
44. **Mora-Escobedo, R., Paredes-López, O. and Gutierrez-Lopez, G.F.,** Effect of germination on the rheological and functional properties of amaranth seeds, *Lebensm.Wiss. Technol.,* 24(3), 241, 1991.
45. **Chavan, J.K. and Kadam, S.S.,** Nutritional improvement of cereals by sprouting, *Crit. Rev. Food Sci. Nutr.,* 28(5), 401, 1989.
46. **Paredes-López, O. and Mora-Escobedo, R.,** Germination of amaranth seeds: effects on nutrient composition and color, *J. Food Sci.,* 54(3), 761, 1989.
47. **Paredes-López, O., Carabez-Trejo, A., Perez-Herrera, S., and Gonzalez-Casteneda, J.,** Influence of germination on physico-chemical properties of amaranth flour and starch microscopic structure, *Starch/Stärke,* 40(8), 290, 1988.
48. **Mendoza, M.C. and Bressani, R.,** Nutritional and functional characteristics of extrusion-cooked amaranth flour, *Cereal Chem.,* 64(4), 218, 1987.
49. **Balasubramanian, T. and Sadasivam, S.,** Changes in carbohydrate and nitrogenous components and amylase activities during germination of grain amaranth, *Plant Food Hum. Nutr.,* 39, 325, 1989.
50. **Paredes-López, O., Barba de la Rosa, A.P. and Carabez-Trejo, A.,** Enzymatic production of high-protein amaranth flour and carbohydrate rich fraction, *J. Food Sci.,* 55(4), 1157, 1990.

51. **Contreras, C.A., Hendershot, L.A., and Chauncey, K.B.,** Knowledge of amaranth and acceptance of an amaranth ready-to-eat breakfast cereal by supermarket and health food store consumers in Lubbock, TX, USA, Proc. 1st Int. Amaranth Congr., Oaxtepec, Mexico, National Autonomous University of Mexico, Mexico City, Mexico, 1991.
52. **Kalinowski, L.S., Navarro, J.P., Concha, A.I.R., Hermoza, G.C., Pacheco, P.A., Choquevilca, Y.C., and Jara, E.V.,** Grain amaranth research in Peru, *Food Rev. Int.*, 8(1), 87, 1992.
53. **Brito, O.,** Nutritional evaluation of two species of amaranths grown locally in Venezuela, Proc. 4th National Amaranth Symp. Amaranth: Perspectives on Production, Processing and Marketing, Minneapolis, MN, 1990, 185.
54. **Lorenz, K.,** *Amaranthus hypochondriacus* — Characteristics of the starch and baking potential of the flour, *Starch/Stärke*, 33(5), 149, 1981.
55. **Pedersen, B., Kalinowski, L.S., and Eggum, B.O.,** The nutritive value of amaranth grain (*Amaranthus caudatus*). I. Protein and minerals of raw and processed grain, *Plant Food Hum. Nutr.*, 36, 309, 1987.
56. **Sanchez-Marroquin, A., Feria-Morales, A., Maya, S., and Ramos-Moreno, V.,** Processing, nutritional quality and sensory evaluation of amaranth enriched corn tortilla, *J. Food Sci.*, 52(5), 1611, 1987.
57. **Sanchez-Marroquin, A. and Maya, S.,** Industrial corn flour enrichment with whole amaranth flour and milling fractions in corn-based products, *Arch. Latinoam. Nutr.*, 35(3), 518, 1985.
58. **Sanchez-Marroquin, A., Domingo, M.V., Maya, S., and Saldana, C.,** Amaranth flour blends and fractions for baking applications, *J. Food Sci.*, 50(3), 789, 1985.
59. **Vargas-Lopez, J.M., Paredes-López, O., and Espitia, E.,** Evaluation of lime heat treatment on some physicochemical properties of amaranth flour by response surface methodology, *Cereal Chem.*, 67(5), 417, 1990.
60. **Sanchez-Marroquin, A., Domingo, M.V., Torres, J.A., and Maya, S.,** Semolina fortification with whole amaranth flour, in Proc. 3rd Amaranth Conf., Rodale Press, Emmaus, PA. 1984, 133.
61. **Slimak, K.M.,** U.S. Patent 4,911,943, 1990.
62. **Jones, M.H.,** *The Allergy Self-help Cookbook*, Rodale Press, Emmaus, PA, 1984.
63. **Irving, D.W., Betschart, A.A., and Saunders, R.M.,** Morphological studies on *Amaranthus cruentus, J. Food Sci.*, 46(4), 1170, 1981.
64. **Irving, D.W. and Becker, R.,** Seed structure and composition of potential new crops, *Food Microst.*, 4, 43, 1985.
65. **Betschart, A.A., Irving, D.W., Shepherd, A.D., and Saunders, R.M.,** *Amaranthus cruentus*: milling characteristics, distribution of nutrients within seed components, and the effects of temperature on nutritional quality, *J. Food Sci.*, 46(4), 1181, 1981.
66. **Becker, R., Irving, D.W., and Saunders, R.M.,** Production of debranned amaranth flour by stone milling. *Lebensm. Wiss. Technol.*, 19, 372, 1986.
67. **Sanchez-Marroquin, A., Maya, S., and Domingo, M.V.,** Milling procedures and air classification of amaranth flours, *Arch. Latinoam. Nutr.*, 35(4), 620, 1985.
68. **Sanchez-Marroquin, A., Del Valle, F.R., Escobedo, M., Avitia, R., Maya, S., and Vega, M.,** Evaluation of whole amaranth (*Amaranthus cruentus*) flour, its air-classified fractions, and blends of these with wheat and oats as possible components for infant formulas, *J. Food Sci.*, 51(5), 1231, 1986.
69. **Wiesner, T.A.,** Milling Evaluation and Protein Isolates of Amaranth Grain, Ph.D. thesis, North Dakota State University, Fargo; *Diss. Abstr. Int.*, 46(10), 1985, Order No. DA8528060.
70. **Saunders, S.R.,** Milling Amaranth with Tangential Abrasive Dehulling Device, M.S. thesis, Kansas State Univeristy, Manhattan, 1987.
71. **Barba de la Rosa, A.P., Paredes-López, O., Cárabez-Trejo, A., and Ordorica-Falomir, C.,** Enzymatic hydrolysis of amaranth flour — Differential scanning calorimetry and scanning electron microscopy studies. *Starch/Stärke*, 41(11), 424, 1989.
72. **Sanchez-Marroquin, A., Maya, S., and Domingo, M.V.,** Effect of heat treatment and milling on the seed, flour, rheology and baking quality of some amaranth ecotypes, *Arch. Latinoam. Nutr.*, 35(4), 603, 1985.
73. **Pant, K.C.,** Effect of heat processing (popping) on protein nutritional quality of grain amaranth, *Nutr. Rep. Int.*, 32(5), 1089, 1985.
74. **Bressani, R., Kalinowski, L.S., Ortiz, M.A. and Elias, L.G.,** Nutritional evaluation of roasted, flaked and popped *A. caudatus*, *Arch. Latinoam. Nutr.*, 37(3), 525, 1987.

75. **Tovar, L.R., Brito, E., Takahashi, T., Miyazawa, T., Soriano, J., and Fujimoto, K.,** Dry heat popping of amaranth seed might damage some of its amino acids, *Plant Food Hum. Nutr.*, 39, 299, 1989.
76. **Imeri, A.G., Flores, R., Elias, L.G., and Bressani, R.,** Effect of processing and of amino acid supplementation on the protein quality of amaranth (*Amaranthus caudatus*), *Arch. Latinoam. Nutr.*, 37(1), 160, 1987. (in Spanish).
77. **Singhal, R.S. and Kulkarni, P.R.,** Puffing effects on functional properties of *Amaranthus paniculatus* (Rajgeera) seed flour, *J. Food Sci.*, 56(4), 1121, 1991.
78. **Singhal, R.S. and Kulkarni, P.R.,** Detection of adulteration of the spice poppy seeds (*Papaver somniferum*) with *Amaranthus paniculatus* (Rajgeera) seeds, *J. Food Qual.*, 13, 375, 1990.
79. **Hubbard, E.S. and Guanella, T.J.,** U.S. Patent 5,069,923, 1991.
80. **Harper, J.M.,** *Extrusion of Foods*, Vol. 1. CRC Press, Boca Raton, FL, 1981.
81. **Schnetzler, K.A.,** Twin-Screw Extrusion Processing of Amaranth, M.S. thesis, University of Minnesota, Minneapolis, 1992.
82. **Hauck, B.W.,** Is the single screw extruder about to become a dinosaur?, *Petfood Ind.*, 30, 16, 1988.
83. **Del Valle, F.R., Sanchez-Marroquin, A., Escobedo, M., Avitia, R., Bourges, H., Maya, S., Vega, M., and Becker, R.,** Development and evaluation of a low cost amaranth (*Amaranthus cruentus*) containing food for preschool children, *Arch. Latinoam. Nutr.*, 37(3), 480, 1987.
84. **Koeppe, S.J., Harris, P.L., Hanna, M.A., Rupnow, J.H., Walker, C.E., and Cuppett, S.L.,** Physical properties and some nutritional characteristics of an extrusion product with defatted amaranth seeds and defatted maize gluten meal (80:20 ratio), *Cereal Chem.*, 64(4), 332, 1987.
85. **Tillman, P.B. and Waldroup, P.W.,** Assessment of extruded grain amaranth as a feed ingredient for broilers. I. Apparent metabolizable energy values, *Poultry Sci.*, 67, 641, 1988.
86. **Vargas-López, J.M., Paredes-López, O., and Ramirez-Wong, B.,** Physicochemical properties of extrusion-cooked amaranth under alkaline conditions, *Cereal Chem.*, 68(6), 610, 1991.
87. **Newman, C.W., Danielson, A.D., Newman, R.K., Pond, W.G., Busch, R.E., and Lewis, B.A.,** Evaluation of the protein quality of raw, boiled and extruded amaranth, *Cereal Foods World*, 37(7) (Abstr.), 566, 1992. (Abstract)
88. **Reddy, N.S., Waghmare, S.Y., and Pande, V.,** Formulation and evaluation of home-made weaning mixtures based on local foods, *Food Nutr. Bull.*, 12(2), 138, 1990.
89. **Del Valle, F.R., Escobedo, M., Sanchez-Marroquin, A., Bourges, H., Bock, M.A., and Biemer, P.,** Nitrogen balance in infants fed formulas containing amaranth or a soy-oats formula, *Cereal Chem.*, 69(2), 156, 1992.
90. **Kulkarni, K.D., Kulkarni, D.N., and Ingle, U.M.,** Sorghum malt-based weaning food formulations: preparation, functional properties, and nutritive value, *Food Nutr. Bull.*, 13(4), 322, 1991.
91. **Harper, J.M. and Jansen, G.R.,** Production of nutritious precooked foods in developing countries by low-cost extrusion technology, *Food Rev. Int.*, 1(1), 27, 1985.
92. **Ketiku, A.O. and Olusanya, J.O.,** Nutrient composition of multimixes for use as weaning foods in Nigeria, *Food Chem.*, 21, 47, 1986.
93. **Imeri, A.G., Elias, L.G., and Bressani, R.,** Amaranth: A technological alternative for child feeding, *Arch. Latinoam. Nutr.*, 37(1), 147, 1987, (in Spanish).
94. **Rathod, P. and Udipi, S.A.,** The nutritional quality and acceptability of weaning foods incorporating amaranth. *Food Nutr. Bull.*, 13(1), 58, 1991.
95. **Gupta, C. and Sehgal, S.,** Development, acceptability and nutritional value of weaning mixtures. *Plant Food Hum. Nutr.*, 41, 107, 1991.
96. **Whittaker, P. and Ologunde, M.O.,** Study of iron bioavailability in a native Nigerian grain amaranth cereal for young children, using a rat model, *Cereal Chem.*, 67(5), 505, 1990.
97. **Ologunde, M.O., Shepard, R.L., Afolabi, O.A., and Oke, O.L.,** Bioavailability to rats of iron from fortified grain amaranth flour, *Int. J. Food Sci. Technol.*, 26, 493, 1991.
98. **Bressani, R., Sanchez-Marroquin, A., and Morales, E.,** Chemical composition of grain amaranth cultivars and effects of processing on their nutritional quality, *Food Rev. Intl*, 8(1), 23, 1992.
99. **Salazar de Buckle, T.,** Composite flours: prospects for their application in developing countries — the case of the Andean pact countries, *Food Rev. Int.*, 4(2), 237, 1988.
100. **Fellers, D.A. and Bean, M.M.,** Composite flours, *Food Rev. Int.*, 4(2), 213, 1988.
101. **Bressani, R., Elias, L.G. and Garcia-Soto, A.,** Limiting amino acids in raw and processed amaranth grain protein from biological tests, *Plant Food Hum. Nutr.*, 39, 223, 1989.

102. **Bressani, R., Elias, L.G., Gonzalez, J.M., and Gomez-Brenes, R.,** The chemical composition and protein quality of amaranth grain germ plasm in Guatemala, *Arch. Latinoam. Nutr.*, 37(2), 364, 1987.
103. **Imeri, A.G., Gonzalez, J.M., Flores, R., Elias, L.G., and Bressani, R.,** Genetic variability and correlations between yield, grain size, chemical composition and protein quality of 25 varieties of amaranth (*Amaranthus caudatus*), *Arch. Latinoam. Nutr.*, 37(1), 132, 1987 (in Spanish).
104. **Sanchez-Marroquin, A.,** Two forgotten crops of agroindustrial importance: amaranth and quinoa, *Arch. Latinoam. Nutr.*, 33(1), 11, 1983 (in Spanish).
105. **Pant, K.C.,** Studies on the nutritional quality of grain amaranths, *Nutr. Rep. Int.*, 28(6), 1445, 1983.
106. **Pedersen, B., Hallgren, L., Hansen, I., and Eggum, B.O.,** The nutritive value of amaranth grain (*Amaranthus caudatus*). II. As a supplement to cereals, *Plant Food Hum. Nutr.*, 36, 325, 1987.
107. **Tovar, L.R. and Carpenter, K.J.,** The effects of alkali-cooking of corn and supplementation with amaranth seed on its deficiencies in lysine and tryptophan, *Arch. Latinoam. Nutr.*, 32(4), 961, 1982.
108. **Pacheco de Delahaye, E. and Portillo, M.,** Enrichment of white corn flour (*Zea mays*) with amaranth (*Amaranthus* sp.) seed flour, *Arch. Latinoam. Nutr.*, 40(3), 360, 1990 (in Spanish).
109. **Morales, E., Lembcke, J. and Graham, G.G.,** Nutritional value for young children of grain amaranth and maize-amaranth mixtures: effect of processing, *J. Nutr.*, 118(1), 78, 1988.
110. **Graham, G.G., Lembcke, J. and Morales, E.,** Protein value for children of soft-endosperm maize, alone and with toasted amaranth flour, *Nutr. Res.*, 9, 859, 1989.
111. **Bressani, R., Gonzalez, J.M., Zuniga, J., Breuner, M., and Elias, L.G.,** Yield, selected chemical composition and nutritive value of 14 selections of amaranth grain representing four species, *J. Sci. Food Agric.*, 38, 347, 1987.
112. **Maeda, E.E.,** Effect of solar dehyration on amino acid pattern and available lysine content in four tropical leafy vegetables, *Ecol. Food Nutr.*, 16(3), 273, 1985.
113. **Makus, D.J.,** Composition and nutritive value of vegetable amaranth as affected by stage of growth, environment and method of preparation, Proc. 4th National Amaranth Symp., Amaranth: Perspectives on Production, Processing and Marketing, August 23 to 25, Minneapolis, MN, 1990, 35.
114. **Abbott, J.A. and Campbell, T.A.,** Sensory evaluation of vegetable amaranth (*Amaranthus* spp.), *HortScience*, 17(3), 409, 1982.
115. **Walters, R.D., Coffey, D.L., and Sams, C.E.,** Fiber, nitrate, and protein content of *Amaranthus* accessions as affected by soil nitrogen application and harvest date, *HortScience*, 23(2), 338, 1988.
116. **Mallory, E., Blair, M., Weber, L., and Purrman, B.,** *Amaranth Bbibliography: Vegetable Amaranth and Leaf Protein Concentrates*, Rodale Press, Emmaus, PA, 1989.
117. **Alfaro, M.A., Martinez, A., Ramirez, R., and Bressani, R.,** Yield and chemical composition of the biomass of amaranth (*A. hypochondriacus* L.) at three different physiological stages, *Arch. Latinoamer. Nutr.*, 37(1), 108, 1987 (in Spanish).
118. **Lazanyi, J., Kapocsi, I., Bene, S., Fazekas, M. and Chrappan, G.,** Agronomic evaluation of some cultivated and wild amaranth species (*Amaranthus spp.*), *Novenytermeles*, 37(3), 219, 1988 (in Hungarian).
119. **Mercadante, A.Z. and Rodriguez-Amaya, D.B.,** Carotenoid composition and vitamin A value of some native Brazilian green leafy vegetables, *Int. J. Food Sci.*, Technol. 25, 213, 1990.
120. **Van der Pol, F., Purnomo, S.U. and van Rosmalen, H.A.,** *Trans-cis* isomerisation of carotenes and its effect on the vitamin A potency of some common Indonesian foods, *Nutr. Rep. Int.*, 37(4), 785, 1988.
121. **Prakash, D. and Pal, M.,** Nutritional and antinutritional composition of vegetable and grain amaranth leaves, *J. Sci. Food Agric.*, 57, 573, 1991.
122. **Latunde-Dada, G.O.,** Effect of processing on iron levels in and availability from some Nigerian vegetables, *J. Sci. Food Agric.*, 53, 355, 1990.
123. **Mallory, E., Blair, M., Weber, L. and Purrman, B.,** *Amaranth Bibliography: Forage and Grain for Livestock*, Rodale Press, Emmaus, PA, 1989.
124. **Tillman, P.B. and Waldroup, P.W.,** Processing grain amaranth for use in broiler diets, *Poultry Sci.*, 65, 1960, 1986.
125. **Tillman, P.B. and Waldroup, P.W.,** Assessment of extruded grain amaranth as a feed ingredient for broilers. II. Apparent amino acid availability values, *Poultry Sci.*, 67, 647, 1988.
126. **Tillman, P.B. and Waldroup, P.W.,** Performance and yields of broilers fed extruded grain amaranth and grown to market weight, *Poultry Sci.*, 67, 743, 1988.
127. **Acar, N., Vohra, P., Becker, R., Hanners, G.D., and Saunders R.M.,** Nutritional evaluation of grain amaranth for growing chickens, *Poultry Sci.*, 67, 1166, 1988.

128. **Laovoravit, N., Kratzer, F.H., and Becker, R.,** The nutritional value of amaranth for feeding chickens, *Poultry Sci.*, 65, 1365, 1986.
129. **Tillman, P.B. and Waldroup, P.W.,** Effects of feeding extruded grain amaranth to laying hens, *Poultry Sci.*, 66, 1697, 1987.
130. **Waldroup, P.W., Hellwig, H.M., Longer, D.E., and Endres, C.S.,** The utilization of grain amaranth by broiler chickens, *Poultry Sci.*, 64, 759, 1985.
131. **Vohra, P., Bersch, S., Acar, N., and Becker, R.,** Use of grain amaranth in diets of Japanese quail and chickens, *Nutr. Rep. Int.*, 39(6), 1227, 1989.
132. **Alfaro, M.A., Ramirez, R., Martinez, A., and Bressani, R.,** Evaluation of different levels of amaranth flour (vegetable parts) in substitution of alfalfa meal for growing rabbits, *Arch. Latinoam. Nutr.*, 37(1), 174, 1987 (in Spanish).
133. **Chairatanayuth, P.,** Inclusion of amaranth crop residues in diet for cattle, *Food Rev. Int.*, 8(1), 159, 1992.
134. **Pond, W.G. and Lehmann, J.W.,** Nutritive value of a vegetable amaranth cultivar for growing lambs, *J. Anim. Sci.*, 67, 3036, 1989.
135. **Pond, W.G., Lehmann, J.W. and Clark, R.,** Utilization of four cultivars of grain amaranth for growth in rats, *Nutr. Rep. Intl.*, 39(5), 1081, 1989.
136. **Lehmann, J.W.,** Anti-nutritional factors in amaranth grain, *Legacy*, 5(1), 6, 1992.
137. **Afolabi, A.O., Oke, O.L., and Umoh, I.B.,** Preliminary studies on the nutritive value of some cereal-like grains, *Nutr. Rep. Int.*, 24(2), 389, 1981.
138. **Makus, D.J.,** Evaluation of amaranth as a potential greens crop in the mid-south, *HortScience*, 19(6), 881, 1984.
139. **Vityakon, P. and Standal, B.R.,** Oxalate in vegetable amaranth (*Amaranthus gangeticus*). Forms, contents, and their possible implications for human health, *J. Sci. Food Agric.*, 48, 469, 1989.
140. **Wohlgemuth, K., Schamber, G.J., Misek, A.R., and Crenshaw, J.D.,** Pigweed is toxic to pigs, *N. D. Farm Res.*, 44(4), 21, 1987.
141. **Flores, H.E.,** Biotechnology and amaranth, Proc. 4th National Amaranth Symp. Amaranth: Perspectives on Production, Processing and Marketing, Minneapolis, MN, 1990, 143.
142. **Flores, H.E., Thier, A., and Galston, A.W.,** *In vitro* culture of grain and vegetable amaranths (*Amaranthus spp.*), *Am. J. Bot.*, 69(7), 1049, 1982.
143. **Teutonico, R.A. and Knorr, D.,** Nondestructive method for determination of water-soluble oxalate in cultured *Amaranthus tricolor* cells, *J. Agric. Food Chem.*, 33(1), 60, 1985.
144. **Van der Krol, A.R., Mol, J.N.M. and Stuitje, A.R.,** Antisense genes in plants: an overview, *Gene*, 72, 45, 1988.
145. **Gupta, K., Barat, G.K., Wagle, D.S., and Chawla, H.K.L.,** Nutrient contents and antinutritional factors in conventional and non-conventional leafy vegetables, *Food Chem.*, 31, 105, 1989.
146. **Correa, A.D., Jokl, L., and Carlsson, R.,** Chemical constituents, *in vitro* protein digestibility, and presence of antinutritional substances in amaranth grains, *Arch. Latinoam. Nutr.*, 36(2), 319, 1986.
147. **Osuntogun, A.B. and Oke, O.L.,** A note on the nutritive value of amaranth seeds, *Food Chem.*, 12, 287, 1983.
148. **Koeppe, S.J. and Rupnow, J.H.,** Purification and characterization of a lectin from the seeds of amaranth (*Amaranthus cruentus*), *J. Food Sci.*, 53(8), 1412, 1988.
149. **Calderon de la Barca, A.M. and Vazquez-Moreno, L.,** *Amaranthus cruentus* lectin: purification, stability, and some biochemical properties, *J. Food Biochem.*, 12, 117, 1988.
150. **Zenteno, E. and Ochoa, J.-L.,** Purification of a lectin from *Amaranthus leucocarpus* by affinity chromatography, *Phytochemistry*, 27(2), 313, 1988.
151. **Calderon de la Barca, A.M., Ochoa, J.L., and Valencia, M.E.,** Effect of the extraction of a hemagglutinin on the nutritive value of *Amaranthus leucocarpus* seeds, *J. Food Sci.*, 50(6), 1700, 1985.
152. **Koeppe, S.J., Rupnow, J.H., Walker, C.E., and Davis, A.,** Isolation and heat stability of trypsin inhibitors in amaranth (*Amaranthus hypochondriacus*), *J. Food Sci.*, 50(5), 1519, 1985.
153. **Fernando, T. and Bean, G.,** Production of aflatoxins on amaranth seeds by *Aspergillus* spp., *Food Chem.*, 17, 33, 1985.
154. **Makus, D.J.,** Aluminum accumulation in vegetable amaranth grown in a soil with adjusted pH values, *HortScience*, 24(3), 460, 1989.
155. **Fernando, T. and Bean, G.,** A comparison of the fatty acids and sterols of weedy and vegetable species of *Amaranthus* spp, *J. Am. Oil Chem. Soc.*, 62(1), 89, 1985.

156. **Pond, W.G., Lehmann, J.W., Elmore, R., Husby, F., Calvert, C.C., Newman, C.W., Lewis, B., Harrold, R.L., and Froseth, J.**, Feeding value of raw or heated grain amaranth germplasm, *Anim. Feed Sci. Technol.*, 33, 221, 1991.
157. **Hill, R.M. and Rawate, P.D.**, Evaluation of food potential, some toxicological aspects, and preparation of a protein isolate from the aerial part of amaranth (pigweed), *J. Agric. Food Chem.*, 30(3), 465, 1982.
158. **Garcia, L.A., Alfaro, M.A., and Bressani, R.**, Digestibility and protein quality of raw and heat-processed defatted and nondefatted flours prepared with three amaranth species, *J. Agric. Food Chem.*, 35(4), 604, 1987.
159. **Bressani, R., Gonzalez, J.M., Elias, L.G., and Melgar, M.**, Effect of fertilizer application on the yield, protein and fat content, and protein quality of raw and cooked grain of three amaranth species, *Plant Food Hum. Nutr.*, 37, 59, 1987.
160. **Cheeke, P.R. and Bronson, J.**, Feeding trials with *Amaranthus* grain, forage and leaf protein concentrates, in: Proc. 2nd Amaranth Conf., Rodale Press, Emmaus, PA, 1980, 5.
161. **Lehmann, J.W.**, Proteins of grain amaranths, *Legacy*, 2(1), 3, 1989.
162. **Lehmann, J.W.**, Carbohydrates of amaranth, *Legacy*, 1(1), 4, 1988.
163. **Tomita, Y., Sugimoto, Y., Sakamoto, S., and Fuwa, H.**, Some properties of starches of grain amaranths and several millets, *J. Nutr. Sci. Vitaminol.*, 27(5), 471, 1981.
164. **Okuno, K. and Sakaguchi, S.**, Differentiation of starch property in perisperm of grain amaranths, JARQ, 18(1), 1, 1984.
165. **Paredes-López, O., Schevenin, M.L., Hernandez-López, D., and Carabez-Trejo, A.**, Amaranth starch—Isolation and partial characterization, *Starch/Stärke*, 41(6), 205, 1989.
166. **Wankhede, D.B., Gunjal, B.B., Sawate, R.A., Patil, H.B., Bhosale, M.B., Gahilod, A.T., and Walde, S.G.**, Studies on isolation and characterization of starch from Rajgeera grains (*Amaranthus paniculatus* Lin.), *Starch/Stärke*, 41(5), 167, 1989.
167. **Yanez, G.A., Messinger, J.K., Walker, C.E., and Rupnow, J.**, *Amaranthus hypochondriacus*: starch isolation and partial characterization, *Cereal Chem.*, 63(3), 273, 1986.
168. **Paredes-López, O. and Hernandez-López, D.**, Application of differential scanning calorimetry to amaranth starch gelatinization—Influence of water, solutes and annealing, *Starch/Stärke*, 43(2), 57, 1991.
169. **Stone, L.A. and Lorenz, K.**, The starch of *Amaranthus*: physico-chemical properties and functional characteristics, *Starch/Stärke*, 36(7), 232, 1984.
170. **Singhal, R.S. and Kulkarni, P.R.**, Some properties of *Amaranthus paniculatus* (Rajgeera) starch pastes, *Starch/Stärke*, 42(1), 5, 1990.
171. **Singhal, R.S. and Kulkarni, P.R.**, Utilisation of *Amaranthus paniculatus* (Rajgeera) starch in salad dressing, *Starch/Stärke*, 42(2), 52, 1990.
172. **Singhal, R.S. and Kulkarni, P.R.**, Studies on applicability of *Amaranthus paniculatus* (Rajgeera) starch for custard preparation, *Starch/Stärke*, 42(3), 102, 1990.
173. **Singhal, R.S. and Kulkarni, P.R.**, Studies on cross-linked *A. paniculatus* (Rajgeera) starch, *Starch/Stärke*, 43(1);15, 1991.
174. **Paredes-López, O., Mora-Escobedo, R., and Ordorica-Falomir, C.**, Isolation of amaranth proteins. *Lebensm. Wiss. Technol.*, 21(1), 59, 1988.
175. **Barba de la Rosa, A.P., Gueguen, J., Paredes-López, O., and Viroben, G.**, Fractionation procedures, electrophoretic characterization, and amino acid composition of amaranth seed proteins, *J. Agric. Food Chem.*, 40(6), 931, 1992.
176. **Barba de la Rosa, A.P., Paredes-López, O., and Gueguen, J.**, Characterization of amaranth globulins by ultracentrifugation and chromatographic techniques, *J. Agric. Food Chem.*, 40(6), 937, 1992.
177. **Gorinstein, S., Arnao de Nue, I., and Arruda, P.**, Alcohol-soluble and total proteins from amaranth seeds and their comparison with other cereals, *J. Agric. Food Chem.*, 39(5), 848, 1991.
178. **Gorinstein, S., Moshe, R., Greene, L.J., and Arruda, P.**, Evaluation of four *Amaranthus* species through protein electrophoretic patterns and their amino acid composition, *J. Agric. Food Chem.*, 39(5), 851, 1991.
179. **Konishi, Y., Fumita, Y., Ikeda, K., Okuno, K., and Fuwa, H.**, Isolation and characterization of globulin from seeds of *Amaranthus hypochondriacus* L., *Agric. Biol. Chem.*, 49(5), 1453, 1985.
180. **Konishi, Y. and Yoshimoto, N.**, Amaranth globulin as a heat-stable emulsifying agent, *Agric. Biol. Chem.*, 53(12), 3327, 1989.

181. **Konishi, Y., Horikawa, K., Oku, Y., Azumaya, J., and Nakatani, N.,** Extraction of two albumin fractions from amaranth grains: comparison of some physicochemical properties and the putative localization in the grains, *Agric. Biol. Chem.*, 55(11), 2745, 1991.
182. **Mora-Escobedo, R., Paredes-López, O., and Ordorica-Falomir, C.,** Characterization of albumins and globulins from amaranth, *Lebensm. Wiss. Technol.*, 23(6), 484, 1990.
183. **Marcone, M.F., and Yada, R.Y.,** Isolation, purification, and characterization of the oligomeric seed globulin from *Amaranthus hypochondriacus*, *Agric. Biol. Chem.*, 55(9), 2281, 1991.
184. **Soriano-Santos, J., Iwabuchi, S., and Fujimoto, K.,** Solubility of amaranth seed proteins in sodium sulphate and sodium chloride: The main factor in quantitative extraction for analysis, *Int. J. Food Sci. Technol.*, 27, 337, 1992.
185. **Bressani, R. and Garcia-Vela, L.A.,** Protein fractions in amaranth grain and their chemical characterization, *J. Agric. Food Chem.*, 38(5), 1205, 1990.
186. **Konishi, Y., Azumaya, J., Horikawa, K. and Nakatani, N.,** A pitfall in determining the globulin/albumin ratio in amaranth grains, *J. Nutr. Sci. Vitaminol.*, 38, 215, 1992.
187. **Rinderle, S.J., Goldstein, I.J., Matta, K.L., and Ratcliffe, R.M.,** Isolation and characterization of amaranthin, a lectin present in the seeds of *Amaranthus cruentus*, that recognizes the T- (or cryptic T)-antigen, *J. Biol. Chem.*, 264(27), 16123, 1989.
188. Anon., Cancer predictor. Herbal screening, *Longevity,* November, 1989, 80.
189. **Lyon, C.K. and Becker, R.,** Extraction and refining of oil from amaranth seed, *J. Am. Oil Chem. Soc.*, 64(2), 233, 1987.
190. **Garcia, L.A., Alfaro, M.A., and Bressani, R.,** Digestibility and nutritional value of crude oil from three amaranth species, *J. Am. Oil Chem. Soc.*, 64(3), 371, 1987.
191. **Lorenz, K. and Hwang, Y.S.,** Lipids in amaranths, *Nutr. Rep. Int.*, 31(1), 83, 1985.
192. **Singhal, R.S. and Kulkarni, P.R.,** Effect of puffing on oil characteristics of amaranth (Rajgeera) seeds, *J. Am. Oil Chem. Soc.*, 67(12), 952, 1990.
193. **Prakash, D. and Pal, M.,** Seed protein, fat and fatty acid profile of *Amaranthus* species, *J. Sci. Food Agric.*, 58, 145, 1992.
194. **Daulatabad, C.M.J.D. and Hosamani, K.M.,** Epoxy and cyclopropenoid fatty acids in *Amaranthus paniculatus* seed oil, *J. Sci. Food Agric.*, 58, 139, 1992.
195. **Singhal, R.S. and Kulkarni, P.R.,** Amaranth oils: a new source of squalene, *Amaranth Newsletter,* No. 3, 7 to 8, September, 1988.
196. **Dailey, Jr., O.D., Severson, R.F., and Arrendale, R.F.,** Nonpolar lipids of *Amaranthus palmeri* S. Wats. I. Fatty alcohols and wax esters (saturated), *J. Agric. Food Chem.*, 37(5), 1317, 1989.
197. **Lehmann, J.W.,** Lipids of grain and feral amaranths, *Legacy,* 4(1), 2, 1991.
198. **Van Sickle, W.A., Angelastro, M.R., Wilson, P., Cooper, J.R., Marquart, A., and Flanagan, M.A.,** Inhibition of cholesterol synthesis by cyclopropylamine derivatives of squalene in human hepatoblastoma cells in culture, *Lipids,* 27(3), 157, 1992.
199. **Malecka, M.,** The effect of squalene on the heat stability of rapeseed oil and model lipids, *Nahrung,* 35(5), 541, 1991.
200. **Kruse, L.,** Oats are a hot commodity, *Successful Farming.*, 1, 3, 1989.
201. **Przybyla, A.E.,** Formulating fiber into foods, *Food Eng.*, 60(10), 77, 1988.
202. **Pedersen, B., Bach Knudsen, K.E., and Eggum, B.O.,** The nutritive value of amaranth grain (*Amaranthus caudatus*). III. Energy and fibre of raw and processed grain, *Plant Food Hum. Nutr.*, 40, 61, 1990.
203. **Prosky, L., Asp, N.-G., Schweizer, T.F., DeVries, J.W., and Furda, I.,** Determination of insoluble, soluble, and total dietary fiber in foods and food products: Interlaboratory study, *J. Assoc. Off. Anal. Chem.*, 71, 1017, 1988.
204. **Osman, H.,** Dietary fiber composition of common vegetables and fruits in Malaysia, *Food Chem.*, 37, 21, 1990.
205. **Danz, R.A. and Lupton, J.R.,** Physiological effects of dietary amaranth (*Amaranthus cruentus*) on rats, *Cereal Foods World.*, 37(7), 489, 1992.
206. **Lehmann, J.W.,** Pigments of grain and feral amaranths, Legacy. 3(1), 3, 1990.
207. **Huang, A.S. and Von Elbe, J.H.,** Stability comparison of two betacyanine pigments-Amaranthine and betanine, *J. Food Sci.*, 51(3), 670, 1986.
208. **Komiyama, K., Iizuka, K., Yamaoka, M., Watanabe, H., Tsuchiya, N. and Umezawa, I.,** Studies on the biological activity of tocotrienols, *Chem. Pharm. Bull.*, 37, 1369, 1989.

209. **Serbinova, E., Kagan, V., Han, D., and Packer, L.,** Free radical recycling and intramembrane mobility in the antioxidant properties of alpha-tocopherol and alpha-tocotrienol, *Free Radical Biol. Med.,* 10, 263, 1991.
210. **Qureshi, A.A., Burger, W.C., Peterson, D.M., and Elson, C.E.,** The structure of an inhibitor of cholesterol biosynthesis isolated from barley, *J. Biol. Chem.,* 261(23), 10544, 1986.
211. **Qureshi, A.A., Becker, K.W., Wells, D.M., and Lane, R.H.,** Processes for recovering tocotrienols, tocopherols and tocotrienol-like compounds, International Publication Number WO 91/17985, 1991.
212. **Qureshi, A.A., Qureshi, N., Wright, J.J.K., Shen, Z., Kramer, G., Gapor, A., Chong Y.H., DeWitt, G., Ong, A.S.H., Peterson, D.M., and Bradlow, B.A.,** Lowering of serum cholesterol in hypercholesterolemic humans by tocotrienols (palmvitee), *Am. J. Clin. Nutr.,* 53(4), 1021S, 1991.
213. **Tan, D.T.S., Khor, H.T., Low, W.H.S., Ali, A., and Gapor, A.,** Effect of a palm-oil-vitamin E concentrate on the serum and lipoprotein lipids in humans, *Am. J. Clin. Nutr.,* 53(4), 1027S, 1991.

Chapter 10

Composition and Nutritional Properties of Amaranth

Ricardo Bressani

CONTENTS

I. INTRODUCTION

The value of a food crop depends on a number of attributes of importance all along the food chain, that is, from production to consumption, on its use in food consumption systems, and on the nutrient contribution to the consumer when ingested alone or as part of a food system. In the case of seed amaranth, a grain apparently very popular in the Aztec, Mayan, and Incan civilizations of the past, it would appear that its recovery as a crop with significant potential depends on several important facts, such as that it can be grown in many areas of the world with relatively good yields; that it is tolerant to drought, high temperatures, and pests; that it is a good solar energy converter, since it has a C_4 metabolism; and that it may be a multiple-purpose crop. Furthermore, the seed has an attractive chemical composition and superior nutritive value when compared on an equal basis to other grains, whether cereals, food legumes, or oil seed crops. The topic of this chapter has been reviewed before;[1-6] however, while presenting data which is already well known, attention will be given to various aspects of its chemical composition and nutritive value. Other sections include the effects of processing on nutritive value, amaranth protein complementary effect to cereal grains and other protein sources, the results of the evaluation of the protein quality of amaranth in human subjects, and its uses in various food products.

0-8493-5374-2/94/$0.00+$.50

Table 1 **Chemical composition of amaranth seed (g/100 g)**

Nutrient	1	2	3	4	5	6	7
Moisture	9.7	10.4	10.7	11.6	10.6	10.3	—
Protein[a]	17.0	15.6	14.9	16.5	15.6	19.3	16.2
Fat[a]	8.1	7.8	9.1	6.9	7.5	8.6	5.8
Ash[a]	3.5	3.5	2.9	4.4	4.0	2.6	3.6
Crude fiber[a]	3.4	3.1	2.8	5.8	7.4	—	2.7
Carbohydrate[a]	67.9	70.0	70.3	66.4	65.5	—	71.7

[a] Moisture-free basis.

Note: (1) *A. cruentus*, (2) *A. cruentus*, (3) *A. caudatus*, (4) *A. edulis*, (5) *A. hybridus*, (6) *A. retroflexus*, (7) *A. mantegazzianus*.

II. CHEMICAL COMPOSITION OF AMARANTH GRAIN

The proximate chemical composition of the seed of various amaranth species is shown in Table 1. The average crude protein content among species varies between 14.9 and 19.3% (N × 6.25); however, there is a high variation between selections within one species. The data in the table suggest that *A. caudatus* is lower in protein content as compared to *A. cruentus* and *A. hypochondriacus*, where various analyses were averaged. The variability between selections within a species must probably be due to genetic effects. For example, Imeri et al.[7] found values ranging from 12.5 to 15.8% (dry weight basis or d.wt.b.) for *A. caudatus* harvested from the same locality, in the same year and cultivated under equal agronomic conditions. A high variation was found for *A. cruentus*, with a range of 16.7 to 18.4% (d.wt.b.), and for *A. hypochondriacus* with a range of 15.2 to 17.8% (d.wt.b.),[8] grown under equal agricultural practices and in the same locality and year. These data suggest the possibility of selecting varieties within species with a protein content above 15%, for example, in order to maintain the concept that amaranth grain indeed has a higher protein content than cereal grains. Although genetic composition may account for the variability reported in the literature, environmental factors and cultural practices as factors influencing protein content should not be discarded. Although various factors to convert total nitrogen to protein nitrogen have been used (5.2 and 6.25), the variability remains if the same value is used for all samples. Very little information is available on the effects of nitrogen fertilizer applications and/or organic matter on the seed protein content. Some results available indicate none or very little influence on the three main *Amaranthus* species.[9,10] Significant increment in crude protein content, however, has been reported from applications of 90 and 180 mg of N per kilogram of soil in four amaranth varieties.[11] Dhan Prakash *et al.*[12] indicated that for a variety of *A. hypochondriacus* (AG-21) which they called genetically stable, there was no year interaction for protein content. Another factor that could influence protein content is the amount of NPN present in the seeds, which may be variable with respect to total N content; however, not much data are available. Balasubramanian and Sadasivan[13] showed *A. hypochondriacus* to contain 0.30% NPN, and Konishi *et al.*[14] indicated the ratio of NPN to albumin-globulins to be 3:7. There is need for a better characterization of NPN in amaranth seed species and varieties. Another factor that could influence protein content in the seed of amaranth is the proportion of germ with respect to the whole grain. Various authors[3,15-17] have shown the germ fraction to contain around 35% protein as compared to the perisperm fraction with about 7 to 9% protein. Likewise, the germ fraction represents a high proportion of the whole grain weight, thus contributing significant amounts of protein to the whole seed.

Fat content is another chemical component of much interest in amaranth grain, since its relatively high amount makes the seed higher in calorie content as compared to cereal grains. Variability is relatively high, with average values apparently higher for *A. caudatus* than for *A. cruentus* and both higher than for *A. hypochondriacus*. Other species show values between these three species. As for protein, there seems to be a genetic influence. The variability for 25 selections of *A. caudatus* was 7.2 to 13.1%;[7] for 7 selections of *A. hypochondriacus* this was 8.8 to 12.1%, and for 3 of *A. cruentus* the variability was 10.6 to 14.6% (d.wt.b.) for materials grown in the same locality.[8] In some 36 selections, the variability in fat content was 6.3 to 11.9% (d.wt.b.).[10] Fatty acid content in amaranth oil is relatively constant among species.[8]

The ash content between species is from 2.6 to 4.4% and the variability within species is relatively small.[1,2,8,15,19] Table 2 summarizes the concentration of various mineral elements in amaranth.[8] The content of the mineral is relatively constant, possibly since all were grown in the same locality and same

Table 2 **Mineral content in the seed of different amaranth species (mg/g)[a]**

	Amaranthus			
Mineral	***Caudatus***	***Hybridus***	***Cruentus***	***Hypochondriacus***
Phosphorus	570	565	556	600
Potassium	532	532	525	563
Calcium	217	303	242	244
Magnesium	319	344	344	342
Sodium	22	26	25	23
Iron	21	104	26	53
Copper	0.86	4.10	1.69	2.40
Manganese	2.9	5.2	3.4	3.5
Zinc	3.4	3.4	4.2	3.8
No. of samples	3	1	3	7

[a] Dry weight basis. All grown in the same locality.

From Bressani, R. González, J.M., Zúñiga, J., Breuner, M., and Elias, L.G., *J. Sci. Food Agric.*, 38, 347, 1987. With permission.

year. The levels of P, Ca, K, and Mg are in general higher than the levels found in common cereal grains. The P/Ca ratio is of interest and varies between 1.9 to 2.6. It is also of nutritional interest to point out the relatively high level of iron, which, again, is higher than those found in cereal grains.

There is a limited amount of information on the dietary fiber of amaranth grain. Values for the three more important grain species are shown in Table 3 from the studies by Pedersen *et al.*[20] and Bressani *et al.*[21] The values are very similar between species, and the percent soluble to the total varies between 18.0 to 48.1%. Pedersen *et al.*[19] established the fiber composition. The non-starch polysaccharides (NSP) made up about 6 and 7% in the light-colored seed and the dark seeds, respectively. The sugars included arabinose, xylose, mannose, galactose, glucose, and uronic acid in very similar amounts in varieties of *A. caudatus*. The dark seeds had a higher content of lignin than the light-colored seeds.

Starch is the most abundant chemical component in amaranth grain. Table 4 shows some values from various authors,[1,2,22] which vary from 48 to 69%. The starch is present mainly as amylopectin, and values from 4.8 to 7.2% of amylose have been reported. Other carbohydrates include sucrose, raffinose, stachyose, and maltose in small variable amounts.[1,23]

The vitamin content of the grain in the three species, *A. caudatus, A. cruentus,* and *A. hypochondriacus* is shown in Table 5. The values are similar between species. The thiamine and niacin content appear in lower concentration than found in cereal grains; riboflavin, however, is present in higher amounts.[1,2,16,24-28]

Various authors have published information on a number of chemical entities which are known to have an antiphysiological effect in animal performance.[16,19,29,30] A summary of values for tannins, phytic acid, nitrates, oxalates, and trypsin inhibitors is presented in Table 6. The ranges shown correspond to various species and crosses between species, since there appears to be little variability. Tannins are present in higher concentration in the dark-colored seed as compared to the light-colored one. As with cereal grains, dehulling of the dark-colored seed decreased tannin content.[30] These authors reported high tannin values in the hulls of the dark-colored seed. Tannins are of importance from the nutritional point of view, since they may affect protein digestibility, reduce mineral bioavailability, and induce a lower palatability. The phytate values in amaranth are higher than those found in rice (0.10 to 0.14%), but lower than values reported for maize and wheat. Lorenz and Wright[30] found phytate to be equally distributed between hulls and the dehulled grain. All authors note only small amounts of heat-labile trypsin and chymotrypsin inhibitor activity.

III. AMINO ACID CONTENT

Various authors have published data on the amino acid content of different amaranth seed species.[1-3,7,8,12,15,16,19,31-33,37-39] The results are summarized in Table 7. Observation of the information indicates that the amino acid content is fairly similar between species. Misra *et al.*,[31] working with wild and cultivated *A. hypochondriacus*, found that protein content varied from 15.3 to 22.1%, but it was not associated with cultural practices. These authors found a significant positive correlation between protein

Table 3 **Dietary fiber content in amaranth seeds**

Species	Dietary Fiber (%)			
	Insoluble	**Soluble**	**Total**	**S/T**
A. caudatus[a,b]	4.6–13.5	2.5–3.6	7.6–16.4	18.0–48.1
A. cruentus[b]	5.5	1.5	7.1	21.5
A. hypochondriacus[b]	7.0	6.5	13.5	41.6

[a] Data from Reference 20; [b] Data from Reference 21.

Table 4 **Starch content in amaranth grain (%)**

Species	Starch	Amylose
Light-colored seed (n = 9)	69 ± 3	—
A. hypochondriacus	62	7.2
A. cruentus	48	4.8–6.4
A. leucosperma		
A. hypochondriacus	62.8	—
Sugars		
Sucrose	1.08–2.26%	—
Raffinose	0.45–1.23	—
Stachyose	0.02–0.15	—
Maltose	0.02–0.36	—

From Saunders, R.M. and Becker, R., Advances in Cereal Science and Technology, Vol. VI, Pomerantz, Y., Ed., *American Association of Cereal Chemists*, Minneapolis, MN, 1984, 357. With permission.

Table 5 **Vitamin content in seed amaranth species (mg/100 g)**

Vitamin	*Amaranthus*		
	Caudatus	***Cruentus***	***Hypochondriacus***
Thiamine	0.10–0.14	0.07–0.10	0.14–0.25
Riboflavin	0.19–0.32	0.19–0.23	0.29–0.32
Niacin	1.00	1.00–1.45	1.00–1.15
Biotin	51.3	42.5	—
Folic acid	42.1	43.8	—
Ascorbic acid	3–7	4.5–4.9	2.8–3.0

Data taken from References 1, 2, 16, 24, 25, 26, 27, and 28.

Table 6 **Content of various compounds with antiphysiological activity**

Compounds	Light-Colored Seed	Dark-Colored Seed
Tannin (cat. equivalent), %	0.08–0.65	0.104–0.116
Phytic acid, %	0.34–0.61	0.52–0.61
Nitrates, mg %	29.0–61.6	—
Oxalic acid, mg %	0–15.6	—
UTI/mg	3.07–5.46	—

Data from References 16, 19, 29, and 30.

Table 7 **Amino acid content in several amaranth seed species (mg/g N)**[a]

Amino Acid	1	2	3	4	5	6	7	8
Isoleucine	226	241	223	216	250	244	244	225
Leucine	352	374	343	347	375	375	394	331
Lysine	421	363	371	328	356	381	387	313
Methionine	120	122	147	126	106	112	112	112
Cystine	127	138	121	124	119	132	132	—
Phenylalanine	249	248	236	250	293	287	300	225
Tyrosine	214	186	217	207	269	269	264	213
Threonine	236	246	225	212	275	269	209	219
Tryptophan	72	65	84	56	75	75	88	—
Valine	263	275	266	253	287	281	207	269
Alanine	227	320	227	212	237	237	244	213
Arginine	416	461	560	541	506	531	587	463
Aspartic acid	499	522	488	500	506	506	537	518
Glutamic acid	982	993	1005	869	1056	1018	1037	962
Glycine	480	565	445	416	537	550	462	544
Histidine	159	161	157	149	175	200	187	137
Proline	244	334	252	225	281	275	281	225
Serine	420	459	393	263	544	537	444	444

[a] Average from two or more samples of the same species.

Note: (1) *A. cruentus*, (2) *A. hypochondriacus*, (3) *A. caudatus*, (4) *A. edulis*, (5) *A. anclancalius*, (6) *A. gangeticus*, (7) *A. mantegazzianus*, (8) *A. hybridus*

content and lysine, and between protein content and methionine. They also reported a negative correlation between protein and threonine content. Various differences between 25 entries of *A. caudatus* grown in the same locality were found by Imeri *et al.*[7] in protein, lysine, leucine, methionine, cystine, and threonine contents. Using statistical analysis, the 25 varieties were classified into low, intermediate, and high protein and amino acid content groups. Pedersen *et al.*,[19] with *A. caudatus*, indicated similar amino acid content between three pale-colored varieties, and all had a high lysine content. In studies with *A. cruentus* and *A. hypochondriacus*, Pant[32] found similar amino acid values between species, and there were no varietal differences in lysine values. Abdi and Sahib[33] noted exceptionally high values for lysine (7.5 to 10.0 g/16 g N) in several varieties of grain amaranth from India. Even in *A. edulis*, identified as a high lysine species of amaranth, the value reported by Downton[34] was not higher than 6.2 g/16 g N. Connor *et al.*[35] established a lysine value of 5.7 g/16 g N for *A. edulis*. Similar results were published by Becker *et al.*[16] for this species, and values of 5.0 to 5.3 g/16 g N for *A. cruentus*. In one study reported by Bressani *et al.*,[8] lysine content in grams per 16 g N varied between 5.55 and 6.44 in *A. caudatus*; from 4.94 to 5.50 in *A. cruentus*, and from 4.68 to 5.62 in *A. hypochondriacus*. For tryptophan, not often reported, the variability found in grams per 16 g N was 1.36 to 1.70 for *A. caudatus*; 1.10 to 1.87 for *A. cruentus*, 1.19 to 1.79 for *A. hypochondriacus*, and 1.45 for *A. hybridus*, suggesting variability between species and within varieties of the same species. On the other hand, Duarte-Correa *et al.*[36] reported that no differences were detected in the seed amino acid composition under different cultivation conditions. Furthermore, fertilizer application level did not affect amino acid values in five different species. Likewise, Dhan Prakash *et al.*[12] indicated that with a variety of *A. hypochondriacus*, protein content and amino acid composition were not affected by year interaction. All authors reporting amino acid values agree that the amino acid content in general and essential amino acid values, in particular, are high and well balanced for a vegetable protein.

In cereal grains, the amino acid content of the whole grain could be influenced by genetic factors, as has been shown for quality-protein maize,[40] sorghum,[41] and barley.[42] Likewise, the contribution to protein content made by the germ of these cereal grains could also increase the levels of certain amino acids, such as lysine.[43,44] Nonetheless, this is not the case for amaranth, based on the results of Betschart *et al.*,[15] who showed somewhat small amino acid values for the germ-seed coat fraction in comparison with amino acid content in the perisperm of *A. cruentus*. However, since 65% of the protein in grain amaranth is provided by the germ and 35% by the endosperm, this relationship may influence the amino acid content in grain amaranth. A larger protein contribution from the perisperm fraction may result in higher lysine in the grain and in other amino acids based on the results of Betschart *et al.*[15]

IV. SEED PHYSICAL STRUCTURE AND CHEMICAL COMPOSITION

Although the subject in this section has been briefly mentioned in a previous one, it is of importance to discuss it further, since the relative quantities of the structural fractions of the grain may very well influence chemical composition, nutritive value, and food products based on amaranth seed. As was very well described by Irving *et al.*[45] and later by Bertoni *et al.*,[29] the very small grain amaranth is covered by the seed coat, a poorly developed endosperm, an annular embryo, and a central zone, the perisperm, the cells of which contain starch. Using a modified barley pearler, Betschart *et al.*[15] found that the seed coat-embryo fraction accounts for approximately 25% of the total seed weight of *A. cruentus*. Hand-dissection by these authors yielded a seed coat-embryo fraction representing 26.3% of the total seed weight, which agrees closely to the milling data. These and other data are shown in Table 8. The data from Bressani[46] obtained with a wheat pearler and using a 30 mesh screen shows a distribution for the tegument and germ of 20.8 to 26.2% and of 73.8 to 79.2% for the perisperm, values also reported by Sánchez-Marroquín *et al.*[17,47] These fractions have been analyzed, and the results from the various authors are shown in Table 9. The tegument and germ fraction differs from the perisperm fraction in all chemical components; however, the perisperm fraction is a rich source of carbohydrate, while the tegument and germ fraction are rich in protein and fat content. No evidence is available to indicate if there is variability between seed selections in the percent distribution of the two fractions, but such variability is likely to exist, as in cereal grains, for example,[48-51] influencing the protein, fat, and the content of other chemical components of the seed. The tegument and germ fraction, according to Betschart *et al.*,[15] contained from 2 to 3 times more vitamins than the whole seed, with around 40 to 60% in the perisperm with respect to the whole seed. The same authors indicate that minerals had a distribution similar to that of other nutrients with respect to the content in the fraction of the whole seed. The seed coat-embryo fraction contained more K, Na, Ca, Fe, Cu, Zn, and Co.

These data suggest, therefore, that the percentage weight distribution of the seed coat-germ fraction has a significant influence on the chemical composition of the whole seed. Although a difficult task due to the small size of amaranth seed, the distribution of the two main fractions among species and varieties within species might provide information on the variability in chemical composition reported for the whole seed.

V. EFFECTS OF PROCESSING

There are various attractive features in amaranth seed which have attracted the attention of food and nutrition scientists throughout the world. Among these, the relatively high protein content in amaranth (as compared to cereal grains) is of particular interest, as well as the essential amino acid pattern the protein contains. Amaranth seed protein has a relatively high concentration of lysine, total sulfur amino acids, and tryptophan, amino acids often deficient in cereal grains. Although not as important as the protein make-up in terms of amino acids, amaranth grain is of interest also due to its relatively high oil content, which contributes to a higher energy level in amaranth in comparison to many other grain staple foods.

Although there are still a number of problems to be solved regarding agricultural productivity of amaranth, its adaption into consumption systems also depends on the use it may have. These uses and the food products to be made from it will be related to successful processing, in terms of functional properties, nutritive value, and the quality and stability of intermediate and/or final products. Actually, the aim of processing would be to improve or develop functional characteristics but not at the expense of nutritive value. As is well established, if not well-controlled, processing may damage the availability of nutrients, lysine in particular, thus diminishing or destroying one of the most important nutritional attributes of amaranth protein, its lysine content.

For purposes of a clearer presentation of this section, the first part will concentrate on the information of the nutritional nature of amaranth seeds, and the second part will deal specifically with the effects of processing on functional properties and nutritive value.

Table 8 **Percent weight distribution of physical fractions of amaranth grain (*A. cruentus*)**

Tegument and germ	Perisperm	Ref.
25a	75a	15
26.3b	73.7b	15
20.8–26.2a	73.8–79.2a	46
22-32a	68–78a	17, 47

[a] Milling; [b] hand-dissection.

Table 9 **Proximate composition of the tegument and germ and of perisperm fractions from amaranth (d.wt.b.)**

	Original Seed		Tegument + Germ		Perisperm	
Protein (%)	18.5	16.9	42.0	37.1	7.7	10.4
Fat (%)	7.4	6.2	19.0	20.9	2.3	5.7
Crude fiber (%)	3.3	5.3	7.7	10.7	0.9	5.1
Ash (%)	3.2	2.8	7.0	6.5	1.2	1.9

Data from References 15, 17, and 47.

A. NUTRITIVE VALUE OF AMARANTH SEED

Possibly, the first suggestion of a high nutritive value of amaranth seeds, particularly of its protein, became apparent when its essential amino acid make-up became available. In this respect, Table 10 shows average essential amino acid values of *A. caudatus*, *A. hypochondriacus*, and *A. cruentus*, as well as the FAO/WHO Reference Amino Acid Pattern (1973)[52] for the purpose of scoring essential amino acid content. This chemical scoring indicates the amino acids valine, isoleucine, and leucine to have lower scores, in particular, leucine; this is the reason why many authors have suggested from this approach that leucine is the first limiting amino acid in amaranth seed protein.[15,16,19] The table also shows threonine to be limiting, but not to the extent of leucine, as was also suggested by Pedersen *et al.*[19] Although essential amino acid balance in amaranth protein is highly acceptable, Bressani *et al.*[53] demonstrated, from multiple biological tests with growing rats, threonine to be the first limiting amino acid in *A. cruentus* for both the dark- and the light-colored seed, when raw or processed. Leucine, as well as methionine, improved the quality of the protein when supplemented with threonine but only to a small extent. Biological confirmation of the high protein quality of amaranth followed the publication of essential amino acid content. These initial reports introduced much controversy since various results showed a very poor animal performance when fed amaranth,[54,55] although good performance was also reported.[56] The results of these studies indicated, however, some benefits in its nutritive value, by processing amaranth. Poor performance was attributed to toxic factors, such as tannins (present in the dark-colored seed), to palatability of amaranth-containing diets, and to heat-labile substances. More recent studies have shown a relatively good growth performance of rats fed the ground raw amaranth grain as the sole source of protein, which is definitively improved by thermal processing.[7-9,15,18,19] This is illustrated in Table 11 for three species.[9] Animal growth for all samples was higher for the cooked grain as compared to the raw material. In all tests, food intake was higher from cooked amaranth; however, this difference in food intake was not similar for all varieties tested, suggesting differences between samples in the concentration of the factor which induces a lower feed intake. Up to the present time, such factors have not been identified, although it has been suggested the effects are due to trypsin inhibitors, tannins, and lectins or palatability.[57,58] Calderón de la Barca[57,58] showed better animal performance on lectin-extracted flour than on the original flour of *A. leucocarpus*. Subchronic toxicity assays with rats conducted for a 90-d period by Castillo Pino *et al.*[59] indicated that amaranth seed does not produce severe pathological effects on the experimental animals. Pedersen *et al.*[19] found a low availability of zinc in rats fed amaranth, as judged by low amounts of zinc in the femur. Likewise, Ca and P in the femur were low in rats fed raw amaranth, as compared to those fed processed seed. Whittaker and Ologunde[60] reported good iron bioavailability from processed

Table 10 **Amino acid content (mg/g N)**

Amino acid	*Amaranth* *Caudatus*	*Hypochondriacus*	*Cruentus*	*Edulis*	FAO Ref.
Lysine	364	374	337	329	340
Threonine	230(92.0)	268	238(95.2)	212(84.8)	250
Methionine	148	106	118	125	
Cystine	116	131	127	123	
Total sulfur	264	237	245	248	220
Valine	264(86.8)	237(78.4)	269(86.8)	254(81.9)	310
Isoleucine	218(87.2)	250	222(88.8)	216(86.4)	250
Leucine	349(79.3)	382(86.8)	344(79.1)	348(79.1)	440
Phenylalanine	238	328	250	250	
Tyrosine	205	269	207	207	
Total aromatic	443	597	457	457	380
Tryptophan	86	84	59	59	60
Histidine	158	169	150	150	—
Arginine	556	506	541	541	—

Note: Values in parentheses indicate percent adequacy.

Table 11 **Feed intake, weight gain and protein quality of rats fed four species of raw and processed amaranth**

Amaranth Species	Raw: Average Food Intake (g)	Raw: Average Weight Gain (g)	Raw: PER	Cooked: Average Food Intake (g)	Cooked: Average Weight Gain (g)	Cooked: PER
cruentus	377 ± 35.0	72 ± 12.7	1.98 ± 0.28	464 ± 34.5	114 ± 17.1	2.48 ± 0.21
cruentus	373 ± 33.9	93 ± 22.5	2.53 ± 0.28	438 ± 34.5	116 ± 13.6	2.66 ± 0.17
hypochondriacus	361 ± 31.7	86 ± 11.3	2.47 ± 0.19	458 ± 46.9	118 ± 19.8	2.56 ± 0.22
caudatus	332 ± 30.8	75 ± 10.8	2.20 ± 0.17	437 ± 43.9	111 ± 12.8	2.60 ± 0.24

From Bressani, R., González, J.M., Elias, L.G., and Melgar, M., *Plant Foods Hum. Nutr.*, 37, 59, 1987.

Table 12 **Nutritional parameters of grain amaranth**

Species	True Protein Digestibility %	Apparent Protein Digestibility %
A. caudatus		
raw	79.2–88.5[a]	76.0 ± 2.8[b]
Toasted	68.1–89.9[a]	62.2 ± 1.9[b]
Popped	85.6–88.4[a]	78.7 ± 2.4[b]
Flaked	89.5[a]	79.5 ± 3.5[b]
	Metabolizable energy[c]	
	Light-Colored Grain (kcal/g)	**Dark-Colored Grain (kcal/g)**
A. cruentus		
Raw	2.91 (2.00–3.70)	2.79 (2.02–3.46)
Atmospheric cooking	3.44 (3.30–3.60)	3.22 (3.37–3.46)
Drum drying	3.67 (3.20–4.05)	3.52 (3.39–3.68)
Extrusion cooking	4.22 (3.90–4.43)	3.36 (2.91–3.90)

[a] Data from Reference 19; [b] data from Reference 67; [c] data from Reference 61.

amaranth. Other nutritional parameters of interest are shown in Table 12 for protein digestibility and energy digestibility.[19,61,67] The values cited do not differentiate between species; dark-colored seed, however, has been shown to have lower protein and digestible energy. García *et al.*[62] found only small differences in organ weight in rats fed raw and processed *A. cruentus*, with the cooked sample giving a linear weight similar to that of the casein control. Likewise, hemoglobulin, hematocrit, and total serum proteins were similar between casein and experimental groups, either raw or processed. It should be indicated that the diets contained iron salts. Some microscopic changes in organs were reported. In the above studies, both raw and cooked amaranth were fed with and without oil removal, concluding that amaranth oil was not responsible for the effects observed when feeding raw amaranth seed.[63]

B. EFFECTS OF PROCESSING

Amaranth grain has been processed by a relatively large number of processes, which can be classified into five groups. These are wet cooking, dry cooking, biological transformation, by physical means, and by chemical techniques.

1. Wet Processing

a. Atmospheric Pressure

Various studies have been carried out with atmospheric cooking on different amaranth grain species. This process merits attention, since it represents a process commonly used at the home level. The results of a study with *A. caudatus*[64] are shown in Table 13. Cooking increased protein quality from the raw seed up to 10 min with a tendency to decrease as cooking time increased. It is of interest to point out that available lysine remained relatively constant. At 60 min the starch was completely gelatinized, as indicated by percentage of damaged starch. Trypsin inhibitors, lectins, and tannins were not present after 30 min of cooking time.

Protein quality improvement by atmospheric pressure cooking from the raw grain has been shown to take place with *A. cruentus*, and *A. hypochondriacus*.[62]

b. Pressure Cooking

Not many studies have been conducted on the effects of pressure cooking, possibly because cooking time under atmospheric pressure conditions takes a short time. Bressani[65] reported on the effects of pressure cooking on functional properties and protein quality with *A. cruentus*. In one case, cooking at 15 psi was done with the grain in water, and in a second case, with no water added to the grain. The protein quality values obtained suggested some decrease after 20 min. In both cases the water absorption index increased, particularly for the sample pressure-cooked with water, and only slightly with vapor. Pressure cooking of the grain in water increased WSI, but a decrease was observed for the seed pressure cooked with vapor only. These differences could have been due to poor cooking on account of poor energy transfer when no water was present. Pressure-cooked amaranth in water gave suspension with little water separation, particularly at 30 min soaking time, which suggests it to be used in cereal *shake-type* products.

c. Drum Drying

After cooking, either at atmospheric or under pressure, the material has to be dried which can be done with hot air, or by means of a drum dryer. Various workers have reported on nutritive quality of amaranth grain made into flakes by drum drying.[19,65-67] The data reveal that for all amaranth selections, there was an increase in weight gain as well as in protein quality. Therefore, the additional thermal treatment during drum drying did not decrease the initial effect of cooking in water. Other workers have soaked the seed in warm water followed by drum drying with similar results.[62] Pedersen *et al.*[19] showed flaked *A. caudatus* to be of better quality than the raw seed. Imeri *et al.*[66] demonstrated that raw *A. caudatus* flour had an amylographic response of 600 B.U. at 16 min and 69°C, while the 10 min cooked/drum dried flour had a peak viscosity of 1420 B.U. at 2 min and at 27°C. The products so prepared have other attractive functional properties, and the authors suggested it could be used in infant cereal grains.

d. Nixtamalization

Lime cooking of maize for tortilla preparation is a process which has been extensively studied by various workers.[68-70] The process induces some interesting changes in maize, both from the chemical, nutritional, and functional points of view. Since maize proteins are of a very low protein quality, various workers have tested amaranth as a supplement to lime-treated maize; however, very few studies have been conducted with lime-cooking of amaranth. Vargas-López *et al.*[71] reported on a study to evaluate the effect of temperature (T), $Ca(OH)_2$ (or lime) concentration, and cooking time on physicochemical properties of

Table 13 **Effect of atmospheric pressure cooking on selected chemical and nutritional characteristics of grain amaranth (*A. caudatus*)**

Cooking Time (min)[a]	True Protein Digestibility (%)	NPR	Damaged Starch (%)	Available Lysine (g/g N)
0	82.0	2.31	0	0.38
10	79.0	3.59	31	0.39
20	81.0	3.38	48	0.34
30	80.0	3.28	54	0.40
40	81.0	3.19	56	0.36
50	79.0	3.30	50	0.44
60	79.0	3.44	100	0.42

[a] Cooking done at 96°C; grain/water ratio = 3/1.

From Imeri, A. Flores, R., Elías, L.G., and Bressani, R., *Arch. Latinoam. Nutr.*, 37, 160, 1987.

amaranth flour for tortilla preparation. It was found that the effects of T and cooking time were significant for pH, WAI, color, chromaticity index, fluid consistency index, and flow behavior index. On the other hand, $Ca(OH)_2$ effects were significant for mass flow properties only. The lime concentration used ranged from 0.8 to 1.0 g/100 g of amaranth, and the pH values obtained were within the reported range for nixtamalized maize flour. These workers concluded that such a nixtamalized amaranth flour may be successfully used to replace maize totally or partially for tortilla preparation. Nutritional evaluations of lime-cooked amaranth flour are also very limited. Bressani[72] found no deleterious effects on protein quality when amaranth grain was processed with 0.6% lime for 10 min. The flour showed WAI and WSI values smaller than those of Vargas-López *et al.*[71] and was quite stable to storage at room temperature as compared to the raw flour.

2. Dry Processing

Dry thermic procedures used to process amaranth seeds include the puffing or popping method, often used to process the grain in the past and today. An extension in terms of time of the puffing process easily leads to roasting. The more recent technology used is extrusion cooking.

a. *Puffing (Popping) and Roasting*

Even though popping of amaranth seems to be the main type of process used by ancient civilizations (and is still used today quite extensively), studies aimed at learning the factors and conditions giving maximum expansion are very few. The process is temperature- and exposure-time dependent, and not all varieties expand to the same extent. As with sorghum and maize,[73,74] perisperm texture, moisture content, and seed coat strength may be important variables affecting the expansion, which increases grain size and introduces attractive functional properties. The objective is to achieve rapid heat transfer into the perisperm to vaporize water which will induce the expansion. The relationships between time and temperature were reported[3,60,70] for *A. cruentus*. Volume increase, percent of expanded grain, and WAI increased up to about 30 s exposure at temperatures between 210 to 230°C. It was indicated that from 0 to 10 s, little expansion took place, which increased significantly from 10 to 30 s, and stabilized at 30 to 40 s. From this moment on, roasting begins to take place. Pant[75] reported little change in proximate composition due to popping; however, lysine decreased from 4.83 to 3.08 g/16 g N representing a loss of 36% in home popped seed. In a commercially popped seed, lysine loss was 25%. The protein quality of the home popped seed was lower than the protein quality of the commercially popped seed, which, in turn, was higher than the protein quality of the raw seed. The home popped seed was obviously roasted. Bressani *et al.*[67] reported a loss of 19% lysine in roasted *A. caudatus*, with a loss in protein quality. Pedersen *et al.*[19] informed lysine losses in popped *A. caudatus* of 7.0 to 17.5%, and in roasted seed, from 7.2 to 32.0%. Lysine losses can therefore be significant, which affects the quality of amaranth protein *per se* and its supplementary effect to cereal grains. It also suggests the need to develop processing control systems capable of increasing grain volume without damaging the quality of the protein. Tovar *et al.*[76] also notified a high lysine loss (55.9%). These authors developed a popping method based on a fluid-bed system and reported losses in lysine, arginine, and cystine. Protein quality was found to decrease in popped seed produced by both the traditional and fluid-bed method. With respect to the effects of puffing on the functional properties of amaranth, Singhal and Kulkarni[77] found that the water- and fat-holding capacities were higher for puffed grain flour, although gelation was not affected and foaming was adversely affected by the process. Flour pastes had a decreased viscosity. In these studies, the grain was processed at 190 to 200°C for 30 to 45 s. Roasted *A. caudatus* had a higher percentage of damaged starch as compared to popped grain and also a higher insoluble dietary fiber content.[67] This was not found by Pedersen *et al.*,[20] who reported that popped seed supplemented cereal grains effectively.[28] Although popping is without doubt a very interesting process to introduce amaranth into consumption systems, there is a need to develop the technology to preserve the nutritional quality of amaranth.

b. Extrusion Cooking

This method has been included under dry-heat processing, although often moisture as water or vapor is added during the procedure. Using a Brady extruder, Mendoza and Bressani[79] reported on the effects of this process on the nutritional and functional properties of extrusion-cooked *A. cruentus* and *A. caudatus*. The process improved protein quality, and available lysine remained the same as in the raw samples. The process increased the WAI more for *A. cruentus* than for *A. caudatus* and increased the percent of damaged starch. The extruded product had higher WA; WR and viscosity decreased significantly. The method was very effective in increasing the protein quality of the seed amaranth by inactivating the factors in raw grain which do not permit a full expression of the nutritional potential of the grain and by introducing attractive functional properties for food product development. Vargas-López *et al.*[80] also studied the extrusion-cooking process using a Brabender extruder. The amaranth grain was processed under alkaline conditions. They found an increase in WAI with barrel temperature and water addition. WA values were about half of those reported by Mendoza and Bressani[79] with similar WSI values. Vargas-López *et al.*[80] concluded that extrusion-cooked amaranth flour under alkaline conditions may be used for tortilla preparation. Tillman and Waldroup[81] evaluated amaranth grain (both pressure-cooked and extrusion-cooked) in poultry diets. They reported that inclusion in the diet of 40% of extrusion-cooked amaranth, either dry or with some moisture added, gave as good poultry growth performance as that observed with the maize-soybean control diet. Likewise, López and Bressani[61] reported that extrusion-cooking of amaranth, (dark and light seed) improved metabolizable energy in poultry as compared to the raw seed. For the light-colored *A. cruentus* seed, the average increase was from 2.91 to 4.22 kcal/g, and for the dark-colored grain, the improvement was from 2.79 to 3.36 kcal/g.

3. Biological Processing

Under this subsection, two methods of processing amaranth will be discussed: germination and enzymatic hydrolysis. Although amaranth mixed with other grains has been subjected to fermentation at the home level in some areas, no published studies are available on amaranth fermentation.

a. Germination

This simple process was studied by Paredes-López *et al.*[82,83] Germination was conducted for 72 h, and they found increases in crude protein, true protein, and crude fiber. Nonetheless, a decrease was observed in the true protein to crude protein ratio and in fat content. Reactive lysine decreased slightly at 72 h, but *in vitro* protein digestibility did not change. They also found that farinography and viscographic performance improved by germination. Their results allow them to conclude that germination might contribute to widen the restricted food uses of amaranth. They also observed differences in the susceptibility of the starch to enzymatic degradation. Germination is a process which has improved the nutrient content of germinated products, particularly in vitamin and other nutrients.[84-86] This possibility was studied by Colmenares de Ruiz and Bressani[27] who found that on a dry weight basis, crude protein, crude fiber, and ash content did not change, whereas lipid and phytic acid content decreased with respect to germination time. Reducing sugars, total sugars, and damaged starch increased, whereas raffinose and stachyose were not detected after 48 h. Thiamine, riboflavin, ascorbic acid, and niacin increased while biotin and folic acid did so only to a small extent. The germinated grain was evaluated for protein quality using raw and cooked seed. With the raw germinated grain, some increase in protein quality due to germination was observed, whereas a significant decrease was measured on the germinated cooked seed. This was explained on the basis of amino acid losses during cooking or of reactions between amino acids and reducing sugars. As concluded by Paredes-López et al.,[82,83] germinated amaranth might be used in food product development.

b. Enzymatic Hydrolysis

The use of amylolytic enzymes or of germinated grains with a high amylolytic activity has been proposed to increase the energy density of cereal foods.[87-89] This approach was studied by Mendoza and Bressani[79] on extruded amaranth, which was suspended in water and to which 10% germinated amaranth grain was added. The partially hydrolyzed product was drum-dried. The product had a high protein quality (net protein retention, or NPR of 3.14 ± 0.34) as compared to the extruded material (NPR, of 3.04 ± 0.35), and it had a high concentration of total sugars. Water absorption did not change, but water retention increased with no change in available lysine. With this product, a drink was prepared with milk and sugar, highly acceptable in sensory analysis. Another approach to enzymatic processing was studied by Paredes-López *et al.*[90] Using amylolytic enzymes and appropriate conditions, they were able to produce a high-protein amaranth flour with 26 to 28% protein, 10 to 16% fat, and 40 to 50% starch, with yields of 38 to 39%. The *in vitro* protein digestibility and reactive lysine were comparable to values for the raw seed. They suggested that the high-protein amaranth flour (HPAF) might be used as a dry milk extender in food products.

4. Physical Processing

Dry grain milling of amaranth was studied by Betschart *et al.*[15] after the studies on the structure of grain amaranth reported by Irving *et al.*[45] Using a modified barley pearler, Betschart *et al.*[15] were able to obtain two physical fractions: one of them a seed-coat germ fraction high in protein content and a second one high in starch and low in protein content. According to the authors, the seed coat-embryo fraction contained 2.3 to 2.6 times as much nitrogen, fat, fiber, and ash; 2.4 to 3.0 times the quantity of thiamine, riboflavin, and niacin; and 1.4 to 2.5 times the concentration of several mineral elements as did the original intact seed and perisperms. A similar approach was studied by Sánchez-Marroquín *et al.*,[17,47] who first ground the seed which was then fractionated by air classification. They tested various mills, different moisture levels, and disc apertures, confirming previous results. Both groups of workers have provided interesting analytical data on the two fractions. Some results are shown in Table 14 for amino acid content and protein quality. Actually, the essential amino acid pattern in the perisperm (low protein fraction) is slightly higher in most essential amino acids than the pattern in the seed coat-germ fraction (high protein-high oil fraction). On the basis of the 1973 FAO/WHO essential amino acid reference pattern,[52] the most limiting amino acid in the protein of both fractions was leucine. The protein and diet digestibilities were similar between fractions; still, the protein quality (protein efficiency ratio or PER) was significantly higher for the testa-embryo fraction than for the perisperm. The authors suggested the difference is due to poor amino acid availability in the perisperm fraction and to its low protein level. Digestibility of the diet and of the nitrogen were essentially alike; therefore, other factors must be involved. Since lysine content was higher in both physical fractions, both could be useful supplements to cereal grains which contain from 160 to 235 mg lysine per gram of N. As was indicated above, Sánchez-Marroquín *et al.*[17,47] tested the efficiency of various mills with air classification. They also found two fractions, one high in protein and one low in protein content, with a gross chemical composition similar to that reported by Betschart *et al.*[15] The same authors conducted rheological studies on these fractions and found that the high protein fraction from either milling system had a lower viscosity peak than the low protein fraction. The temperature of initial gelatinization was higher for the low protein fraction than for the high protein fraction. They also concluded these fractions to have applications in food product development based on wheat flour.[15,17,47,91]

5. Protein Extraction

In another chapter of this book, the protein distribution, characterization, and amino acid content of grain amaranth is presented. In this section, the studies reported to concentrate the protein by chemical means will be discussed. Already mentioned are the studies conducted by Paredes-López *et al.*[90] and Barba de la Rosa and Paredes López[92] in which enzymes were used to concentrate the proteins of amaranth. A more chemical approach was reported by Pacheco de Delahaye and Gómez.[93] These authors extracted the proteins at pH 9.0 to 9.5 with a 0.02 $\underline{N}$ NaOH solution. With a flour: solvent ratio of 1:15 g/ml, 5 min of extraction time at 45°C was found optimum for maximum extraction. Precipitation of the protein was accomplished at pH 4.6. A similar extraction was done at pH 3.5 with succinic anhydride. The alkaline protein isolate contained 66.6% protein, while the succinylated isolate was 67.1%, with 4.1 to 4.3 g of available lysine per 100 g protein. The protein solubility at pH 3 to 6, water absorption, oil absorption, emulsion activity, and stability were higher for the succinylated protein. They suggested this preparation would be useful as protein-rich foods and as emulsifying agents.

Table 14 **Amino acid content (mg/g n) and protein quality of physical fractions of grain amaranth**

Amino Acid	Grain	Perisperm	Tegument and Germ	FAO/WHO Ref. Pattern
Isoleucine	224	267	222	250
Leucine	320	388	316	440
Lysine	318	401	306	340
Methionine	121	131	124	220
Cystine	134	136	142	
Phenylalanine	213	256	217	380
Tyrosine	164	191	183	
Threonine	211	244	209	250
Valine	261	304	266	310
Protein quality (% of casein)	60.3	30.0	78.3	—

From Betschart, A.A., Irving, D.W., Shepherd, A.D., and Saunders, R.M., *J. Food Sci.*, 46, 1181, 1981.

6. Human Studies on Processed Amaranth

Some studies have been published on the protein quality of processed amaranth in children and adults. Morales *et al.*[94] tested toasted, popped, and flaked products from *A. caudatus* in convalescent malnourished infants and young children. Protein digestibility was similar between samples (69.0 to 69.6%), and nitrogen retention (as percent of casein used as control) was 70.9 ± 8.8, 65.9 ± 14.1, and 59.0 ± 9.3% for toasted, popped, and flaked amaranth, respectively. The above authors indicated that apparent N retention from amaranth was superior to that from most cereals studied at isonitrogenous levels and was similar to that from rice or high-lysine maize. They further indicated that in isoenergetic amounts, the higher protein content of the grain amaranths (13 to 14% of energy) should make them superior to rice and high-lysine maize as sources of dietary protein.

Bressani *et al.*[95] conducted nitrogen balance studies with young adult human subjects fed popped and extrusion-cooked amaranth from *A. cruentus*. The true protein digestibility was 89.8 and 85.5% for extruded and popped amaranth, respectively. The protein quality relative to cheese protein, used as control, was 88.6% for extrusion-cooked amaranth and 81.4% for popped amaranth. When the protein quality was corrected for protein digestibility, both products were statistically equal to cheese protein. In this study, popping was conducted at 250°C for 15 to 20 s, and followed by removal of popped grains which became roasted. The extruded product was produced from a Brady extruder, as described previously.[79]

VI. SUPPLEMENTARY AND COMPLEMENTARY PROTEIN EFFECTS OF AMARANTH SEEDS TO OTHER PROTEIN SOURCES

As already indicated elsewhere in this chapter, from the nutritional point of view, the attractive features in amaranth grain are its relatively high protein content and essential amino acid pattern, particularly lysine, sulfur amino acids, and tryptophan content. Two approaches have been used to study the improvement effects of amaranth protein to other food proteins: the supplementary and the complementary effects. The two approaches are essentially alike; the difference is the way the evaluation is conducted.

A. SUPPLEMENTARY EFFECTS

One of the first attempts to measure the supplementary effects of amaranth protein to maize was reported by Tovar and Carpenter.[96] Since lime-treated maize protein is deficient in lysine and tryptophan,[97] and amaranth is a good source of these two amino acids, the experimental design used was to make maize lysine- or tryptophan-limiting by adding either lysine or tryptophan in the presence of 25% amaranth seed flour which had been popped. The approach was attractive; however, they were able to demonstrate an effect from amaranth only when tryptophan was made limiting. These studies show the high biological availability of the amino acid from amaranth protein. The failure to demonstrate an effect for lysine was due to the low levels of this amino acid in the popped product. A loss of around 30% was reported to be due to popping processing. Furthermore, the level of amaranth flour used was only 25% which provided minimal levels of lysine (about 0.15%) and around 0.05% of tryptophan. The study proved the importance of correct processing to preserve the nutritional qualities of amaranth protein. Using popped *A. caudatus*, Pedersen *et al.*[78] studied its supplementary effect to maize, wheat, and low-tannin sorghum at levels of 0, 10, 25, and 50%. The popped seed contained 5.19 g lysine per 16 g N, so it provided 0.075, 0.19, and 0.38 g of lysine. The protein digestibility was, in general, not affected by amaranth substitution with any of the raw or cooked cereal grains. The biological value of the protein, as well as the net protein utilization, increased considerably with an increasing level of amaranth inclusion. They concluded that amaranth is an effective source of protein to combine with cereal proteins. These studies confirm previous results by Sánchez-Marroquín *et al.*[98] on the supplementary effect of amaranth flour and fractions on wheat flour. Similar results were obtained with oats.[99] Pant[75] also showed an increase in protein quality of wheat with an NPR of 2.68 when supplemented with amaranth (W/A 6:3 protein ratio) with an NPR of 3.48. This was further improved in a blend of wheat, amaranth, skim milk (5:3:1 protein ratio). Sánchez-Marroquín and Maya[98] also found improvement in protein quality of maize supplemented with the amaranth milling fractions.

Some studies conducted in children have been published on the protein quality of amaranth/maize blends. Morales *et al.*[94] found nitrogen retention values of 28.0, 29.0, and 36.6% of intake from diets of 80% maize and 20% amaranth, 70% maize and 30% amaranth, and casein used as control. They concluded that ordinary maize meal in which as little as 12.7% by weight has been replaced with toasted amaranth flour (20% of the protein) becomes a nutritionally superior source of protein and fat that can, when providing approximately 70% of diet energy, satisfy a good part of the protein requirement of young children. Nonetheless, Graham *et al.*[100] indicated that, in these cases, when protein supplementation was not evident in human studies, it may be due to damage of the amaranth protein quality during processing. Using post-prandial measurements of free-plasma amino acids of children fed amaranth grain products and mixtures with maize, Graham *et al.*[101] established that tryptophan levels did not fall when the children were fed maize plus amaranth; leucine elevation which took place with maize was blunted, but there was a fall in lysine. This suggested not only small levels of addition through amaranth supplementation, but a partial destruction of lysine due to processing.

B. COMPLEMENTATION

Protein complementation is the evaluation of the efficiency with which the essential amino acid content of two or more protein foods balance each other, that is make up a much more efficient essential amino acid pattern. Some results for the complementation of amaranth protein to the protein of oats, wheat flour, maize and sorghum have been described.[3,46] With oats, as increasing amounts of protein are supplied by amaranth, there is an increase in protein quality up to a point in which amaranth proteins make up 75% of the dietary protein and oats 25%. This improvement is brought about by the contribution of lysine and threonine from amaranth to oats. There is also an increase in leucine in the mixture provided by oat protein.

For wheat flour, as the amount of protein from amaranth increases in the amaranth wheat flour diet, protein quality increases in a straight line relationship. Wheat flour is known to be low in lysine, threonine, and sulfur amino acids,[102] which are provided by amaranth protein as its contribution in the mixture increases.

The optimum complementation between maize and amaranth takes place when maize contributes about 46% of the dietary protein and amaranth about 54%. This response is due to the correction of the tryptophan and lysine deficiencies in maize by amaranth protein and the apparent deficiency of leucine in amaranth by maize protein, although threonine remains low. The same type of response is observed with sorghum, in which both protein sources contribute about the same amount of protein to the diet, 50%. Lysine deficiency in sorghum in corrected by amaranth, and the leucine excess in sorghum is diluted by amaranth. In the case of rice, the optimum complementary effect would take place when amaranth provides 25% of the dietary protein and rice 75%. The effect is primarily due to the higher lysine content in amaranth.

Complementary studies with foods of a protein content higher than that found in cereal grains have also been reported.[46] With common beans, the optimum mixture seems to occur when amaranth provides 60% of the protein in the diet and beans around 40%. The effect based on amino acid content is due to the sulfur amino acid contribution made by amaranth protein. Some minimal effect also takes place with mixtures of soybean and amaranth, explained on the basis of the sulfur amino acid contribution from amaranth. In the case of peanut protein, as more amaranth protein makes up the diet, the higher the response in protein quality, attributed to the deficiencies in lysine, threonine, and sulfur amino acids in peanuts. The last case reported[46] is that of amaranth and milk protein combinations, which show some optimum combination at 25% amaranth protein to 75% milk protein, effect which could be due to the sulfur-containing amino acids.

These mixtures of proteins, from various sources and from amaranth, can be used in food product formulation so as to maximize the protein quality of the ingredient of poorer protein quality.

VII. NUTRITIONAL QUALITY OF AMARANTH GRAIN-BASED FOOD PRODUCTS

Amaranth, as already indicated, was an important staple food crop of the Aztec civilization and was also cultivated by the Maya, Inca, and other pre-Columbian people. Through its dissemination, amaranth grain became relatively important for populations in Nepal, India, and the Himalayas. All these civilizations were responsible for the development of the traditional amaranth foods listed in Table 15, taken from Teutonico and Knorr[2] and others.[103-108]

These foods continue to be produced and marketed in the countries or regions indicated. Since the rediscovery of amaranth as a high-quality protein source, various efforts have been made to introduce it into the market mainly as a food ingredient, because of its higher price as compared to cereal grains. In these formulations, the purpose of amaranth grain is to increase the protein quality of the cereal grain as shown in a previous section to levels usually not higher than 20%. According to Breene,[103] amaranth-based consumed food products currently available in the U.S. market include those listed in Table 16.

In recent years, some activity has evolved regarding food product development using amaranth because of its particular nutritional characteristics. The types of foods include processed whole grain flours, fractionated flours, weaning and infant foods, infant cereal grains, and as supplement to cereal grains, applications which do not differ much from those already mentioned. Bressani and Elías[104] indicated that due to the relatively high protein and oil content of whole amaranth flour, and to its amino acid make-up and nutritive value, 100% amaranth flours already were a high-quality food to be used for weaning purposes. The addition of 12% skim milk increased protein quality to 98% of the value of casein. The 100% amaranth flour concept was also expressed by Imeri *et al.*[66] and Whittaker and Ologunde.[60] For such a purpose and to retain as much as possible the high protein quality of whole grain amaranth flour, it is necessary to use processes which do not destroy the quality of the protein. A 100% whole amaranth flour is very similar in chemical composition as available infant cereal grains which contain small amounts of milk.[46] Whittaker and Ologunde[60] and Ologunde *et al.*[109] studied Fe availability from nonsupplemented or iron-supplemented amaranth flour, since its phytate content (around 0.35 g%) or tannin content (around 80 mg/100 g) could interfere with its availability. In the study of Whittaker and Ologunde,[60] cooked amaranth flour was fortified with NaFe EDTA, ferrous fumarate, or ferrous sulfate. The protein quality of the product was found to be favorable, and the normal body weight gain of rats and the high RBV measured indicated good iron absorption from the amaranth alone, which was improved by Fe supplementation. Ologunde *et al.*[109] in a follow-up study reported RBV of 0.40 for unfortified amaranth flour, 1.55, 1.75, and 1.67 for NaFe EDTA, ferrous fumarate, and ferrous sulfate, respectively. These authors recommended ferrous fumarate as an iron supplement.

Another area of food product development with amaranth is the area of weaning foods. Sánchez-Marroquín *et al.*[99,100] evaluated whole amaranth and its air-classified products and blends of these with wheat and oats as possible components for infant formulas. A low-cost amaranth preschool-children food was formulated and tested by del Valle *et al.*[111] using whole amaranth flour, 16.1% dehulled soybean flour, 28.7% pearled oats, 18.3% sucrose, 5.9% fat, minerals, vitamins, and flavoring. The mixtures contained 17.2% protein, 15.2% fat; it was prepared by extrusion cooking and had a protein quality as PER of 2.9. In a short-term feeding trial with 1 to 3-year old children, the percent weight gain reported was 4.1% for the amaranth-containing food, which was similar to the weight gain of a product with soybeans and oats of 3.9%, and both significantly superior to the control group of 0.3%. Using a high and a low protein-containing amaranth flour fractions as developed by Sánchez-Marroquín *et al.*,[110] del Valle and coworkers[112,113] tested two formulations in nitrogen balance studies with children. Formula iR, using the high-protein air classified flour contained 42.2% fraction iR, 28.5% oats, 21.1% sucrose, 6.3% vegetable oil and vitamins and minerals. The second formula contained 14.5% amaranth flour fraction

Table 15 **Some traditional amaranth foods**

Food	Country	Brief Description of the Food
Alegría	México	Popped seeds with molasses or syrup
Alboroto	Guatemala	Popped seeds with molasses or syrup
Atole	México and Guatemala	Gruel of roasted amaranth grain, sugar, water
Bollos	Perú	Ground popped seeds and syrup
Millmi	Bolivia	Popped and toasted seed
Lodoos	India	A confection
Satoo	Nepal	Gruel of roasted ground seeds, water
Chappati-like pancakes	Himalaya	Amaranth flour

From Teutonico, R.A. and Knorr, D., *Food Technol.*, 39, 49, 1985.

Table 16 **Amaranth based consumer food products available in the U.S.**

Product	Observation
Whole amaranth products	Include whole grain, popped, puffed and toasted flour, flakes, high or low bran
Baking mixes	Flour, flakes, fractionated flour in specialty mixes for pancakes, muffins and other baked items
Malted flours	Same product application as regular flours
Breakfast cereals	In hot types and cold ready-to-eat types; granola, granola bars; for cold type breakfast cereals, amaranth in the form of flakes, puffed, popped, or extended
Snacks	As corn curl types or "nacho" type chips popped amaranth is often used
Baked goods	Special market for multigrain breads, muffins, crackers, and cookies
Miscellaneous food products	Pasta products and in a large number of recipes

From Breene, W.M., Contribution of amaranth to consumer products in Proc. 4th Natl. Amaranth Symp. Amaranth: Perspectives on Production, Processing and Marketing, Minneapolis, MN, 1990, 75.

1R (low protein), 25.0% oats, 25.0% dehulled soybeans, 16.4% sucrose, 10.8% corn starch, 6.3% vegetable oil, and vitamins and minerals. Both contained 0.1% artificial flavoring. These two formulations were compared to one made of soybean flour and oats in nitrogen balance studies with children. They reported no significant differences between the three formulations in nitrogen absorption or nitrogen retention.

Similar types of formulas were evaluated by Gupta and Sehgal.[114] One contained 60 g of malted pearled millet, 20 g of roasted amaranth grain, 40 g of roasted green gram, and 45 g of sorghum. The second was made up of 6.0 g of malted barley, 20 g of roasted amaranth grain, 40 g of roasted green gram, and 45 g of sorghum. These formulae contained 9.8 to 9.9% protein, 416 to 441 kcal/100 g. Upon processing, the levels of phytic acid, polyphenolic compounds, and saponins decreased, with an increase in protein digestibility. These authors reported that both mixtures were acceptable to trained panelists and to children. Rathod and Upidi[115] developed weaning mixes using malted, roasted, and popped amaranth and leafy vegetables, together with a cereal in one of four forms: as malted wheat, raw milled rice, puffed rice, and rice flakes. All blends tested were mixed in a 7:2:1 ratio of cereal, amaranth, and dehydrated spinach or drumstick leaves (*Moringa olcifera*). Intake and acceptability trials conducted with children with four of the most acceptable formulae in laboratory sensory evaluation indicated such mixes to be used effectively at the community or village level. Lazcano *et al.*[116] reported studies on the development of infant foods based on germinated amaranth, whole amaranth flour, and popped amaranth, with defatted soybean flour, wheat, and a fruit concentrate. The product packed in retort pouches had a high protein content, a high microbiological quality, and a high acceptability. Other products include a food bar for school children,[107] a macaroni product,[106] and cookies,[117,118] all the latter products made from amaranth flour and wheat flour.

Although there is still much ground to cover in food product development, all these studies indicate that amaranth can be well-utilized in improving the nutritional quality of wheat and maize-based products. The question at the present time is availability of the grain, through an increased productivity.

ACKNOWLEDGMENT

The author acknowledges the technical support provided by Ms. María Antonieta Rottmann in the preparation of this document.

REFERENCES

1. **Saunders, R. M. and Becker, R.,** Amaranthus: a potential food and feed resource, in *Advances in Cereal Science and Technology*, Vol. VI, Pomeranz, Y., Ed., American Association of Cereal Chemists, Minneapolis, MN, 1984, 357.
2. **Teutonico, R. A. and Knorr, D.,** Amaranth: composition, properties and applications of a rediscovered food crop, *Food Technol.*, 39, 49, 1985.
3. **Bressani, R.,** The proteins of grain amaranth, *Food Rev. Int.*, 5, 13, 1989.
4. **Paredes-López, O., Barba de la Rosa, A. P., Hernández-López, D., and Cárabez-Trejo., A.,** *Amaranto — características alimentarias y aprovechamiento agroindustrial*, Laboratorio de Biotecnología de Alimentos, Centro de Investigación y de Estudios Avanzados del Instituto Politécnico Nacional, Irapuato, México y Secretaría de la OEA, Washington, D.C., 1990.
5. **Joshi, B. D. and Rana., R. S.,** *Grain Amaranths: The Future Food Crop*, Shimla Sci-Monogr. No. 3, National Bureau of Plant Genetic Resources, Regional Station, Phagli, Shimla, India, 1991.
6. **Sinhal, R. S. and Kulkarni, P. R.,** Review: amaranths: an underutilized resource, *Int. J. Food Sci. Technol.*, 23, 125, 1988.
7. **Imeri, A., González, J. M., Flores, R., Elías, L. G., and Bressani, R.,** Variabilidad genética y correlaciones entre rendimiento, tamaño del grano, composición química y calidad de la proteína de 25 variedades de amaranto (*A. caudatus*), *Arch. Latinoam. Nutr.*, 37, 132, 1987.
8. **Bressani, R., González, J. M., Zúñiga, J., Breuner M., and Elías, L. G.,** Yield, selected chemical composition and nutritive value of 14 selections of amaranth grain representing four species, *J. Sci. Food Agric.*, 38, 347, 1987.
9. **Bressani, R., González, J. M., Elías, L. G., and Melgar, M.,** Effect of fertilizer application on the yield, protein and fat content, and protein quality of raw and cooked grain of three amaranth species, *Plant Foods Hum. Nutr.*, 37, 59, 1987.
10. **Pulido, M. L. and Trinidad, S. A.,** Efecto de nitrógeno y humedad en el rendimiento de grano, materia seca y contenido de proteína de *Amaranthus hypochondriacus* y *Amaranthus cruentus*, in *Memorias Coloquio Nacional del Amaranto*, Instituto de Desarrollo Estatal para la Acción Social, Querétaro, México, 1987, 99.
11. **Abidin, Z. and Hilman, Y.,** Effect of nitrogen fertilization on the nutritional content of four amaranth varieties, *Amaranth Newsl.*, 4, 6, 1986.
12. **Dhan Prakash, Misra, P. S., Pandey, R. M., and Pal, M.,** Genetic stability for protein and amino acids composition in *Amaranthus*, *Amaranth Newsl.*, 3, 1989.
13. **Balasubramanian, T. and Sadasivan, S.,** Changes in carbohydrate and nitrogenous components and amylase activities during germination of grain amaranth, *Plant Foods Hum. Nutr.*, 39, 325, 1989.
14. **Konishi, Y., Fumita, Y., Ikeda, K., Ohono, K., and Fuwa, H.,** Isolation and characterization of globulin from seeds of *Amaranthus hypochondriacus*, *Agric. Biol. Chem.*, 49, 1453, 1985.
15. **Betschart, A. A., Irving, D. W., Shepherd, A. D., and Saunders, R. M.,** *Amaranthus cruentus*: milling characteristics, distribution of nutrients within seed components and the effects of temperature on nutritional quality, *J. Food Sci.*, 46, 1181, 1981.
16. **Becker, R., Whealer, E. L., Lorenz, K., Stafford, A. E., Grosjean, O. K., Betschart, A. A., and Saunders, R. M.,** A compositional study of amaranth grain, *J. Food Sci.*, 46, 1175, 1981.
17. **Sánchez-Marroquín, A., Maya, S., and Domingo, M. V.,** Milling procedures and air classification of amaranth flours, *Arch. Latinoam. Nutr.*, 35, 620, 1985.
18. **Bressani, R., Elías, L. G., González, J. M., and Gómez-Brenes, R.,** The chemical composition and protein quality of the amaranth grain germ plasm from Guatemala, *Arch. Latinoam. Nutr.*, 37, 364, 1987.

19. **Pedersen, B., Kalinowski, L. S., and Eggum, B. O.,** The nutritive value of amaranth grain (*Amaranthus caudatus*). I. Protein and mineral of raw and processed grain, *Plant Foods Hum. Nutr.*, 36, 309, 1987.
20. **Pedersen, B., Bach Knudsen, K. E., and Eggum, B. O.,** The nutritive value of amaranth grain (*Amaranthus caudatus*). III. Energy and fibre of raw and processed grain, *Plant Foods Hum. Nutr.*, 40, 61, 1990.
21. **Bressani, R., Velásquez, L., and Acevedo, E.,** Dietary fiber content in various grain amaranth species and effect of processing, *Amaranth Newsl.*, 1, 5, 1990.
22. **Paredes-López, O., Schevenin, M. L., Hernández-López, D., and Cárabez-Trejo, A.,** Amaranth starch — isolation and partial characterization, *Starch/Stärke*, 41, 205, 1989.
23. **Lorenz, K. and Gross, M.,** Saccharides of amaranth, *Nutr. Rep. Int.*, 29, 721, 1984.
24. **Elías, J.,** Food composition tables for comparative nutrient composition of amaranth greens and seeds, in *Proc. First Amaranth Sem.*, Rodale Press, Emmaus, PA, 1977.
25. **Sánchez-Marroquín, A.,** Dos cultivos olvidados de importancia agroindustrial: el amaranto y la quinua, *Arch. Latinoam. Nutr.*, 33, 11, 1983.
26. **Watt, B. K. and Merrill, A. L.,** *Composition of Foods*, Agriculture Handbook 8, U.S. Department of Agriculture, Washington, D.C.
27. **Colmenares de Ruiz, A. S. and Bressani, R.,** Effect of germination on the chemical composition and nutritive value of amaranth grain, *Cereal Chem.*, 67, 519, 1990.
28. **INCAP-ICNND,** *Tabla de composición de alimentos para uso de América Latina*, INCAP, Guatemala, 1961.
29. **Bertoni, M. H., Gómez, R. G., Cattaneo, P., and Covas, G.,** Estudios sobre semilla de especies americanas de *Amaranthus*. II. Harinas de extracción de *A. caudatus*, *A. cruentus* y *A. mantegazzianus*, *An. Asoc. Quím. Argent.*, 72, 597, 1984.
30. **Lorenz, K. and Wright, B.,** Phytate and tannin content of amaranth, *Food Chem.*, 14, 27, 1984.
31. **Misra, P. S., Pandey, R. M., and Pal, M.,** Amino acids composition in *Amaranthus*, *Fitoterapia*, 54, 135, 1983.
32. **Pant, K. C.,** Studies on the nutritional quality of grain *Amaranthus*, *Nutr. Rep. Int.*, 28, 1445, 1983.
33. **Abdi, H. and Sahib, M. K.,** Distribution of lysine in different legumes and some species of amaranth seeds, *J. Food Sci. Tech. (Mysore)*, 13, 237, 1976.
34. **Downton, W. J. S.,** *Amaranthus edulis*: a high lysine grain amaranth, *World Crops*, 25, 20, 1973.
35. **Connor, J. K., Gartner, R. J. W., Runge, B. M., and Amos, R. N.,** *Amaranthus edulis*: an ancient food source reexamined, *Aust. J. Exp. Agric. Anim. Husb.*, 20, 156, 1980.
36. **Duarte-Correa, A., Joke, L., and Carlsson, R.,** Amino acid composition of some *Amaranthus sp.* grain proteins and of its fractions, *Arch. Latinoam. Nutr.*, 36, 466, 1986.
37. **Carlsson, R.,** Quantity and quality of *Amaranthus* grain from plants in temperate, cold and hot, and subtropical climates. A review, in *Proc. 2nd Amaranth Conf.*, Rodale Press, Emmaus, PA, 1979, 48.
38. **Harrold, R. L. and Nalewaja, J. D.,** Proximate, mineral and amino acid composition of 15 weed seeds, *J. Anim. Sci.*, 44, 389, 1977.
39. **Uzo, J. O. and Okorie, A. V.,** *Amaranthus hybridus*: a potential grain crop for West Africa, *Nutr. Rep. Int.*, 27, 519, 1983.
40. **Mertz, E. T., Bates, L. S., and Nelson, O. E.,** Mutant gene that changes protein composition and increases lysine content of maize endosperm, *Science*, 145, 279, 1964.
41. **Singh, R. and Axtell, J. D.,** High lysine mutant gene (hl) that improves protein quality and biological value of grain sorghum, *Crop Sci.*, 13, 535, 1973.
42. **Munck, L., Karlsson, K. E., Hagberg, A., and Eggum, B. O.,** Gene for improved nutritional value in barley seed protein, *Science*, 168, 985, 1970.
43. **Bjarnason, M. and Pollmer, W. G.,** The maize germ: its role as a contributing factor to protein quality and quality, *Z. Pflanzenzech.*, 68, 83, 1972.
44. **Poey, F. R., Bressani, R., García, A. A., García, M. A., and Elías, L. G.,** Germ-endosperm relationship in the nutritional improvement of maize grain, in *Seed Protein Improvement in Cereals and Grain Legumes*, Vol. 1, IAEA, Viena, 1979, 230.
45. **Irving, D. W., Betschart, A. A., and Saunders, R. M.,** Morphological studies on *Amaranthus cruentus*, *J. Food Sci.*, 40, 1170, 1981.
46. **Bressani, R.,** Amaranto. Composición química y valor nutritivo del grano, in *Primer Congreso Internacional del Amaranto*, Oaxtepec, México, 1991.

47. **Sánchez-Marroquín, A., Maya, S., and Domingo, M. V.,** Effect of heat treatment and milling on the seed, flour, rheological and baking quality of some amaranth ecotypes, *Arch. Latinoam. Nutr.*, 35, 603, 1985.
48. **Earle, F. R., Curtis, J. J., and Hubbard, J. E.,** Composition of the component parts of the corn kernel, *Cereal Chem.*, 23, 504, 1946.
49. **Bradsbury, D., Cull, I. M., and MacMasters, M. M.,** Structure of the mature wheat kernel. I. Gross anatomy and relationship of parts, *Cereal Chem.*, 33, 329, 1956.
50. **Bressani, R. and Mertz, E. T.,** Studies on corn proteins. IV. Protein and amino acid content of different corn varieties, *Cereal Chem.*, 35, 227, 1958.
51. **Rooney, L. W. and Sullins, R. D.,** Chemical, physical and morphological properties of diploid and tetraploid *Sorghum bicolor* (L) Moench kernels, *Crop Sci.*, 10, 97, 1970.
52. **FAO/WHO Ad Hoc Expert Committee,** *Energy and Protein Requirements*, WHO Technical Report Series No. 522, FAO Nutrition Meeting Report Series No. 52, WHO, Geneva, FAO, Rome, 1973.
53. **Bressani, R., Elías, L. G., and García-Soto, A.,** Limiting amino acids in raw and processed amaranth grain protein from biological tests, *Plant Foods Hum. Nutr.*, 39, 223, 1989.
54. **Cheeke, P. R. and Bronson, J.,** Feeding trials with Amaranthus grain, forage and leaf protein concentrates, in *Proc. 2nd Amaranth Conf.*, Rodale Press, Emmaus, PA, 1979, 5.
55. **Afolabi, O. A., Oke, O. L., and Umoh, I. B.,** Preliminary studies on the nutritive value of some cereal-like grains, *Nutr. Rep. Int.*, 24, 389, 1981.
56. **Osuntogun, A. B. and Oke, O. L.,** A note on the nutritive value of amaranth seeds, *Food Chem.*, 12, 287, 1983.
57. **Calderón de la Barca, A. M., Ochoa, J. L., and Valencia, M. E.,** Effect of the extraction of a hemagglutinin on the nutritive value of *Amaranthus leucocarpus* seeds, *J. Food Sci.*, 50, 1700, 1985.
58. **Calderón de la Barca, A. M. and Vázquez-Moreno, L.,** *Amaranthus cruentus* lectin: purification, stability and some biochemical properties, *J. Food Biochem.*, 12, 117, 1988.
59. **Castillo Pino, A., Abreo Peñate, M., González Pérez, T., Rodríguez, N., Rion, J., and González, J.,** Evaluación toxicológica de una variedad de amaranto, *Rev. Cubana Alimtn. Nutr.*, 5, 18, 1991.
60. **Whittaker, P. and Ologunde, M. O.,** Study of iron bioavailability in a native Nigerian grain amaranth cereal for young children, using a rat model, *Cereal Chem.*, 67, 505, 1990.
61. **López, M. and Bressani, R.,** The true metabolizable energy of amaranth (*Amaranthus sp.*) seed in poultry, *Amaranth Newsletter*, 3, 7, 1987.
62. **García, L. A., Alfaro, M. A., and Bressani, R.,** Digestibility and protein quality of raw and heat-processed defatted and non-defatted flours prepared with three amaranth species, *J. Agric. Food Chem.*, 35, 604, 1987.
63. **García, L. A., Alfaro, M. A., and Bressani, R.,** Digestibility and nutritional value of crude oil from three amaranth species, *J. Am. Oil Chem. Soc.*, 64, 371, 1987.
64. **Imeri, A., Flores, R., Elías, L. G., and Bressani, R.,** Efecto del procesamiento y de la suplementación con aminoácidos sobre la calidad proteínica del amaranto (A. caudatus), *Arch. Latinoam. Nutr.*, 37, 160, 1987.
65. **Bressani, R.,** Effects of processing on the nutritional qualities and functional properties of amaranth, in *Amaranth: Perspectives on Production, Processing and Marketing, Proc. 4th National Amaranth Symp.*, Minneapolis, MN, 1990, 55.
66. **Imeri, A. G., Elías, L. G., and Bressani, R.,** Amaranto: una alternativa tecnológica para alimentación infantil, *Arch. Latinoam. Nutr.*, 37, 147, 1987.
67. **Bressani, R., Kalinowski, L. S., Ortiz, M. A., and Elías, L. G.,** Nutritional evaluation of roasted, flaked and popped *A. caudatus, Arch. Latinoamer. Nutr.*, 37, 525, 1987.
68. **Trejo-González, A., Feria-Morales, A., and Wild-Altamirano, C.,** The role of lime in the alkaline treatment of corn for tortilla preparation, *Adv. Chem. Series*, 198, 245, 1982.
69. **Serna-Saldívar, S. O., Gómez, M. H., and Rooney, L. W.,** The chemistry, technology and nutritive value of alkaline cooked corn products, *Adv. Cereal Chem. Technol.*, 10, 243, 1990.
70. **Bressani, R.,** Chemistry, technology and nutritive value of maize tortillas, *Food Rev. Int.*, 6, 225, 1990.
71. **Vargas-López, J. M., Paredes-López, O., and Espitia, E.,** Evaluation of lime heat treatment on some physico-chemical properties of amaranth flour by response surface methodology, *Cereal Chem.*, 67, 417, 1990.
72. **Bressani, R.,** Nixtamalization of maize and nutritive value, submitted.

73. **Hoseney, R. C., Zeleznak, K., and Abdelrahmam, A.,** Mechanism of popcorn popping, *J. Cereal Sci.*, 1, 43, 1983.
74. **Chike, A. V.,** *Effect of Dry-Heat Processing on the Expansion of Different Sorghum Cultivars*, Thesis, Texas A & M University, College Station, 1980.
75. **Pant, K. C.,** Effect of heat processing (popping) on protein nutritional quality of grain amaranth, *Nutr. Rep. Int.*, 32, 1089, 1985.
76. **Tovar, L. R., Brito, E., Tahahashi, T., Miyazawa, T., Soriano, J., and Fujimoto, K.,** Dry heat popping of amaranth seed might damage some of its essential amino acids, *Plant Foods Hum. Nutr.*, 39, 299, 1989.
77. **Singhal, R. S. and Kulkarni, P. R.,** Puffing effects on functional properties of *Amaranthus paniculatus* (Rajgeera) seed flour, *J. Food Sci.*, 56, 1121, 1991.
78. **Pedersen, B., Hallgren, L., Hansen, I., and Eggum, B. O.,** The nutritive value of amaranth grain (*Amaranthus caudatus*). II. As a supplement to cereals, *Plant Foods Hum. Nutr.*, 36, 325, 1987.
79. **Mendoza, C. and Bressani, R.,** Nutritional and functional characteristics of extrusion-cooked amaranth flour, *Cereal Chem.*, 64, 218, 1987.
80. **Vargas-López, J. M., Paredes-López, O., and Ramírez-Wong, B.,** Physico-chemical properties of extrusion-cooked amaranth under alkaline conditions, *Cereal Chem.*, 68, 610, 1991.
81. **Tillman, P. B. and Waldroup, P. W.,** Processing grain amaranth for use in broiler diets, *Poultry Sci.*, 65, 1960, 1986.
82. **Paredes-López, O., Cárabez-Trejo, A., Pérez-Herrera, S., and González-Castañeda, J.,** Influence of germination on physico-chemical properties of amaranth flour and starch microscopic structure, *Starch/Stärke*, 40, 290, 1988.
83. **Paredes-López, O. and Mora-Escobedo, R.,** Germination of amaranth seeds: effects on nutrient composition and color, *J. Food Sci.*, 54,761, 1989.
84. **Chen, L. H., Wells, C. E., and Fordham, J. R.,** Germinated seeds for human consumption, *J. Food Sci.*, 40, 1290, 1975.
85. **Martínez, A. B., Gómez-Brenes, R. A., and Bressani, R.,** Relación del contenido de lisina y triptofano con el de zeína, durante la germinación del grano de maíz y su posible vinculación con el ciclo vegetativo de la planta, *Arch. Latinoam. Nutr.*, 30, 607, 1980.
86. **Lorenz, K.,** Cereal sprouts: composition, nutritive value and food applications, *CRC Crit. Rev. Food Sci. Nutr.*, 13, 353, 1980.
87. **Hansen, M., Pedersen, B., Munck, L., and Eggum, B. O.,** Weaning foods with improved energy and nutrient density prepared from germinated cereals. I. Preparation and dietary bulk of gruels based on barley, *Food Nutr. Bull.*, 11, 40, 1989.
88. **Pedersen, B., Hansen, M., Munck L., and Eggum, B. O.,** Weaning foods with improved energy and nutrient density prepared from germinated cereals. II. Nutritional evaluation of gruels based on barley, *Food Nutr. Bull.*, 11, 46, 1989.
89. **Kulkarni, K. D., Kulkarni, D. N., and Ingle, U. M.,** Sorghum malt-based weaning food formulations: preparation, functional properties and nutritive value, *Food Nutr. Bull.*, 13, 322, 1991.
90. **Paredes-López, O., Barba de la Rosa, A. P., and Cárabez-Trejo, A.,** Enzymatic production of high-protein amaranth flour and carbohydrate rich fraction, *J. Food Sci.*, 55, 1157, 1990.
91. **Sánchez-Marroquín, A., Domingo, M. V., Maya, S., and Saldaña, C.,** Amaranth flour blends and fractions for baking applications, *J. Food Sci.*, 50, 789, 1985.
92. **Barba de la Rosa, A. and Paredes-López, O.,** Development of an enzymatic procedure to produce high-protein amaranth flour, *Biotechnol. Lett.*, 11, 417, 1989.
93. **Pacheco de Delahaye, E. and Gómez, V.,** Procesos de elaboración de aislados proteínicos de harina de semilla integral de amaranto: extracción alcalina y succinilación de las proteínas. Propiedades funcionales de los aislados proteínicos, *El Amaranto y su Potencial*, 4, 6, 1987.
94. **Morales, E., Lembcke, J., and Graham, G. G.,** Nutritional value for young children of grain amaranth and maize-amaranth mixtures: effect of processing, *J. Nutr.*, 118, 78, 1988.
95. **Bressani, R., De Martell, E. C. M., and De Godínez, C. M.,** Protein quality evaluation of amaranth in adult humans, *Plant Foods Hum. Nutr.*, 43, 123, 1993.
96. **Tovar, L. R. and Carpenter, K. J.,** The effect of alkali cooking of corn and supplementation with amaranth seed on its deficiencies in lysine and tryptophan, *Arch. Latinoam. Nutr.*, 32, 961, 1982.
97. **Bressani, R., Elías, L. G., and Braham, J. E.,** Suplementación con aminoácidos del maíz y de la tortilla, *Arch. Latinoa. Nutr.*, 18, 123, 1968.

98. **Sánchez-Marroquín, A. and Maya, S.,** Industrial corn flour enrichment with whole amaranth flour and milling fractions in corn-based products, *Arch. Latinoamer. Nutr.*, 35, 518, 1985.
99. **Sánchez-Marroquín, A., del Valle, F. R., Escobedo, M., Avilés, R., Maya, S., and Vega, M.,** Evaluation of whole amaranth (*A. cruentus*) flour. Its air-classified fractions, and blends of these with wheat and oats as possible components for infant formulas, *J. Food Sci.*, 51, 1231, 1986.
100. **Graham, G. G., Lembcke, J., and Morales, E.,** Protein value for children of soft-endosperm maize, alone and with toasted amaranth flour, *Nutr. Res.*, 9, 859, 1989.
101. **Graham, G. G., Lembcke, J., and Morales, E.,** Post-prandial plasma aminograms in the assessment of protein quality for young children: maize and grain amaranth, alone and combined, *Eur. J. Clin. Nutr.*, 44, 35, 1990.
102. **Howe, E. E., Jansen, G. R., and Gilfillan, E. W.,** Amino acid supplementation of cereal grains as related to the world food supply, *Am. J. Clin. Nutr.*, 16, 315, 1965.
103. **Breene, W. M.,** Contribution of amaranth to consumer products, in *Proc. 4th National Amaranth Symp., Amaranth: Perspectives on Production, Processing and Marketing*, Minneapolis, MN, 1990, 75.
104. **Bressani, R. and Elías, L. G.,** Development of 100% amaranth food, in *Proc. 3rd Amaranth Conf., Grain Amaranth: Expanding Consumption Through Improved Cropping, Marketing and Crop Development*, Rodale Press, Emmaus, PA, 1984, 8.
105. **Necoechea, M. H., Camacho, C. J. L., and Pérez Gil, F.,** Elaboración de una pasta para sopa a base de alegría (*Amaranthus leucocarpus* S. Wats), *Tecnol. Aliment. (Méx.)*, 17, 12, 1982.
106. **Santin, H. C., Morales, L. J., and Ibañez, G. L.,** Desarrollo a nivel piloto de una pasta para sopa con base en una mezcla de amaranto y trigo, *Tecnol. Aliment. (Méx.)*, 22, 13, 1987.
107. **Rico, N. N., Morales, L. J., and Doria, V. R.,** Barras troqueladas: un producto para el niño en edad escolar, *Tecnol. Aliment. (Méx.)*, 23, 5, 1988.
108. **Koeppe, S. J., Harris, P. L., Hanna, M. A., Rupnow, J. H., Walker, C. E., and Cuppett, S. L.,** Physical properties and some nutritional characteristics of an extrusion product with defatted amaranth seeds and defatted maize gluten meal (80:20 ratio), *Cereal Chem.*, 64, 332, 1987.
109. **Ologunde, M. O., Shepard, R. L., Afolabi, O. A., and Oke, O. L.,** Bioavailability to rats of iron from fortified grain amaranth flour, *Int. J. Food Sci. Technol.*, 26, 493, 1991.
110. **Sánchez-Marroquín, A., del Valle, F. R., Escobedo, M., Avitia, R., Maya, S., and Vega, M.,** Evaluation of whole amaranth (*Amaranthus cruentus*) flour, its air-classified fractions and blends of these with wheat and oats as possible components for infant formulas, *J. Food Sci.*, 51, 1231, 1986.
111. **Del Valle, F. R., Sánchez-Marroquín, A., Escobedo, M., Avitia, R., Bourges, H., Maya, S., Vega, M., and Becker, R.,** Development and evaluation of a low-cost amaranth (*Amaranthus cruentus*) containing food for preschool children, *Arch. Latinoam. Nutr.*, 37, 480, 1987.
112. **Del Valle, F. R., Escobedo, M., Sánchez-Marroquín, A., Bourges, H., Bock, M. A., and Biemer, P.,** Nitrogen balance in infant fed formulas containing amaranth or a soy-oat formula, *Cereal Chem.*, 69, 156, 1992.
113. **Del Valle, F. R., Escobedo, M., Sánchez-Marroquín, A., Bourges, H., Bock, M. A., and Biemer, P.,** Chemical and nutritional evaluation of two amaranth (*Amaranthus cruentus*) based infant formulas, *Plant Foods Hum. Nutr.*, 43, 145, 1993.
114. **Gupta, C. and Sehgal, S.,** Development, acceptability and nutritional value of weaning mixtures, *Plant Foods Hum. Nutr.*, 41, 107, 1991.
115. **Rathod, P. and Udipi, S. A.,** The nutritional quality and acceptability of weaning foods incorporating amaranth, *Food Nutr. Bull.*, 13, 58, 1991.
116. **Lazcano, S. M., Morales, L. J., and Rico, N. N.,** Utilización del amaranto en la elaboración de papillas para infantes, *Tecnol. Aliment. (Méx.)*, 23, 13, 1988.
117. **Rico, N. N., Morales, L. J., and Suárez, N. L.,** Elaboración de una galleta con base en mezclas trigo-amaranto, *Tecnol. Aliment. (Méx.)*, 22, 19, 1987.
118. **Bressani, R. and Benavides, V.,** Nutritive value of wheat and amaranth flour mixtures, *Amaranth Newsl.*, 4, 14, 1990.

Chapter 11

Amaranth: Commercialization and Industrialization

James W. Lehmann

CONTENTS

I. INTRODUCTION

Amaranth has been perceived as lacking uniqueness among modern grains. However, the crop's unusual seed architecture, unique processing history, and microgranular starch argue to the contrary. To understand the fabric of tomorrow's industrialized amaranth, threads of the crop's cultural and evolutionary history must be interwoven. Grain amaranth has been a food crop for at least seven centuries. Its roles in future commerce will be determined, in part, by its genetic and cultural legacies, its amenability to modern technology, and its niches in world agriculture. The blueprint for amaranth commerce has not been fully developed, but a decade or so of scientific investigation has revealed glimpses of a future crop. To shed light on the potential routes for development, the focus of this chapter will be (1) the unusual seed architecture of amaranth and its relationship to genetic, historical, and technological functions; (2) the historical eras of amaranth development; and (3) crop development models and projected needs for the crop.

II. AMARANTH — A RING-EMBRYO GRAIN

Botanically, amaranth is a dicotyledonous herb rather than a monocotyledonous herb such as corn, oats, rice, wheat, etc. Like buckwheat and quinoa, grain amaranth is nongramineous plant (Table 1), but unlike these other pseudocereals, it belongs to a separate plant family, the Amaranthaceae of about 60 species.

0-8493-5374-2/94/$0.00+$.50

Table 1 **Characteristics of amaranth and other dicotyledonous pseudocereals**

Crop	Scientific Name	Plant Family	Starch Storage Tissue	Presence of Ring Embryo
Amaranth	*Amaranthus* spp.	Amaranthaceae	Perisperm	Yes
Quinoa	*Chenopodium* spp.	Chenopodiaceae	Perisperm	Yes
Buckwheat	*Fagopyrum* spp.	Polygonaceae	Endosperm	No

These plants possess little endosperm and have a starchy portion derived from the diploid nucellus, the perisperm. The ring-like embryo of amaranth nearly circumscribes the starchy perisperm. At first glance, these differences might appear to be arbitrary designations by plant taxonomists, but they may be central to amaranth's identity and utilization.

The chemistry of any grain is interrelated with the structure and physiology of the plant that produced it. Appropriate technologies for fractionalizing, extracting, and processing a grain must inherently deal with the grain's component structures and the ease of manipulating those structures.[1] These generalizations have been applied to the gramineous cereals such as wheat,[2] rice, barley, etc. For example, milling quality of wheat and rice grains is affected by deep creases or uneven aleurone layers, while malting quality of barley is affected by kernel plumpness and hardness and attachment of germ to the caryopsis.[1,3]

Also, we might consider grain structures such as the amaranths' ring-embryo (Figure 1) from different functional viewpoints.

A. GENETIC OR EVOLUTIONARY FUNCTIONS

Genetically, the ring-like embryo of amaranth (Figure 1) is a reproductive structure nourished by meager food reserves compared to larger-seeded crops. The preformed seed leaves (cotyledons), apical meristem, and radicle (root) are essentially dormant until the seed imbibes water and begins to grow. Amaranth's protein and oil are mostly contained in the ring-like embryo, while the readily mobilized, microgranular starch (1 to 3 μm in diameter) is in the central perisperm or portion. Although we view amaranth oil, protein, and starch as nutrients or industrial ingredients, they are, first and foremost, chemical compounds and the structure of a life form!

B. CULTURAL OR HISTORICAL FUNCTIONS

Culturally, the ring-embryo or germ appears to have been a sexual structure for classifying the crop. In 1788, a European botanist named Gärtner first drew the ring embryo of *Amaranthus hypochondriacus* L. as he sorted out early Linnean concepts of plant taxonomy (Figure 2).[4] There appears to be no mention of the ring-embryo per se in pre-Linnean times; rather the entire grain with its fish egg-like appearance was considered to have food and curative properties.[5]

C. INDUSTRIAL OR TECHNOLOGICAL FUNCTIONS

Industrially, the ring-embryo presents unusual opportunites for pearl-like milling, for physical separation from the starchy perisperm, and for an intact germ after popping or milling. For instance, preliminary evidence suggests that wet milling might remove the seed coat and allow centrifugal separation of the protein-rich embryo from the starchy perisperm.[6] This situation resembles germ separation in rice, where the germ adsorbs water and swells more quickly than its starchy endosperm.[1]

III. HISTORICAL ERAS IN AMARANTH DEVELOPMENT

A. PRE-CONQUEST AMARANTH (B.C. TO 1519)

As a long-domesticated crop, amaranth's prehistoric and pre-Conquest uses give insight into early trial and error technologies and into artificial selection. Ancient peoples viewed amaranths as both an edible, storable grain and as a vegetable, the latter being gathered before anthesis. Wild amaranths were probably encountered along silty riverbanks, ditches, and lakeshores; even today, taxonomists describe amaranths as riverbank pioneers or desert wash plants. The hardiness of the plants and their vibrant colors probably attracted early gatherers, just as in recent centuries, people selected colorful variants for ornamentals. Remarkably, one wild relative of the grain amaranths, *A. palmeri* S. Wats, is capable of completing its entire life cycle with the moisture of one rain.[7] Archaelogical records document amaranths at early North

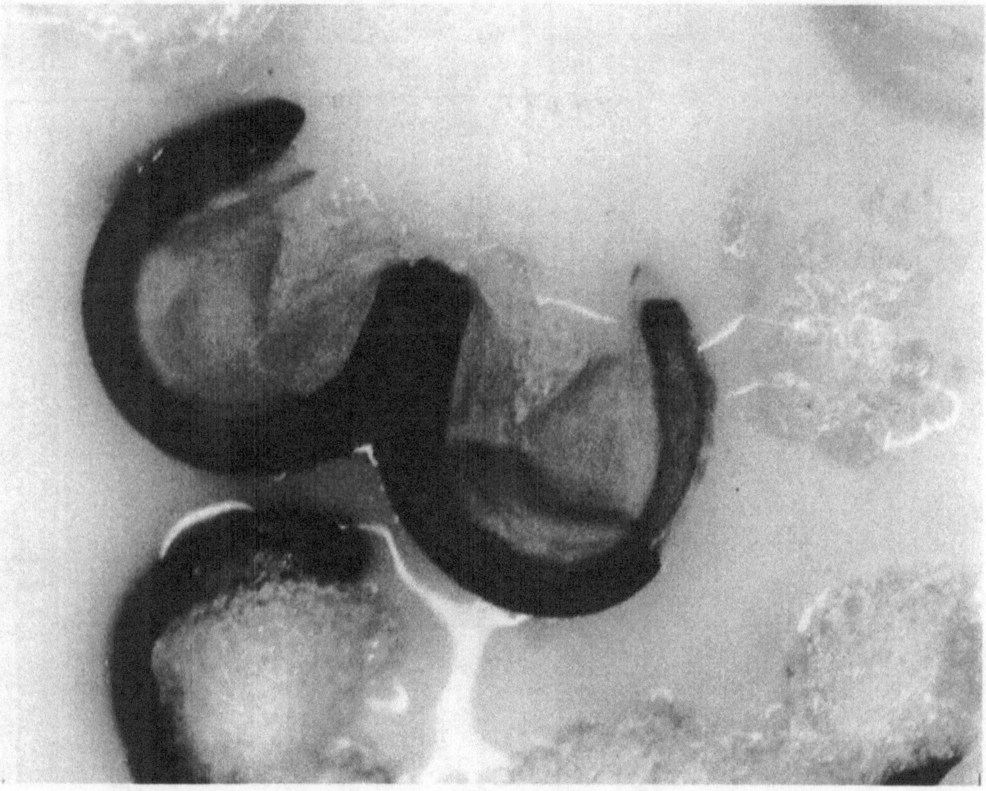

Figure 1 The darkened ring embryo that stays intact after popping of amaranth grain. The grain was intentionally burnt during popping to allow greater visiblility of the otherwise cream-colored germ (magnification × 40). (From Lehmann, J. W., *Int. Food Ingredients,* 3, 26, 1992. With permission.)

American campsites[8] and in Mexican caves.[9] The crop's adventive nature in campsite refuse prompted the late maize geneticist, Edgar Anderson, to describe amaranth as a "dump heap plant of civilization".

Crop evolutionists generally agree that grain amaranths originated in the New World and probably in the Mesoamerican, southwestern U.S. or northwestern Mexico area. Wild amaranths were sought by early hunters and gatherers, such as the aboriginal Cocopans and Yumans along the Colorado River delta and → other early Baja California tribes.[10] The historical abundance of crops and their wild relatives can often be reconstructed from pollen deposits in lakebeds or alluvial sediments. Unfortunately, the pollen record of amaranth throughout much of its center of origin is obscured by its similarity to chenopod pollen, although some differentiation is possible.[11]

Amaranths, with their tiny seeds, transplanted seedlings, and hard seed coats, might seem an unlikely companion staple crop for maize during the pre-Conquest era. Why amaranth attained prominence before and during the Aztec era and why it declined so drastically after the Conquest remains unclear, although various theories exist about its role in Mesoamerican agriculture.[12] Certainly, the physical grandeur of 10 to 12 ft grain amaranths (such as the Mixtec race of *Amaranthus hypochondriacus*), resplendent in their 3 to 4 ft tall, golden or amaranthine seedheads, might evoke animism in prescientific societies. Modern evidence suggests that amaranths complemented the amino acids of maize and thus formed a complete vegetable protein. Whether the Aztecs resorted to cannabalistic rituals to terrorize their subjects or due to a lack of protein in their diets is controversial, but Ortiz de Montellano[13] presents evidence suggesting that, with amaranth and other foods, the Aztec diet contained adequate protein and calories.

B. PRE-INDUSTRIAL AMARANTH (1519 TO 1973)

After the Spanish Conquest in Mexico, cultivation and usage of amaranth was either discouraged, destroyed, or the crop fell into disuse. Amaranths were likely carried to Africa and India, although there

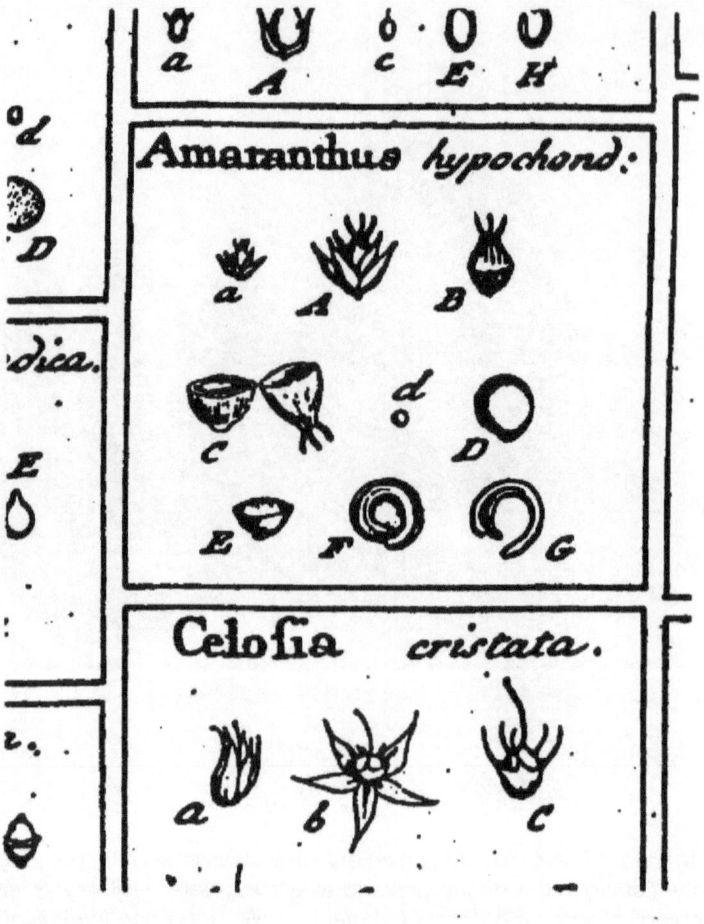

Figure 2 Taxonomic illustrations by an 18th century European botanist which first show the ring-embryo (letters "F" and "G") of grain amaranth (*Amaranthus hypochondriacus* L.) (From Gärtner, J., De fructibus et seminibus plantarum, Stvtgardiae, typis Academiae Carolinae, 1788.)

is uncertainty as to how and when these events occurred. Nevertheless, wherever light- or white-seeded amaranths were grown, the technology of popping or roasting was applied.[5,14] Both technologies probably reflect the need to break the hard seed coat and to negate or lessen any antinutrients present.[15] Also, seed coat thickness and hardness distinguish domesticated grain amaranth from its thicker-coated wild progenitor.[16]

C INDUSTRIAL AMARANTH (1973+)

Wild amaranths have competed well with monoculture crops during the 20th century. The impact of *A. retroflexus* L. and *A. hybridus* L. on crop production, development of atrazine herbicides, and elucidation of amaranth's atrazine resistance are topics onto themselves. Ironically, modern agriculture tends to view amaranths as weeds rather than as vegetables and grains. Separation of weedy and feral amaranths by herbicides is as yet undeveloped.[17]

Grain and vegetable amaranths as crops per se might be modeled after other crops. Predicting the development of crop amaranths is a pseudoscience, even though Delphi techniques have been applied.[18] Still, many elements of crop development have been associated with successful commercialization and industrialization:[19-21]

1. Breakthroughs
2. Crop champions
3. Public support
4. Industry support
5. Timelines of the production-marketing-consumption (PMC) chain

Narrative reviews of these elements relative to amaranth are available.[18,22] Commercial applications for the crop, included projected ones, are listed in Figure 3.

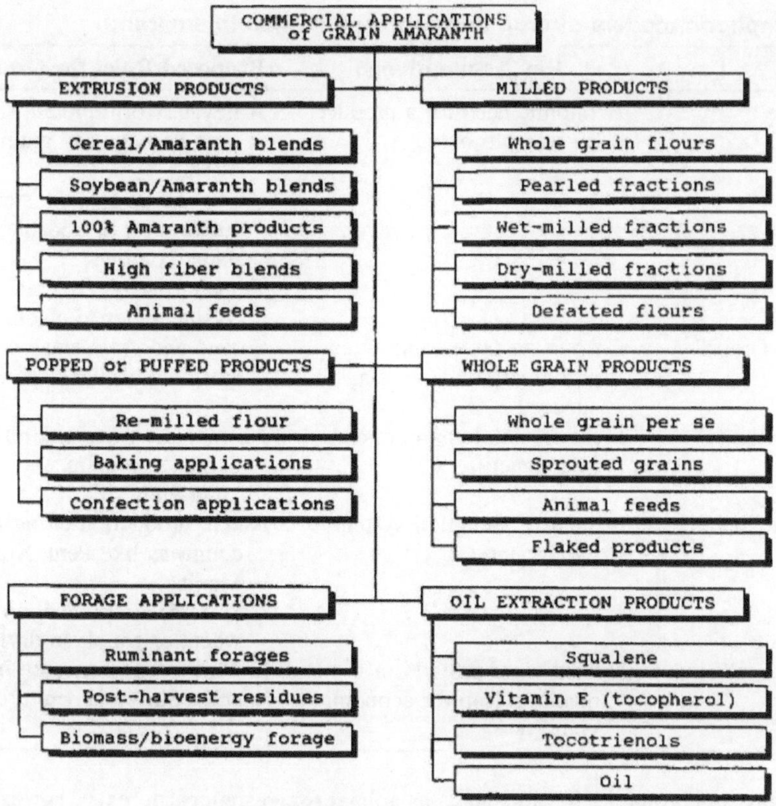

Figure 3 Categories of projected products from grain amaranth.

IV. METAPHORIC MODELS OF CROP DEVELOPMENT

Metaphoric models of crop development might be applied to amaranth (Table 2); but, they invariably overlap or often fit in a hierarchy of influence.

A. THE "DRIVING FORCE" MODEL

In the case of amaranth, some researchers subscribe to the "driving force theory of crop development." A unique trait, commercial advantage, or economic breakthrough for a crop overcomes the energy or monetary barrier(s) needed to industrialize the crop. For instance, rapeseed bred for low glucosinolate levels gave rise to canola oil, which was both economical and suited to human consumption. For amaranth, Saunders[23] suggested that if its price per pound were less than $.10, then the grain would become more competitive as a protein source.

B. THE "CREAM OF THE CROPS" MODEL

Another name for this model might be the ecologically inspired "limiting factor(s)" model. Some observers suggest that minor crops as amaranths or quinoas (*Chenopodium* spp.) have serious defects: seed size, yield, mechanical harvest capability, limited production and area, etc. These defects, in turn, limit production and economic viability and, thus, the crops become inferior to staple crops. Considering the successful plant improvements seen in major crops using plant breeding and biotechnology, this model has limited appeal.

C. THE "GOLDEN OPPORTUNITY" MODEL

Another perspective is that a "Cinderella crop" such as soybeans blossomed into a major commercial crop because of protein and oil needs, particularly those generated by World War II. Another way to phrase this school of thought is that trends and societal needs eventually dictate what is developed. Some elements of the "driving force" model are also present.

Table 2 **Metaphoric models of crop development applied to amaranth**

Model	Key Assumption(s)	Proposed Roles for Grain Amaranth
"Driving Force"	Economic need for a product or crop attribute	Allievate world protein shortage with both herbage and grain (Reference 57) Extract microgranular starch for fat substitutes, degradable plastics, or paper coatings. Transfer drought tolerance or high lysine genes to other crops
"Cream of the Crops"	Crops with fewest liabilities and greatest adaptability succeed	Forage and grain amaranth in marginal, dryland agriculture
"Golden Opportunity"	Meet crucial, temporal needs in a society	Nutraceutic transmission of anti-oxidants, protein, and iron via the "ring-embryo"
"Diversify and Conquer"	Influence of cultural system paramount	Native production of oat-like grain in countries like Peru, Kenya, and Mexico U.S. domestic production coupled with wheat or canola in dryland systems
"The Hybrid Leap"	Hybrid vigor provides a quantum leap for economic viability	Hybrid forage amaranth in Russian and Chinese agriculture

Internationally, opportunities for amaranths as animal forages appear to exist. For example, Kazan State University (Russia) and a private company have asked the U.S. Department of Agriculture for technical and economic assistance to plant 9 to 12 million acres of amaranth for animal fodder. Although this expansion may seem premature, Chinese researchers have reported 50,000 ha devoted to amaranth fodder production in 1991.[24] It is unclear whether prevailing economic, political, and commodity interests in the U.S. will cooperate by transferring amaranth technology or, instead, view these expressed foreign needs or interests as competitive.

D. THE "DIVERSIFY AND CONQUER" MODEL

As in the development of canola in Canada, research sponsorship and aid by wheat organizations enabled a concerted development of low-glucosinolate repeseed.[20] In the U.S., amaranth, too, might be viewed also as a rotation crop in wheat or canola production zones. Furthermore, amaranth could bolster the protein quality of the canola oil by-product, canola meal.

E. THE "HYBRID LEAP" MODEL

Economic incentive for a seed industry and quantum yield leaps are both hallmarks of the hybrid corn industry. Amaranth, as the former companion crop of maize, could be developed as a hybrid crop, but no systematic program of inbred population development and hybrid evaluation has been applied to the three grain amaranth species. Amaranth is a mostly self-pollinated crop with variable outcrossing. Its breeding system resembles that of cotton. Preliminary breeding evidence within *A. cruentus* suggests a 10 to 30% increase in grain yield might be anticipated by hybrids with this species.[25] Interspecific crosses within the *Amaranthus* gene pool (e.g., *A. hypochondriacus* × *A. cruentus*) tend to be late flowering, vegetative, and high producers of biomass at middle latitudes.[26] Comprehensive evaluation of hybrid vigor could be facilitated by male sterility sources.[27-29]

F. SPECIFIC CROP MODELS

Are models for commerialization or industrialization of other crops pertinent to amaranth? For instance, assuming that amaranth follows the developmental model of soybean, then its early development as a forage is reminiscent of forage soybean in the early 1900s. Other parallels within the successful soybean model include the need for crusaders, missionaries, and pioneers[30] and for high quality livestock feeds.

Or, will amaranth instead model a minor, specialty niche crop like buckwheat? In some respects, this model is already being followed in the organic amaranth market in the U.S..

Amaranth industrialization will likely borrow aspects of numerous crop development models, whether they be metaphorical (Table 2) or other crop models. A generalized framework for agroindustrial development could be applied to amaranth.[31] Whatever potpourri or blend of development occurs, the genes and germplasm[32] precede the technologies that will be applied to the crop. Curious indigenous, processing technologies[33] have been applied to the crop such as popping, followed by milling into flour. Because such technologies may have evolved during the crop's halcyon years and were apparently adopted by divergent peoples, they should be evaluated as to their effectiveness in retaining, modifying, and transmitting nutrients.

V. PROJECTED NEEDS FOR AMARANTH

The overriding and broader questions of industrialization, though, concern the economic, ecological, and nutritional needs for amaranth. For example, Dr. Shoaxian Yue[24] stated that Chinese "amaranth industrialization should be guided by the market," which he viewed as the food and feed industry. On the other hand, Bressani[34] viewed integrated research and knowledge of the crop's value in foods, forages, and by-products as crucial to commercialization. What then are the developing market needs which may influence, by either impeding or accelerating, amaranth development? Again, these needs can be only partially identified as domestic and worldwide marketplaces evolve.

A. HUMAN NUTRITION: COMPLETE VEGETABLE PROTEINS WITH HIGH ENERGY VALUE

The need for protein sources to fight malnutrition has given way to a need for foods with adequate energy.[35] Other high quality vegetable proteins, such as quinoa and quality protein maize, have been ineffectively introduced or advanced as crops which would nutritionally aid the poor.[35] Amaranth, too, has been viewed by some developers as a protein source for infant formulas.[36] Inevitably, amaranth must be viewed as a cash crop in rural development and modern commerce. Overemphasis on amaranth as a self-reliance crop obscures the crop's potential for processing by both indigenous and modern food technologies. Composite flours of locally grown amaranth and imported cereals may be feasible in some countries; for example, DeBuckle[37] has summarized similar composite flours in the Andean Pact countries. Incidentally, Peruvian amaranth production gained public support and promotion, but the effort was thwarted by political upheaval.

B. PROTECTIVE AND REGULATORY NUTRIENTS: THE AGE OF THE NUTRACEUTICAL

Nutraceuticals, nutratherapy, phytotherapy, and phytomedicine are new words to describe an age-old situation — foods are more than nutrients. Grains and many other botanicals can possess at least three functions or dimensions: (1) nutrition, (2) regulation, and (3) protection. Separation of life-nourishing or vital compounds into simplistic categories sometimes diminishes the complexity and interrelatedness of these roles in living systems. For example, some forms of vitamin E, tocopherols and tocotrienols, function as antioxidants or protectors, while others function as inhibitors of cholesterol biosynthesis or regulators. Even essential amino acids are involved in a multiplicity of roles — in specific enzymes and structural proteins in the human body. We simply call lysine a part of protein, but its roles are slightly known.

Is amaranth a nutraceutical or "functional food"? By definition, nutraceuticals are food substances that provide medical or health benefits, including the prevention and treatment of disease.[38] The Aztecs certainly credited amaranth and many other foods with medicinal and sometimes magical properties.[5,13] In this vein, this author has advanced a ring hypothesis for grain amaranth and quinoa: upon popping, the conserved ring-embryo retains released vitamin E and harbors oil-components, minerals, and high quality protein. The interaction of the seed structure and thermal processing intimates a prescientific nutraceutical. A recent patent confirms that a thermal release mechanism for vitamin E is indeed present in plant oils, including amaranth oil.[39] Today's food distributors will require an extended shelf life for thermally processed amaranth, but it is unknown if the vitamins released in amaranth will persist long enough to meet these needs. Altogether, the ring hypothesis may account for some of the crop's magical or sacred qualities attributed by prescientific cultures.[40,41]

Food technologists, druggists, and consumers show a growing interest in nutraceuticals.[42] Cholesterol-lowering attributes of amaranth are emerging, with either squalene, fiber components,[43] or vitamin E isomers[44] as possible agents. Another dimension of nutraceuticals concerns bioavailability vs. the mere presence of analytically detected nutrients.[45] For instance, calcium levels in amaranth may surpass those in milk, but bioavailability of the mineral may be limited by the presence of phytates and by interactions with other minerals. Bioavailability studies that test the effect of processing on the release of calcium are needed. Comparatively, amaranth's iron content has a high biological value when measured in animal tests.[46]

For centuries, grains and cereals have been the "staffs of life" for civilizations. In today's societies, tensions prevail between marketing a "healthy food" or nutraceutical, satisfying the taste needs of the consumer, and meeting the requirements of the food distribution. Other issues could be raised about highly processed vs. freshly processed foods and the merits of indigenous wisdom vs. industrial knowledge, but insufficent data exists to relate them to amaranth. Indigenous methods of amaranth usage merely give clues as to palatablity and processing.[14]

C. SUSTAINABLE AGRICULTURAL SYSTEMS

Sustainable agriculture systems are needed that increase or maintain soil tilth and fertility, while utilizing nitrogen and non-nitrogen fixing crops in economical cropping sequences. Additionally, crop rotation and options are highly desirable in thwarting many soil-borne diseases and in preventing build-up of crop pests. Amaranth's roles in sustaining agriculture are poorly known, although Chinese researchers[47] suggest that the crop's dense fibrous root system and modest fertility requirements make it a candidate for saline and sandy soils.

D. POPCORN ECONOMICS AND MEDIA FOODS

The proprietor of any cinema can tell you why the air in popcorn is economically important, while advertisers using popular persona can identify the association of pleasure with products. Amaranth, like popcorn, can be popped or puffed to increase volume and profitability while its nutritive qualities are often related to pleasure. Historically, sweeteners such as crude sugar, honey, and molasses made the otherwise bland or bitter grain attractive while pleasureable words such as the Spanish alegria or "joy" were associated. Most likely the modern "sweetners" of amaranth will entail health imagery or slogans, the identity of its colorful history, and unrelated associations with popular culture.

E. INDUSTRIAL FEEDSTOCKS

Because of the luxurious growth of feral and domestic amaranths, their biomass output may be useful as industrial feedstocks. Murray[48] first documented the striking biomass of interspecific amaranth hybrids, while Lehmann and others[26] determined dry matter gains in some hybrids ranging from 21.6 to 23.9 Mg/ha. Whether amaranths possess any usuable paper fibers, hemicelluloses, or other specialty botanochemicals has not been determined. At present, amaranth is a lignocellusic feedstock which could potentially act as a seasonal supplement to energy or chemical production, thus shoring up or buffering the lignocellulosics generated by municipal solid, pulp mill, and agricultural wastes.[49]

VI. TECHNOLOGIES APPROPRIATE TO AMARANTH DEVELOPMENT

To probe the future of a crop like amaranth, the nature of technologies that could be applied should be considered. Details of these technologies have been reviewed in other parts of this publication and elsewhere:

1. Extrusion[50,51]
2. Biotechnology[52]
3. Antioxidant-releasing and stabilization processes[39]
4. Fractionation into oil, protein, and starch[6,50]
5. Thermal expansion and processing[53,54]

VII. FOUNDATIONS FOR A 21ST CENTURY CROP

Whether amaranth will fulfill some of the perceived economic and societal needs mentioned above is unknown. However, substantial data on the crop's development have been assembled: genetic resources,[32] nutritional evaluation,[55] and bibliographic summaries.[56] Integration of technologies and marketing appropriate to amaranth seem to be lacking as yet.

Since time immemorial on this continent, peoples have popped amaranths and inadvertently conserved their nutritive "embryonic rings". To increase commercial value, amaranth processing should retain major nutrients and therapeutic agents such as antioxidant vitamins and valuable minerals. Amaranth's past "magic" and today's science may merge when we understand how its nutrients are released, conserved, or transmitted, especially during thermal processes such as popping or extrusion.

Commercialization of amaranth is a complex process, involving political, technological, marketing, and economic dimensions. Societal needs for nutraceuticals, high energy foods, industrial feedstocks, etc. may also influence the crop's development. Industrialization of the grain will be based on the seed's architecture and processing chemistry as well as on the markets for microgranular starch, proteins, and other by-products. Genetic improvement of the crop has started with the forage or fodder, a trend resembling soybean introduction on this continent. Given the models of other successful crops, concerted improvement of crop or grain yield and of specialty attributes will hasten amaranth's commercialization. Harnessing the industrial and staple crop aspects of the old regal crop, though, is only beginning.

REFERENCES

1. **Pomeranz, Y.,** Grain endosperm structure and end-use properties, in Proc. 7th World Cereal and Bread Congress, Prague (1982), *Developments in Food Science 5A*, 19-48, 1983.
2. **Saunders, R. M.,** Cereal by-products: wheat and rice, In *CRC Handbook of Nutritional Supplements*, Vol. I, Human Use, Rechcigl, Jr., M., Ed., CRC Press, Boca Raton, FL, 1983.
3. **Hockett, E. A. and Nilan, R. A.,** Genetics, In *Barley*, Rasmusson, D.C., Ed., American Society of Agronomy, Madison, WI, 1985.
4. **Gärtner, J.,** De fructibus et seminibus plantarum, Stvtgardiae, typis Academiae Carolinae, 1788.
5. **Sauer, J. D.,** The grain amaranths: a survey of their history and classification, *Ann. Mo. Bot. Gard.*, 37, 561, 1950.
6. **Fox, S. R., and Myers, D. J.,** A comparative study of the wet milling characteristics between pearled and unpearled amaranth seed, *Cereal Foods World*, (Abstr.), 1991.
7. **Ehleringer, J.,** Ecophysiology of *Amaranthus palmeri*, a Sonoran desert summer annual, *Oecologia*, 57, 107, 1983.
8. **Agogino, G. A.,** Pigweed seeds dated oldest U.S. food grain, *Sci. Newsl.*, 72, 345, 1957.
9. **MacNeish, R. S., Folwer, M. L., Cook, A. G., Peterson, F. A., Nelken-Terner, A., and Neely, J. A.,** *The Prehistory of the Tehuacan Valley.* Vol. 5, *Excavations and Reconnaissance*, University of Texas Press, Austin, 1975.
10. **Castetter, E. F. and Bell, W. H.,** *Yuman Indian Agriculture*, University of New Mexico Press, Tucson, AZ, 1951.
11. **Tsukada, M.,** Chenopod and amaranth pollen: electron-microscopic identification, *Science*, 157, 80, 1967.
12. **Gordon, C. L.,** An Anthropo-Geographic View of an Economy: the Late Pre-Conquest Basin of Mexico, Ph.D. thesis, Columbia University, New York, DAI 8023497, 1980.
13. **Ortiz de Montellano, B. R.,** *Aztec Medicine, Health, and Nutrition*, Rutgers University Press, New Brunswick, NJ, 1990.
14. **Early, D. K.,** Amaranth production in Mexico and Peru, in *Advances in New Crops: Proc. 1st National Symp. New Crops-Research, Development, Economics*, Janick, J. and Simon, J. E., Eds., Indianapolis, IN, Oct 23 to 26, 1988, Timber Press, Portland, OR, 1990.
15. **Lehmann, J. W.,** Anti-nutritional factors in amaranth grain, *Legacy* 5, 6, 1992.
16. **Pal, A., Singh, R. P., and Pal, M.,** Development and structure of seeds in *Amaranthus hypochondriacus* and its wild progenitor, *A. hybridus* L., *Phytomorphology*, 40, 145, 1990.
17. **Endres, C. S. and Longer, D. E.,** Herbicide selectivity among grain and weedy *Amaranthus* species, *Agron. J.*, 79, 824, 1987.
18. **Lehmann, J. W.,** The potential of grain amaranths in the 1990s and beyond, in *Proc. 4th National Amaranth Symp.*, Minneapolis, MN, August 23 to 25, 1990, Minnesota Extension Service, St. Paul, 1990.
19. CAST, Development of New Crops: Needs, Procedures, Strategies, and Options, Council for Agricultural Science and Technology, Rep. No. 102, Ames, IA, 1984.

20. **Joliff, G. D. and Snapp, S. S.,** New crop development: opportunity and challenges, *J. Prod. Agric.*, 1, 83, 1988.
21. U.S. Senate, New uses for farm and forest products, S. Hrg. 100-471, June 26, 1987.
22. **Lehmann, J. W.,** Crop development of grain amaranth: an interface of science and culture, in *Proc. Strategies for Alternative Crop Development: Case Histories*, University of Minnesota, Extension Special Publications, St. Paul, MN, 1989, 11.
23. **Saunders, R. M.,** Comments re post-harvest processing and nutrition, and marketing, *Amaranth Newsletter*, No. 4, INCAP, Guatemala City, Guatemala, 1985.
24. **Yue, S. X,** personal communication, 1991.
25. **Lehmann, J. W.,** unpublished data, 1987.
26. **Lehmann, J. W., Clark, R. L., and Frey, K. J.,** Biomass heterosis and combining ability in interspecific and intraspecific matings of grain amaranths, *Crop Sci.*, 31, 1111, 1991.
27. **Peters, I., and Jain, S.,** Genetic variation of sex expression in cultivated *Amaranthus* spp. — solution to many puzzles, *Am. J. Bot.*, 72, 878, 1985.
28. **Peters, I. and Jain, S.,** Genetics of grain amaranths. III. Gene-cytoplasmic sterility, *J. Hered.*, 78, 251, 1987.
29. **Gudu, S. and Gupta, V. K.,** Male sterility in the grain amaranth (*Amaranthus hypochondriacus* ex-Nepal) variety *Jumla*, *Euphytica*, 37, 23, 1988.
30. **Windish, L. G.,** The soybean pioneers, trailblazers ...crusaders...missionaries, Privately published by author. Henry, IL, 1981.
31. **Austin, J. E.,** *Agroindustrial Project Analysis*, John Hopkins University Press, Baltimore, MD, 1981.
32. **Kauffman, S.,** The status of grain amaranth for the 1990s, *Food Rev. Int.*, 8, 165, 1992.
33. **Early, D. K.,** Seed capital — a long-forgotten grain may have the potential to feed the world's hungriest nations, *Earthwatch*, September/October, 12, 1992.
34. **Bressani, R.,** Amaranth: the nutritive value and potential uses of the grain and by-products, *Food Nutr. Bull.*, 10, 49, 1988.
35. **Tripp, R.,** Does nutrition have a place in agricultural research?, *Food Policy*, 50, 467, 1990.
36. **Sanchez-Marroquin, A., Delvalle, F.R., Escobedo, M., Avitia, R., Maya S., and Vega, M.,** Evaluation of whole amaranth (*Amaranthus cruentus*) flour, its air classified fractions and blends of these with wheat and oats as possible components for infant formulas, *J. Food Sci.*, 51, 1231, 1986.
37. **DeBuckle, T. S.,** Composite flours: prospects for their application in developing countries-the case of the Andean pact countries, *Food Rev. Int.*, 4, 237, 1988.
38. FIM, The nutraceutical initiative: a proposal for economic and regulatory reform, Foundation for Innovation in Medicine, Cranford, NJ, 1991.
39. **Qureshi, A., Becker, K., Wells, D., and Lane, R. H.,** International Patent No. WO 91/17985, 1991.
40. **Lehmann, J. W.,** Grain amaranth — developing 21st century products from the legendary sister crop of maize, *Int. Food Ingredients*, 3, 26, 1992.
41. **Lehmann, J. W.,** The industrialization and commercialization of amaranth, in *Proc. 1st Int. Amaranth Congr.*, Oaxtepec, Morelos, Mexico, Universidad Nacional Autonoma de Mexico, Mexico City, Mexico, September 22 to 26, 1991.
42. **Duxbury, D. D. and Swientek, R. J.,** "Eating healthy" trend spurs therapeutic foods, *Food Process.*, 53, 22, 1992.
43. **Danz, R. A. and Lupton, J. R.,** Physiological effects of dietary amaranth (*Amaranthus cruentus*) on rats, *Cereal Foods World*, 37, 489, 1992.
44. **Qureshi, A. A. and Lehmann, J.W.,** Cholesterol regulation by grain amaranth and its oil in 6-week old female chickens, *Cereal Foods World* 36, (Abstr.), 710, 1991.
45. **Ferrando, R.,** From analysis to reality: bioavailability in nutrition and toxicology - a misunderstood concept, *World Rev. Nutr. Diet*, 53, 28, 19XX.
46. **Whittaker, P. and Ologunde, M. O.,** Study of iron bioavailability in a native Nigerian grain amaranth cereal for young children, using a rat model, *Cereal Chem.*, 67, 505, 1990.
47. **Yue, S. X. and Sun, H. L.,** The nutritional composition of grain amaranth and its potential of utilization, *Acta Agron. Sinica*, 13, 151, 1987.
48. **Murray, M. J.,** Environmental vegetative heterosis, *Agron. J.*, 52, 609, 1960.
49. **Rowell, R. M.,** Opportunities for lignocellulosic materials and composites, in *Emerging Technologies for Materials and Chemicals from Biomass*, Rowell, R. M., Schultz, T. P., and Narayan, R., Eds., American Chemical Society, Washington, D.C., 1992, 12.

50. **Breene, W. M.,** Food uses of grain amaranth, *Cereal Foods World*, 35, 426, 1991.
51. **Mendoza, C. and Bressani, R.,** Nutritional and functional characteristics of extrusion-cooked amaranth flour, *Cereal Chem.*, 64, 218, 1987.
52. **Flores, H. E.,** Biotechnology and amaranth, in *Proc. 4th National Amaranth Symp.*, Minneapolis, MN, August 23-25, 1990, Minnesota Extension Service, St. Paul, 1990.
53. **Tovar, L. R., Brito, E., Takahashi, T., Miyazawa, T., Soriano, J., and Fujimoto, K.,** Dry heat popping of amaranth seed might damage some of its essential amino acids, *Plant Foods Hum. Nutr.*, 39, 299, 1989.
54. **Hubbard, E. S. and Guanella, T.,** U.S. Patent 5069923, 1991.
55. **Bressani, R., Sanchez-Marroquin, A., and Morales, E.,** Chemical composition of grain amaranth cultivars and effects of processing on their nutritional quality, *Food Rev. Int.*, 8, 23, 1992.
56. **Mallory, E., Blair, M., and Weber, L.,** *Amaranth Bibliographies*, Rodale Press, Emmaus, PA, 1989.
57. **Rawate, P. D.,** Amaranth (pigweed): a crop to help solve the world protein shortage, in *Environmentally Sound Agriculture*, Lockeretz, W., Ed., Praeger Scientific, New York, 1983.

INDEX

G

H

I

L

M